Vibrational Spectroscopy Applications in Biomedical, Pharmaceutical and Food Sciences

AF713741

Vibrational Spectroscopy Applications in Biomedical, Pharmaceutical and Food Sciences

Andrei A. Bunaciu
Hassan Y. Aboul-Enein
Vu Dang Hoang

ELSEVIER

Elsevier
Radarweg 29, PO Box 211, 1000 AE Amsterdam, Netherlands
The Boulevard, Langford Lane, Kidlington, Oxford OX5 1GB, United Kingdom
50 Hampshire Street, 5th Floor, Cambridge, MA 02139, United States

© 2020 Elsevier Inc. All rights reserved.

No part of this publication may be reproduced or transmitted in any form or by any means, electronic or mechanical, including photocopying, recording, or any information storage and retrieval system, without permission in writing from the publisher. Details on how to seek permission, further information about the Publisher's permissions policies and our arrangements with organizations such as the Copyright Clearance Center and the Copyright Licensing Agency, can be found at our website: www.elsevier.com/permissions.

This book and the individual contributions contained in it are protected under copyright by the Publisher (other than as may be noted herein).

Notices
Knowledge and best practice in this field are constantly changing. As new research and experience broaden our understanding, changes in research methods, professional practices, or medical treatment may become necessary.

Practitioners and researchers must always rely on their own experience and knowledge in evaluating and using any information, methods, compounds, or experiments described herein. In using such information or methods they should be mindful of their own safety and the safety of others, including parties for whom they have a professional responsibility.

To the fullest extent of the law, neither the Publisher nor the authors, contributors, or editors, assume any liability for any injury and/or damage to persons or property as a matter of products liability, negligence or otherwise, or from any use or operation of any methods, products, instructions, or ideas contained in the material herein.

Library of Congress Cataloging-in-Publication Data
A catalog record for this book is available from the Library of Congress

British Library Cataloguing-in-Publication Data
A catalogue record for this book is available from the British Library

ISBN: 978-0-12-818827-9

For information on all Elsevier publications
visit our website at https://www.elsevier.com/books-and-journals

Publisher: Susan Deans
Acquisitions Editor: Kathryn Eryilmaz
Editorial Project Manager: Lena Sparks
Production Project Manager: Debasish Ghosh
Designer: Greg Harris

Typeset by SPi Global, India

ELSEVIER Book Aid International Working together to grow libraries in developing countries
www.elsevier.com • www.bookaid.org

Contents

Part IV
Food analysis applications

Preface

Vibrational spectroscopy, comprising infrared absorption and Raman scattering spectroscopy, is being currently and widely used in different branches of natural science such as chemistry, physics, astronomy, biology, medicine, geology, and mineralogy.

These spectroscopic techniques have been unceasingly matured since the historical discovery of infrared radiation by Sir Frederick William Herschel (1738–1822) and Raman scattering by Sir Venkata Raman (1888–1970).

Typically, they are used in connection with each other so as to get a more complete picture of molecular structure that is extensively useful for characterizing and identifying compounds.

The application of vibrational spectroscopy is ever expanding, due to its nondestructive and versatile nature. Historically speaking, it started with pioneer works in the field of infrared spectroscopy by Coblentz in 1913 and Raman spectroscopy by Garfinkel and Edsall in 1958. Since then, its development has been undeniably evidenced by an enormous number of review and research papers published every year, especially in biomedical, pharmaceutical, and food analysis.

Bearing this in mind, this book specifically aims at providing readers with up-to-date applications of vibrational spectroscopy in biomedical, pharmaceutical, and food analysis. It is suitable for both graduate students and experienced researchers in academia and industry. It contains 10 chapters, being organized into four main sections. The first section deals with the theoretical aspects of vibrational spectroscopy, sampling methods, and instrumentation used by infrared and Raman spectroscopy. The last three sections focus on describing studies selectively and illustratively related to biomedical, pharmaceutical, and food analysis. An appendix is also provided to highlight the importance of the chemometric tools used in vibrational spectroscopy data analysis.

Andrei A. Bunaciu
Hassan Y. Aboul-Enein
Vu Dang Hoang

Chapter 1

Introduction

Vibrational spectroscopy is one of the classical instrumental methods of chemical analysis that can shed light on molecular chemical composition and architecture of molecules. Since the discovery, at the end of the 19th century, this branch of molecular spectroscopy could be used as an approach to understand the bond lengths/bond angles/bond distortion relationship, measured with picometer precision. It is also suitable for studying redox state, interactions with the environment—like hydrogen bonding and electric fields—as well as conformational degrees of freedom.

Nowadays, it has become one of the most commonly used techniques in the field of biomedical, pharmaceutical, and food sciences for identification, structural elucidation, characterization, reaction monitoring, quality control, and quality assurance. This standing is due to the fact that any kind of substances (i.e., liquids, solutions, powders, pastes, films, fibers, gaseous, and different surfaces) can be investigated by using vibrational spectroscopy with a thoughtful choice of sampling techniques. With modernized machines and informatics information development, more sensitive analytical procedures have been increasingly developed in order to examine samples previously intractable.

As a collective term, vibrational spectroscopy encompasses several techniques, i.e., infrared (IR) and Raman spectroscopy. It involves the study of changes in molecular vibrational state caused by photon energy transfer in the interaction of electromagnetic radiation with the molecule. While IR bands arise from an electric dipole-mediated transition between vibrational energy levels by cause of absorbing mid-IR radiation (a resonance condition), Raman bands arise from a change in polarizability of the molecule (an off-resonance condition).

In principle, mid-IR and Raman spectroscopy yield characteristic fundamental vibrations, which is useful for the interpretation of molecular structure. On the other hand, near-IR spectra are suitable for rapid and accurate quantitation because they are generated by two processes: broad overtone and combination bands of some fundamental vibrations transitions (only the higher frequency modes). To fully assess molecular vibrational modes of a molecule, Raman and mid-IR spectroscopy are commonly requested for symmetric vibrations of nonpolar groups and asymmetric vibrations of polar groups, respectively.

Vibrational Spectroscopy Applications in Biomedical, Pharmaceutical and Food Sciences
https://doi.org/10.1016/B978-0-12-818827-9.00001-9

© 2020 Elsevier Inc. All rights reserved.

To really appreciate the importance of vibrational spectroscopy in analytical sciences, its brief history is introduced at first in this chapter with the most pioneering works in biomedical, pharmaceutical, and food application.

For more details on the theoretical knowledge that are beyond the scope of this book, readers may refer to other textbooks edited by Chalmers and Griffiths [1–3] on vibrational spectroscopy.

Brief history of vibrational spectroscopy

Although Sir Isaac Newton was the first to understand the visible spectrum of light by using a prism, in 1666, when refracting white light into various colors [4], vibrational spectra experiments started with the first studies of the astronomer Sir Frederich William Herschel in 1800 [5, 6], in the IR region. In a systematic study of the heating power of colored rays, he created the spectrum—a rainbow, by directing the sunlight through a glass prism placed in front of a thin slit made in a window shutter. The temperature of each color was measured by using three mercury-in-glass thermometers with blackened bulbs (i.e., placing one bulb in a visible color and the other two beyond the spectrum as control). His findings were published in the first article [5] speculating on the fact that the maximum heat effect lies beyond the red edge of the visible spectrum, now named as IR radiation.

Sir Frederich W. Herschel
(1738–1822)

In the second article [6], he stated the detection of IR radiation with the apparatus presented in Fig. 1.1: "*the four last experiments prove that the maximum of the heating power is vested among the invisible rays.*"

It must be noted that when IR radiation was discovered with Herschel's glass prism, most scientists did not approve the wave theory of light proposed by Christiaan Huygens [7] stating that wavelength is what determines the color of light. This did not change for more than a decade, even though shortly afterwards an English physicist, Thomas Young, determined the wavelengths of the colors of visible light by using narrowly separated slits to isolate the interference fringes [8, 9].

Thomas J. Seebeck
(1770–1831)

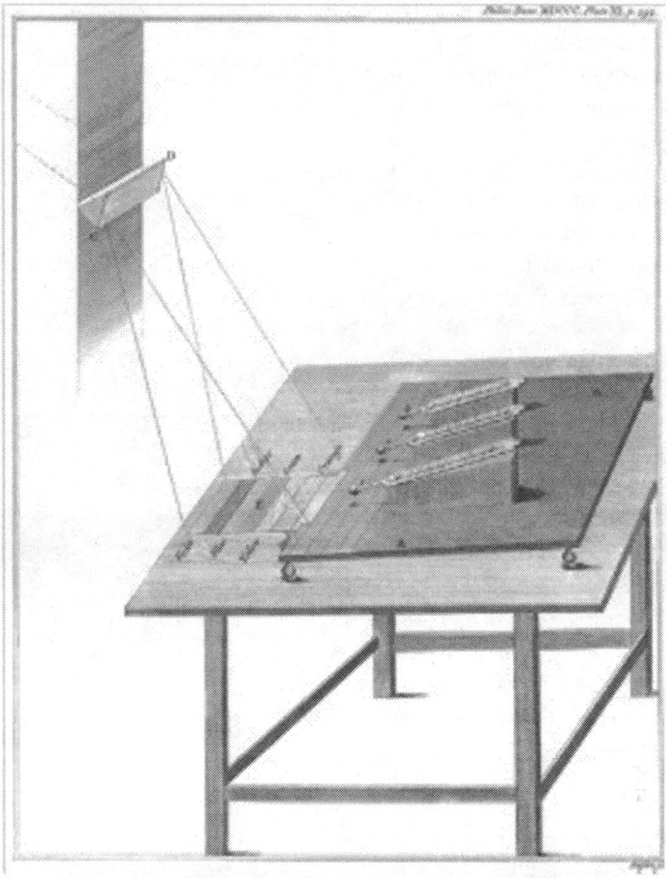

FIG. 1.1 Experimental setup for the discovery of IR radiation in 1800. A prism dispersed sunlight; the spectrum fell on a table and a movable stand with mounted thermometers. Thermometers 1 and 2 were exposed to the radiation, whereas thermometer 3 served as a control. *(Reproduced from N. Sheppard, The historical development of experimental techniques in vibrational spectroscopy, in: J.M. Chalmers, P.R. Griffits (Eds.), Handbook of Vibrational Spectroscopy, John Wiley & Sons, Ltd., 2002, pp. 1–32 with permission.)*

At the same time, further advances on IR spectroscopy depended on the possibility to replace the mercury-in-glass thermometer by more sensitive temperature-measurement methods to find out better IR optical materials and sources of heat radiation for laboratory work other than sunlight. These objectives were accomplished by the discovery [10] of the thermoelectric effect by Thomas Johann Seebeck and the discovery of the thermocoupling effect by Jean Claude Athanase Peltier and Thomas Balt-Johann Seebeck in 1822.

Nobili, in 1825, developed the astatic galvanometer; and together with his younger colleague Melloni, in 1833, initially used a thermopile (a series-connected array of thermocouples) and a galvanometer to increase the sensitivity for measuring temperature and IR radiation.

Using thermopile technology, in the 1850s, John Tyndall, at the Royal Institution of Great Britain in London, was the first who measured correctly the relative infrared absorptive powers of a wide variety of liquids and gases [11]. He was a pioneer in attributing IR absorption bands to vibrational degrees of freedom of the molecules concerned, and demonstrating that visually transparent gaseous elements (e.g., O_2, N_2, and H_2) were IR emitters.

John Tyndall (1820–1893)

In 1881, Sir William de Wiveleslie Abney and Edward Robert Festing employed photographic means to record the first NIR spectra for about 48 organic substances [12] up to 1.3 μm showing that some molecules (e.g., CCl_4 and CS_2) did not absorb in this spectral region. They postulated the correlation of observed absorption bands to different types of bonds involving the light hydrogen atoms (CH, NH, OH, etc.) available in the molecules under study.

Sir William de Wiveleslie Abney (1843–1920)

The association of individual NIR absorption bands with smaller functional groups was also done for complex organic molecules. His work was further developed by Julius, who extended the range of measurements to 10 μm for 20 organic liquids and assigned the maxima at 3.45 and 6–7 μm to the existence of methyl groups ($-CH_3$) in the molecule.

In this direction, William Weber Coblentz studied a very broad range of compounds mostly in IR absorption but also in IR reflection or emission spectroscopy, under the guidance of professor Edward Leamington Nichols at Cornell University. His collected data were later published by the Smithsonian Institution of Washington, DC [13], and some examples are displayed in Fig. 1.2. Coblentz listed 15 group-characteristic bands for aromatic rings, most polar groupings (NO_2, CN, $-SCN$, and $-NCS$), and types of XH groups (CH_3, CH_2, NH_2, OH, etc.). Surprisingly, the specificity of the strong C=O bond-stretching absorptions was not recognized, possibly due to their (still systematic) variations in position in aldehydes, ketones, carboxylic acids, esters, etc.

William W. Coblentz (1873–1962)

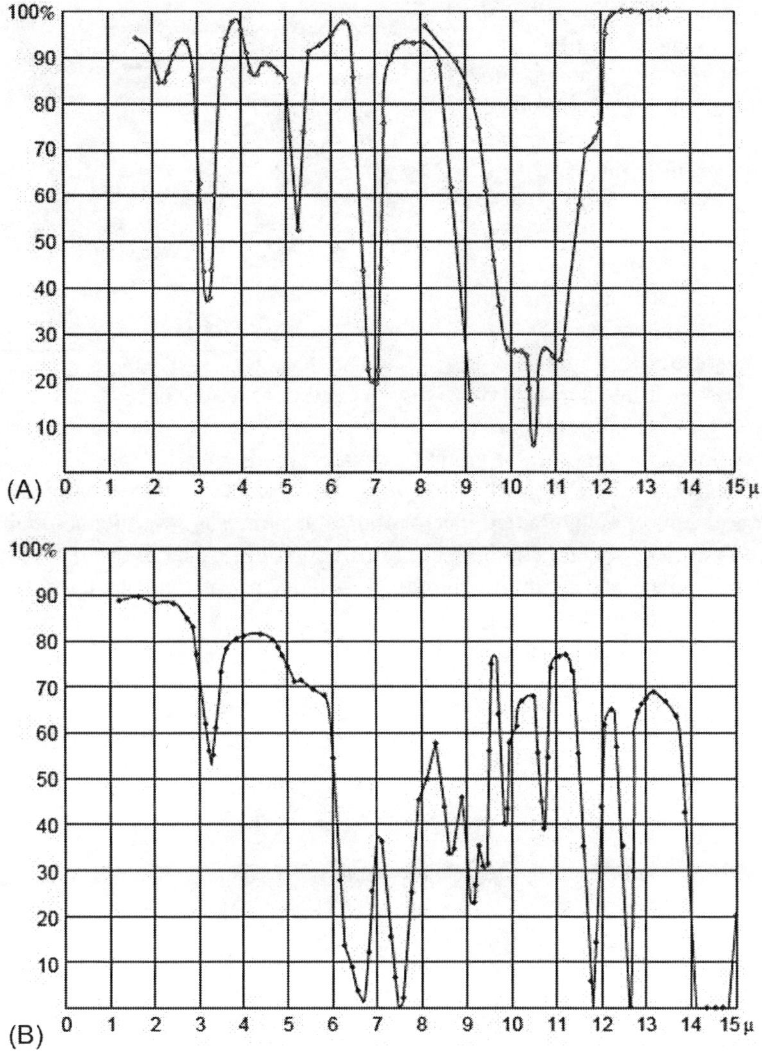

FIG. 1.2 Coblentz's IR spectra of (A) ethylene (ethene) and (B) nitrobenzene. *(Reproduced N. Sheppard, The historical development of experimental techniques in vibrational spectroscopy, in: J.M. Chalmers, P.R. Griffits (Eds.), Handbook of Vibrational Spectroscopy, John Wiley & Sons, Ltd., 2002, pp. 1–32 with permission.)*

Raman scattering or the Raman effect is the inelastic scattering of photons of light upon the interaction with matter. The effect was discovered by Sir Chandrasekhara Venkata Raman and his student Kariamanickam Srinivas Krishnan [14], in Calcutta in 1928 while trying to use a green filter to intercept the scattered light emerged at right angles to the original beam (i.e., the violet portion of the sunlight) after it had passed through a liquid. In fact, this effect had been predicted by Smekal [15] in 1925. Some of the first spectra, obtained by Raman and Krishnan [14], are presented in Fig. 1.3 (the left photograph shows the incident light from a mercury arc lamp after passing through a blue filter, while the right photograph shows the same spectrum after passing through liquid benzene). The interest in Raman discovery blossomed into some 70 papers by the end of that year because this spectroscopic technique could provide a second method for studying the frequency ranges linked with molecular vibrations and rotations.

Sir C.V. Raman (1888–1970)   Sir K.S. Krishnan (1898–1961)

As early as in 1931, Karl Wilhelm Friedrich Kohlrausch summarized the measurements of many Raman spectra of organic liquids in several monographs [16, 17]. Raman spectra could be used in addition to IR data to enable at least nearly complete and accurate vibrational assignments of fundamental normal modes, i.e., chemical grouping that shows weak or missing bands in the IR region, but often gives strong Raman features.

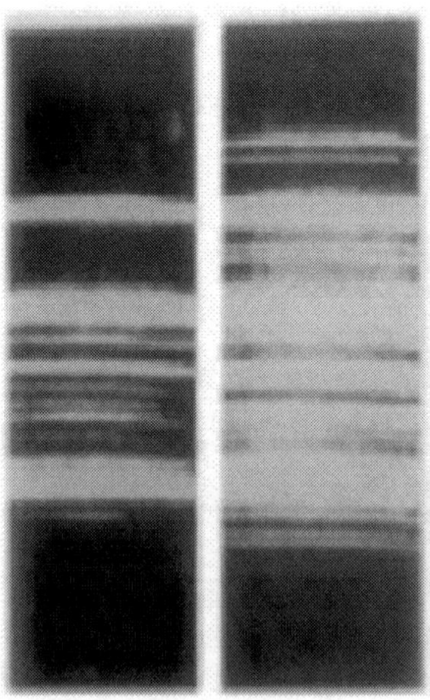

FIG. 1.3 The first Raman spectra obtained by photography. *(Reproduced from The Raman Effect—75 years, Curr. Sci. 84(5) (2003) 627, https://www.jstor.org/stable/24108480 (Accessed 10 January 2020) with permission.)*

Because the experimental setup could be much more easily established in the visible light than in the IR region, more than 1800 papers were published on the Raman effect by 1939.

By the late 1930s, Raman spectroscopy was principally chosen for nondestructive chemical analysis for both organic and inorganic compounds, identification was made by referring to the unique spectrum of Raman scattered light of any particular substance served as a "fingerprint" for its identification, whereas the intensity of the spectral lines was related to the amount of the substance.

In 1942, a remarkable progress happened in Raman measurement as the first photoelectric Raman spectrograph (using a cooled cascade-type RCA IP21 photomultiplier detector) was introduced by Rank and Wiegand for quantitative hydrocarbon analysis. This Raman instrument improved the limited photometric accuracy as compared to the use of photographic plates as detector [18]. Photographic and photoelectric spectra of CCl_4 are presented for comparison in Fig. 1.4.

During the World War II, IR measurements quickly became routine operations due to the availability of sensitive detectors and advances in electronics. At that time, however, the requirement of skilled operators and darkroom facilities

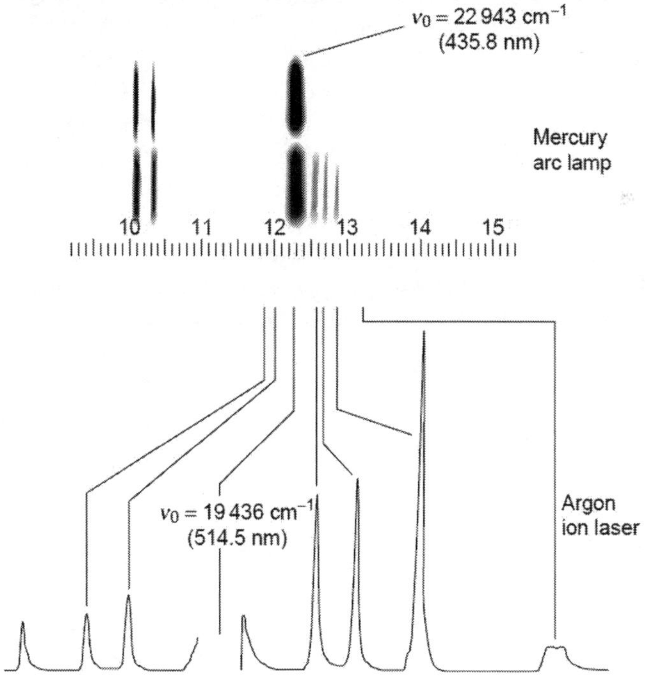

FIG. 1.4 The Raman spectrum of carbon tetrachloride, CCl_4, taken by photographic recording (Hg 435.8 nm excitation) and by photoelectric recording (Ar 514.5 nm excitation). *(Reproduced from N. Sheppard, The historical development of experimental techniques in vibrational spectroscopy, in: J.M. Chalmers, P.R. Griffits (Eds.), Handbook of Vibrational Spectroscopy, John Wiley & Sons, Ltd., 2002, pp. 1–32 with permission.)*

made Raman spectroscopy less competitive than IR spectroscopy. Moreover, Raman scattering was a relatively weak process, so it needed more intense light sources for effect amplification. This problem was solved by Charles Hard Townes, who developed the laser [19] (Light Amplification Stimulated Emission of Radiation), a much more powerful light source to serve as a probe exploring properties to generate dramatically new effects. Basically, a laser is a maser (i.e., Microwave Amplification by Stimulated Emission of Radiation) that works with higher-frequency photons in the UV-Vis spectrum. The first papers about the maser were coauthored by C.H. Townes, J.P. Gordon, and H.J. Zeiger, who created the first ammonia-beam maser at Columbia University to produce amplification of microwaves at a frequency of about 24.0 GHz.

C.W. Townes
(1915–2015)

The late 1980s experienced a resurgence in the use of the original Raman effect by virtue of commercially available Fourier Transform (FT) Raman spectrophotometers. It was also realized that the use of a NIR laser in place of a visible laser as the excitation source could circumvent fluorescence and photodecomposition, but reduce sensitivity in Raman experiments. Fortunately, a successful approach to overcome the latter has engaged with the adoption of interferometry and FT techniques for signal processing.

Jean-Baptiste
Joseph Fourier
(1768–1830)

In practice, Fourier transform infrared (FTIR) is the preferred technique of IR spectroscopy. An FTIR spectrum arises from the mathematical method of Fourier-transformation of an interferogram being yielded by the interference of radiation between two beams. An FTIR instrument is better than a dispersive IR one with reference to shorter analysis time (no energy separation into individual frequency), less reflection loss (no individual frequency limit and fewer mirror surfaces), and spectral comparison with confidence (the laser is available as a source of wavelength calibration within the instrument).

J. von Neumann
(1903–1957)

Vibrational spectroscopy experienced a final impetus and advance through its digital measuring devices, fathered by the computer scientist J. von Neumann (real name Neumann János Lajos) [20].

Table 1.1 briefly summarizes some of the differences between the techniques of vibrational spectroscopy.

IR spectroscopy was probably first exploited in the field of biomedical analysis by Coblentz [21], in 1911 for radiometric investigation of water of crystallization, light filters, and standard absorption bands, Stair and Coblentz [22], in 1936, for measuring IR absorption spectra of plant, animal tissue, and various other substances.

The pioneers of Raman spectroscopy utilization in biomedical analysis were probably Garfinkel and Edsall [23] in 1958. These authors used a high-pressure mercury lamp for scattering excitation and photographic plates for detection to record the first Raman spectrum of a protein, lysozyme.

The first paper that used MIR spectroscopy in order to characterize fats and oils dates back to 1905, when Coblentz published the first compilation of IR spectra of several vegetable oils and fatty acids [13].

In conclusion, a short family tree of vibrational spectroscopy can be presented in Fig. 1.5.

TABLE 1.1 Comparison of Raman, mid-IR, and near-IR spectroscopy.

	Raman	Mid-infrared	Near-infrared
Ease of sample preparation	Very simple	Variable	Simple
Liquids	Very simple	Very simple	Very simple
Powders	Very simple	Simple	Simple
Polymers	Very simple	Simple	Simple
Gases	Simple	Very simple	Simple
Fingerprinting	Excellent	Excellent	Very good
Best vibrations	Symmetric	Asymmetric	Comb/overtone
Group frequencies	Excellent	Excellent	Fair
Aqueous solutions	Very good	Very difficult	Fair
Quantitative analysis	Good	Good	Excellent
Low-frequency modes	Excellent	Difficult	No

Reproduced from P. Larkin, Infrared and Raman Spectroscopy; Principles and Spectral Interpretation, Elsevier Science, Oxford, 2011 with permission.

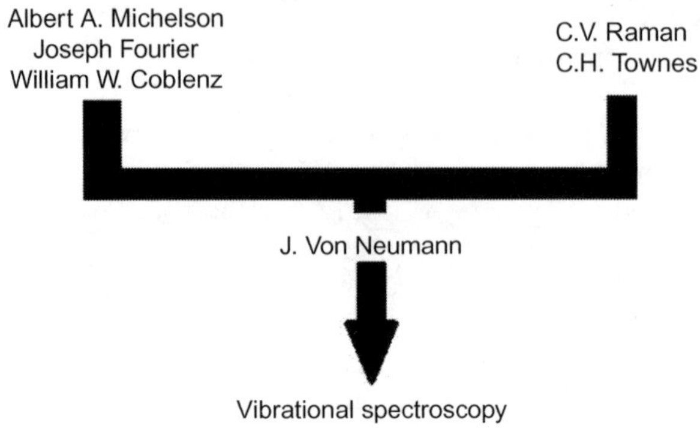

FIG. 1.5 Chronology of major contribution to vibrational spectroscopy. *(Reproduced from J.E. Katon, G.E. Pacey, J.F. O'Keefe, Vibrational molecular microspectroscopy, Anal. Chem. 58(3) 1986, 465A–478A with permission.)*

References

[1] J.M. Chalmers, P.R. Griffiths (Eds.), Handbook of Vibrational Spectroscopy. Volume 1: Theory and Instrumentation, John Wiley & Sons, Ltd., 2002.

[2] J.M. Chalmers, P.R. Griffiths (Eds.), Handbook of Vibrational Spectroscopy. Volume 2: Sampling Techniques for Vibrational Spectroscopy, Wiley& Sons, Ltd., 2002.

[3] J.M. Chalmers, P.R. Griffiths (Eds.), Handbook of Vibrational Spectroscopy. Volume 3: Sample Characterization and Spectral Data Processing, John Wiley & Sons, Ltd., 2002.

[4] W.W. Rouse Ball, A Short Account of the History of Mathematics, Dover, New York, 1908, p.325.

[5] F.W. Herschel, Experiments on the solar, and on the terrestrial rays that occasion heat; with a comparative view of the laws to which light and heat, or rather the rays which occasion them, are subject, in order to determine, whether they are the same, or different. Part I, Philos. Trans. R. Soc. Lond. 90 (1800) 255–283.

[6] F.W. Herschel, Experiments on the solar, and on the terrestrial rays that occasion heat; with a comparative view of the laws to which light and heat, or rather the rays which occasion them, are subject, in order to determine, whether they are the same, or different. Part II, Philos. Trans. R. Soc. Lond. 90 (1800) 284–292.

[7] C. Huygens, Treatise on light containing the explanation of reflection and of refraction and especially of the remarkable refraction which occurs in Iceland Spar, in: J.S. Ames (Ed.), Scientific Memoirs—X. The Wave-Theory of Light, American Book Company, 1900, pp. 1–42.

[8] E.S. Barr, Historical survey of the early development of the infrared spectral region, Am. J. Phys. 28 (1960) 42–54.

[9] E.S. Barr, Men and milestones in optics. II: Thomas Young, Phys. Teach. 5 (1967) 53–60. (reprint of Appl. Opt. 2, 639, (1963)).

[10] T.J. Seebeck, Magnetic polarization of metals and minerals by temperature differences, Treatises R. Acad. Sci. Berl. (1825) 265–373.

[11] J. Tyndall, Heat, a Mode of Motion, sixth ed., Longmans, Green and Co., London, 1880.

[12] W. Abney, R.E. Festing, On the influence of atomic grouping in the molecules of organic bodies on their absorption in the infra-red region of the Spectrum, Philos. Trans. R. Soc. Lond. 172 (1881) 887–918.

[13] W.W. Coblentz, Investigations of Infrared Spectra, Parts I to V, Carnegie Institution, 1905–1908. Publication No. 35, 65 and 97. (Republished by Coblentz Society and the Perkin-Elmer Corp., 1962, Norwalk, CT).

[14] C.V. Raman, K.S. Krishnan, A new type of secondary radiation, Nature 121 (1928) 501–502.

[15] A. Smekal, Zur quantentheorie der dispersion, Naturwissenschaften 11 (43) (1923) S873–S875.

[16] K.W.F. Kohlrausch, Der Smekal-Raman-Effekt, in: Struktur und Eigenschaften der Materie, vols. XII and XIX, Springer Verlag, Berlin, 1931/1938.

[17] K.W.F. Kohlrausch, Ramanspektren, in: Hand- und Jahrbuch der Chemischen Physik, Vol. 9, Part VI, Akad. Verlagsgesellschaft, Leipzig, 1943.

[18] D.H. Rank, R.V. Wiegand, A photoelectric Raman spectrograph for quantitative analysis, J. Opt. Soc. Am. 36 (6) (1946) 325–334.

[19] J. Gordon, H. Zeiger, C.H. Townes, The maser—new type of microwave amplifier, frequency standard, and spectrometer, Phys. Rev. 99 (4) (1955) 1264–1274.

[20] D. Knuth, Von Neumann's first computer program, in: W. Aspray, A. Burks (Eds.), Papers of John von Neumann on Computing and Computer Theory, MIT Press, Cambridge, ISBN: 978-0-262-22030-9, 1987, pp. 89–95.

[21] W.W. Coblentz, Radiometric investigation of water of crystallization, light filters and standard absorption bands, Bull. Natl. Bur. Stand. (U.S.) 7 (1911) 619–663.

[22] R. Stair, W.W. Coblentz, Infrared absorption spectra of plant and animal tissue and of various other substances, J. Res. Natl. Bur. Stand. 15 (1935) 295–316.

[23] D. Garfinkel, J.T. Edsall, Raman spectra of amino acids and related compounds. X. The Raman spectra of certain peptides and of lysozyme[1–3], J. Am. Chem. Soc. 80 (15) (1958) 3818–3823.

Further reading

J.E. Katon, G.E. Pacey, J.F. O'Keefe, Vibrational molecular microspectroscopy, Anal. Chem. 58 (3) (1986) 465A–478A.

P. Larkin, Infrared and Raman Spectroscopy; Principles and Spectral Interpretation, Elsevier Science, Oxford, 2011.

N. Sheppard, The historical development of experimental techniques in vibrational spectroscopy, in: J.M. Chalmers, P.R. Griffits (Eds.), Handbook of Vibrational Spectroscopy, John Wiley & Sons, Ltd., 2002, pp. 1–32.

The Raman Effect—75 years, Curr. Sci. 84 (5) (2003) 627. https://www.jstor.org/stable/24108480. (Accessed 10 January 2020).

Part I

Fundamental aspects of vibrational spectroscopy

Chapter 2

Basic theory, sampling techniques, and instrumentation

Basic theory

Spectroscopy is defined to be a branch of natural sciences concerned with the study of the absorption and emission of light and other electromagnetic radiation by matter as well as the interactions between particles (e.g., electrons, protons, and ions) as a function of their collision energy [1].

In molecular spectroscopy, more specifically speaking, molecules can undergo three types of quantized transitions (electronic, rotational, and vibrational) when being excited by ultraviolet (UV), visible (Vis), and infrared (IR) radiation [2]. For electronic transition, an electron residing in a low-energy orbital is pushed forward to a higher-energy orbital as the energy $h\nu$ of the photon in the UV-Vis region exactly matches the energy gap between the two orbitals. Unlike UV-Vis rays, IR radiation (from 1 to 15 kcal/mol) is not energetic enough to induce electronic transitions, but it can generate transitions in the rotational and vibrational states involved in the ground-state electronic energy of a molecule given its being in resonance with a vibrating bond. In other words, absorbing IR radiation is characteristic of molecular species having a small energy discrepancy between the rotational and vibrational states.

Provided that n is the number of atoms in a molecule and each atom has three degrees of freedom of motion (corresponding to its position in three-dimensional space described by the Cartesian coordinate system), the internal degrees of freedom (that describes the vibrational motion of a molecule, i.e., change in the distance between atoms (stretching) or the angle between bonds (bending)) will be $3n-6$ and $3n-5$ for nonlinear and linear molecules, respectively. The two main modes of vibration (i.e., stretching and bending) may be further given descriptive names as shown in Fig. 2.1 [3]. It is mentioned that some general trends are applicable to vibrational modes as follows: (i) the frequencies of stretching are higher than those of bending because it is much easier to bend than to stretch or compress a bond; (ii) the frequencies of stretching are higher for bonds to hydrogen than those to heavier atoms; (iii) the decreasing order for the frequencies of stretching is: triple bonds > double bonds > single bonds (except for bonds to hydrogen).

Vibrational Spectroscopy Applications in Biomedical, Pharmaceutical and Food Sciences
https://doi.org/10.1016/B978-0-12-818827-9.00003-2

© 2020 Elsevier Inc. All rights reserved.

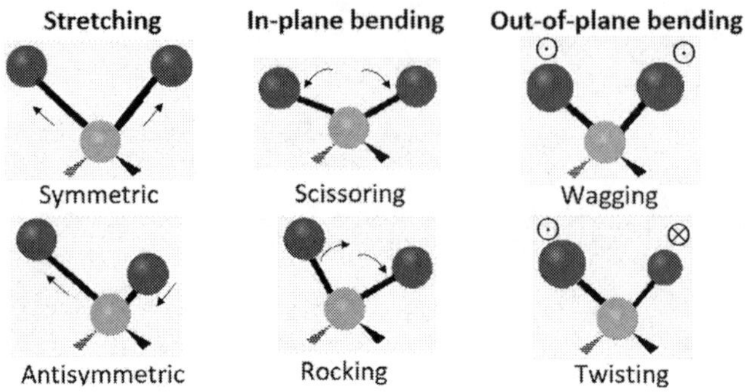

FIG. 2.1 Possible vibrational modes of a molecule.

Conventionally, the IR region covers the electromagnetic spectrum range from 13,000 to 100 cm^{-1}, which can be subdivided into three regions (near-IR, mid-IR, and far-IR) as follows [4]:

- The mid-IR region (4000–400 cm^{-1}) is often generalized as the X-H stretching region [4000–400 cm^{-1}: O–H stretching observed as a broad band in the range 3700–3600 cm^{-1}; N–H stretching (usually much sharper than O–H stretching) seen between 3400 and 3300 cm^{-1}; C–H stretching occurred in the range 3800–2850 cm^{-1} (from aliphatic compounds) or between 3100 and 3000 cm^{-1} (if adjacent to a double bond or aromatic ring)], the triple-bond region [2500–2000 cm^{-1}: C≡C (normally very weak intensity) and C≡N (medium intensity) stretching occurred in the ranges 2300–2050 and 2300–2200 cm^{-1}, respectively], the double-bond region [2000–1500 cm^{-1}: C=N, C=C and C=O (usually the most intense band) stretching], and the fingerprint region [1500–600 cm^{-1}: uniquely found for most bending and skeletal vibrations making it difficult to assign all the absorption bands].
- The near-IR region (13,000–4000 cm^{-1}) includes weak and overlapping absorption bands because they arise from overtones (i.e., a vibrational mode is excited from the ground state to a higher state and the quantum number $v \geq 2$) and combinations (i.e., two molecular vibrations are simultaneously excited) of C–H, N–H, or O–H stretching bands. This region is less useful than mid-IR region for qualitative analysis, but it can be often exploited for quantitative analysis because of important differences existed between different functional groups.
- The far-IR region (400–100 cm^{-1}) is rarely used for structural elucidation, but does provide information on the intramolecular stretching modes involving heavy atoms, skeleton bending modes involving an entire molecule containing heavier atoms, torsional modes (i.e., certain small groups bonded to a large group undergo a motion with regard to the heavier "anchor" group), and crystal lattice vibrations (i.e., the movement of the whole molecular chains with regard to each other in crystalline solids).

When one of the electrons of a molecule is excited to a higher energy level, the molecule almost instantaneously relaxes to the lowest level in the excited electronic state without emitting radiation by collision with other molecules (i.e., internal conversion). It is followed by emitting fluorescence light in the de-excitation process when the excited molecule goes back to one of the vibrational levels of the ground state.

Conversely, if a molecule is shined with the light (being more energetic to excite any vibrational or rotational states, but less energetic to bring it out of the ground state), it is excited to a virtual state (i.e., a very short-lived, unobservable quantum state) and decays back down to lower energy states. In this case, Raman or Rayleigh scattering may occur [5]. For Rayleigh scattering, the scattered photon has its energy preserved because the molecule decays back to the initial state (i.e., elastically scattered radiation); whereas for Raman scattering (i.e., inelastically scattered radiation), the shifted photons can be of either higher (anti-Stokes radiation) or lower (Stokes radiation) energy as compared to Rayleigh radiation, depending upon the vibrational state of the molecule under study (Fig. 2.2). The Stokes line is much more intense than the anti-Stokes line since only molecules vibrationally excited prior to irradiation may give rise to the anti-Stokes line. Hence, in Raman spectroscopy only the more intense Stokes line is normally measured and the Raman effect is relatively weak with an observed intensity of ca. 10^{-6} times that of the incident light for strong Raman scattering. However, the intensity of Raman-active vibrations (associated with

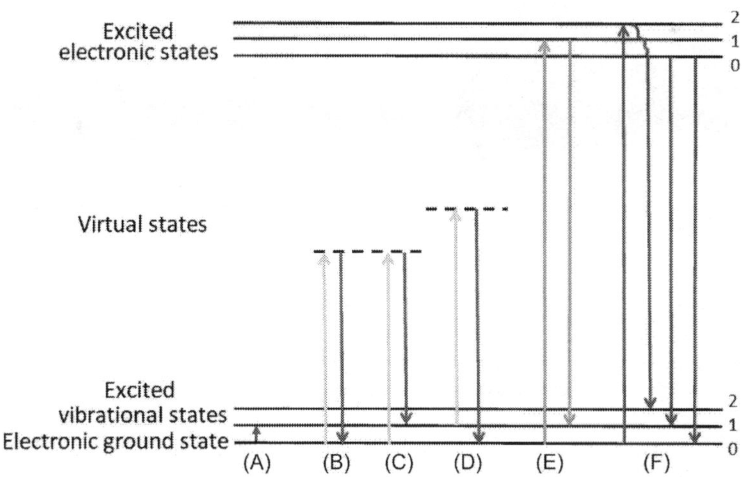

FIG. 2.2 Energy diagram of IR and Raman processes: IR absorption (A), Rayleigh scattering (B), Stokes Raman scattering (C), anti-Stokes Raman scattering (D), resonance Raman scattering (E), and fluorescence (F). The numbers represent different vibrational levels within each electronic state. *(Modified from H. Baranska, An introduction to Raman scattering, in: H. Baranska, A. Labudzinska, J. Terpinski, (Eds.), Laser Raman Spectrometry: Analytical Applications, Ellis Horwood, Chichester, 1987, pp. 9–31.)*

the absorbing chromophore) could be enhanced by a factor of 10^2–10^4 (resonance Raman effect) if the incident laser line in a Raman experiment is tuned near, and finally through, the electronic transition of a molecule [6]. Moreover, the Raman scattering from a molecule (or ion) absorbed on or even within a few Angstroms of the surface of suitably nanostructured metallic substrates can be strongly amplified (i.e., 10^3–10^6 times greater than in solution). This surface-enhanced Raman scattering (aka. SERS) is related to both the electromagnetic and chemical effects as illustrated in Fig. 2.3 [7]. Alternatively stated, SERS may arise from two mechanisms: (i) an enhanced electromagnetic field produced at the metal surface (conduction electrons in the metal surface are excited to a state called a surface plasmon resonance when the wavelength of the incident light

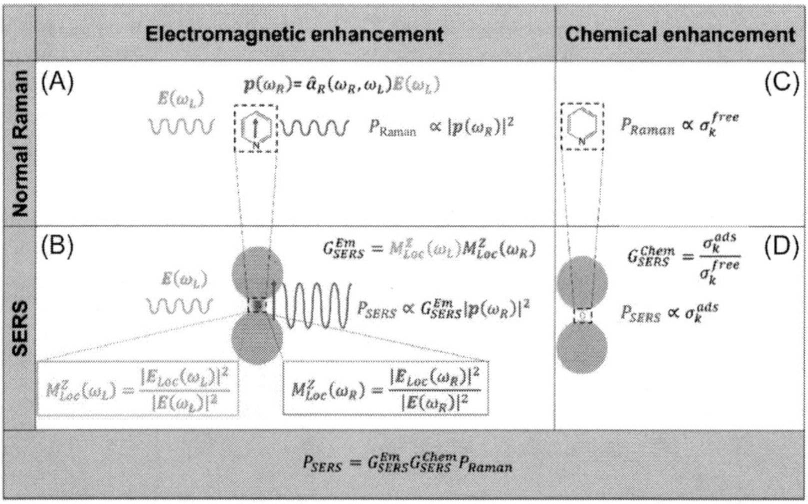

FIG. 2.3 Electromagnetic enhancement. (A) Normal Raman. A laser radiation, with electric field $E(\omega_L)$ oscillating at (angular) frequency ω_L impinges on a molecule, characterized by a Raman polarizability tensor $\hat{\alpha}_R$ (ω_R, ω_L). The laser induces a dipole oscillating at the Raman frequency (vertical *red arrow* (ω_R)); the Raman power radiated by this dipole is proportional to the square modulus of the dipole itself. (B) Surface-enhanced Raman scattering (SERS) electromagnetic enhancement. When the molecule is placed near a plasmonic substrate, the electric field experienced by the molecule is E_{Loc} (ω_L), normally much stronger than the input laser $E(\omega_L)$; this local field enhancement is quantified by M_{Loc}^Z (ω_L). Moreover, the presence of the plasmonic substrate also enhances the efficiency with which the dipole emits Raman radiation; this reradiation enhancement is quantified by M_{Loc}^Z (ω_R). The total electromagnetic enhancement factor, within the $|E|^4$ approximation, is defined as: $G_{SERS}^{Em} = M_{Loc}^Z$ (ω_L) M_{Loc}^Z (ω_R). Chemical enhancement. (C) Normal Raman. The vibrational modes of a molecule in free space are characterized by the cross-section(s) σ_k^{free}; (D) SERS chemical enhancement. The interaction with the plasmonic substrate modifies the structure of the molecule and consequently also the cross-section(s) of its modes (σ_k^{ads}). The chemical enhancement is quantified as $G_{SERS}^{Em} = \frac{\sigma_k^{ads}}{\sigma_k^{free}}$.*(Reproduced with permission from R. Pilot, R. Signorini, C. Durante, L. Orian, M. Bhamidipati, Laura Fabris, A review on surface-enhanced Raman scattering, Biosensors 9 (2019) 57.)*

is close to the plasma wavelength of the metal. It makes molecules absorbed or in close proximity experience an exceptionally large electromagnetic field, most strongly enhancing vibrational modes normal to the surface); (ii) the formation of a charge-transfer complex between the surface and analyte molecule (the resonance enhancement happens as a result of the electronic transition of many charge-transfer complexes in the Vis region. The strongest SERS effect is observed for molecules with lone pair electrons or pi clouds).

Theoretically, the symmetry of a molecule, or the lack of it, will determine what vibrations are IR and Raman active. In general, Raman spectra could be most easily recorded for symmetric or in-phase vibrations and nonpolar groups; on the other hand, IR spectra could be most conveniently ascribed for asymmetric or out-of-phase vibrations and polar groups. It was suggested that the mathematical theory of group could be applied for predicting the number of vibrational bands, their shape and polarization, and the qualitative description of their associated normal modes [8]. It is based on the fact that a molecule may have at least one symmetry element, allowing it to be classified by a point group (i.e., a set of compactible symmetry operations). For small molecules, the IR and Raman activities may be often defined by simply inspecting vibrational forms, i.e., according to the rule of mutual exclusion, no vibration can be active in both the IR and Raman spectra of molecules having a center of symmetry. For instance, vibrations (retaining the center of symmetry) are IR inactive and may be Raman active, and vice versa for vibrations (not retaining the center of symmetry). Conversely, some vibrations can be active in both the IR and Raman spectra (for molecules without any center of symmetry) or in only either one of the IR and Raman spectra (for molecules having other suitable symmetry elements other than a center of symmetry).

For spectral interpretation, IR and Raman frequencies of common functional groups are displayed in Tables 2.1 and 2.2 [4].

Sampling techniques and instrumentation

Basically, IR absorption and Raman scattering differ from each other with respect to the underlying principle by which molecular vibrations occur, i.e., the molecule must be subjected to a change in polarizability in Raman spectroscopy, while there is a change in the net molecular dipole in IR spectroscopy. As a consequence, each technique requires very different instrumentation for spectral registration, i.e., an IR spectrum is obtained by projecting the image of the IR source through a sample onto a detector, by contrast a Raman measurement is performed by imaging the focused laser beam in a sample [9].

In practice, there have been so far two basic types of vibrational instrumentation: (i) dispersive instruments and (ii) Fourier transform instruments. For the former (sometimes called grating or scanning spectrometers, emerged in the 1940s), a diffraction grating is used to sort polychromatic radiation spatially into monochromatic components and direct the dispersed radiation through

TABLE 2.1 IR and Raman frequencies of common functional organic groups.

Functional group	Position	Assignment	Intensity	
			IR	Raman
OH for water	3700–3300		s	w
H bonded OH	3550–3230	str	br	
OH group	3670–3680	str	ms	
R–CH_3	2975–2950	asym str	vs	vs
R–CH_3	2885–2860	sym str	vs	vs
R–CH_3	1470–1440	asym bend	ms	ms
R–CH_3	1380–1370	sym bend	m	vw
R–$CH(CH_3)_2$	1385–1380	Bend-bend	m	vw
R–$CH(CH_3)_2$	1373–1365	Bend-open	m	vw
Aryl–CH_3	2935–2915	Sym str + bend overtone	ms	ms
	2875–2855		m	m
$R(CH_3)_3$	1395–1385	Bend-bend	m	vw
$R(CH_3)_3$	1373–1365	Bend-open	ms	ms
Aliphatic CH_2	2936–2915	asym str	vs	vs
Aliphatic CH_2	2895–2833	sym str	vs	vs
Aliphatic CH_2	2920–2890	Fermi resonance	w	m
Aliphatic CH_2	1475–1445	Bend	ms	ms
$(CH_2)_{>3}$	1305–1295	In-phase twist	–	m
$(CH_2)_{>3}$	736–720	In-phase rock	m	–
R_3CH	1360–1320	CH bend	m	m
C–C≡C–H	3340–3267	CH str	s	w
C–C≡C–H	2140–2100	C≡C str	w	vs
C–C≡C–H	710–578	CH wag	sbr	w
>C=C< trans, tri, tetra	1600–1665	C=C str	w-0	s
>C=CH–R mono, cis, trans	3020–2995	CH str	m	m
C=C mono, cis 1,1	1660–1630	C=C str	m	s
>C=CH_2 mono 1,1	3090–3075	CH_2 asym str	m	m
>C=CH_2 mono 1,1	3000–2980	CH_2 sym str	m	s

TABLE 2.1 IR and Raman frequencies of common functional organic groups—cont'd

Functional group	Position	Assignment	Intensity	
			IR	Raman
>C=CH_2 mono 1,1	1420–1400	CH_2 bend	w	m
R–CH=CH_2	995–985	trans CH_2 in-phase wag	s	w
R–CH=CH_2	910–905	>CH_2 wag	s	w
Aryl CH	3100–3000	CH str	mw	s
Aromatic ring	1620–1585	Quadrant str	var	m
Aromatic ring	1590–1565	Quadrant str	var	m
Aromatic ring	1525–1470	Semicircle str	var	vw
Aromatic ring	1465–1400	Semicircle str	m	vw
Mono, meta, (1,3,5), (2,4,6)	1010–990	In-phase str	vw	vs
Meta, (1,2,4), (1,3,5)	9365–810	Lone H wag	m	–
Para, (1,2,4)	880–795	2 adj. H wag	s	–
Meta, (1,2,3)	825–750	3 adj. H wag	s	–
Ortho, meta	800–725	4 and 5 adj. H wag	s	–
Mono, meta, (1,3,5)	710–665	Ring out-of-plane bend	s	–
Para	650–630	Ring in-plane bend	–	m
Mono	630–605	Ring in-plane bend	w	m
R–CO–H	1740–1720	C=O str	s	m
Conj –CO–H	1710–1685	C=O str	s	w
R–CO–R	1725–1705	C=O str	s	m
Conj –CO–R	1700–1670	C=O str	s	m
H–CO–O–R	1725–1720	C=O str	s	m
R–CO–O–R	1750–1735	C=O str	s	m
R–CO–OH dimer	1720–1680	C=O out-of-phase str	s	–
R–CO–OH dimer	1670–1630	C=O in-phase str	–	m
R–COO^-	1650–1540	C=O out-of-phase str	s	w

Continued

TABLE 2.1 IR and Raman frequencies of common functional organic groups—cont'd

Functional group	Position	Assignment	Intensity	
			IR	Raman
R–COO^-	1450–1360	C=O in-phase str	ms	s
R–CO–O–CO–R	1755–1745	C=O out-of-phase str	mw	m
R–CO–O–CO–R	1825–1815	C=O in-phase str	s	m
R_2CH–OH	1150–1075	C–O str	m	mw
R_2CH–OH	900–800	C–O str	mw	s
R_3C–OH	1210–1180	C–O str	s	mw
R_3C–OH	800–750	C–O str	mw	s
Ar–OH	1260–1180	C–O str	s	w
O=C–O–C	1300–1140	C–O str	s	w
O=C–OH	1300–1200	C–O str	s	w
CH_2–NH_2	3500–3300	NH_2 out-of-phase str	m	vw
CH_2–NH_2	3400–3200	NH_2 in-phase str	m	m
CH_2–NH_2	1630–1590	NH_2 bend	m	vw
CH_2–NH_2	900–600	NH_2 wag	sbr	w
–CH_2–NH–CH_2–	3450–3250	NH str	vw	w
–CH_2–NH–CH_2–	1150–1125	C–N–C out-of-phase str	m	mw
O=C–NH–	About 3300	NH str	s	w
O=C–NH–	Near 3100 (overtone of 1550)	NH str	w	w
C–C≡C–C	2245–2100	C≡C str	–	s
CH_2–C≡N	2260–2240	C≡N str	m	vs
	1440–1405	CH_2 bend	m	m
Conj –C≡N	2235–2185	C≡N str	var	s
Ar–NH_2	1380–1260	C–N str	sbr	m
C–NH_3^+...X^-	3200–2700	NH_3 str	s	vw

TABLE 2.1 IR and Raman frequencies of common functional organic groups—cont'd

Functional group	Position	Assignment	Intensity	
			IR	Raman
$C{-}NH_3^+...X^-$	1625–1560	NH_3 out-of-phase str	mw	vw
$C{-}NH_3^+...X^-$	1550–1505	NH_3 in-phase str	w	vw
$C_2NH_2^+...X^-$	1620–1560	NH_2 bend	mw	w
$C_2NH_2^+...X^-$	3000–2700	NH_2 str	sbr	w
$C_3NH^+...X^-$	2700–2300	NH_2 str	s	w
$CH_2{-}NO_2$	1600–1530	NO_2 out-of-phase str	s	mw
$CH_2{-}NO_2$	1380–1310	NO_2 in-phase str	s	vs
$Ar{-}NO_2$	1555–1485	NO_2 out-of-phase str	s	–
$Ar{-}NO_2$	1357–1318	NO_2 in-phase str	s	vs
$-CH_2{-}Cl$	830–560	C–Cl str	s	s
$-CH_2{-}Br$	700–515	C–Br str	s	vs
$-CH_2{-}F$	1100–1000	C–F str	s	w
Pyridine	3100–3300	Aryl CH str	m	m
Pyridine	1615–1570	Quadrant str	s	m
Pyridine	1400–1440	Semicircle str	s	mw
Pyridine	1035–1025	2,4,6 carbon radial str	m	vs
Pyridine	995–985	Ring breath/str	m	s
Pyridine	660–600	Quadrant in-plane bend	–	m
Pyrrole	3500–3000	NH str	s	m-w
Pyrrole	3135–3103	=CH str	m	s
Pyrrole	1530	Quadrant str + CH rock	s	–
Pyrrole	1468	Quadrant str + CH rock	m	s

Continued

TABLE 2.1 IR and Raman frequencies of common functional organic groups—cont'd

Functional group	Position	Assignment	Intensity	
			IR	Raman
Pyrrole	1418	Semicircle str + CH rock	m	–
Pyrrole	1380	Semicircle str + CH rock	s	m
Pyrrole	1143	Ring in-phase str	m	vs
Furan	3156, 3121, 3092	=CH str	m	s-m
Furan	1590	Quadrant str + CH rk	s	w
Furan	1483		s	vs
Furan	1378	Semicircle str + CH rock	s	s
Furan	1140	Ring in-phase str	–	vs
γ-Lactones	1795–1760	C=O str	s	m
Cyclic anhydride	1870–1845	C=O sym str	m	s
Cyclic anhydride	1755–1745	C=O asym str	s	mw
Epoxy	1270–1245	Ring sym str	m	s
Epoxy	935–880	Ring asym str	s	m
Epoxy	880–830	Ring asym str	s	m

s, strong; *m*, medium; *w*, weak; *v*, very; *br*, broad; *var*, variable; –, zero.
Data referenced from P. Larkin, Infrared and Raman spectroscopy: principles and spectral interpretation, second ed., Elsevier, 2018.

a slit to isolate a frequency range reaching the detector (Fig. 2.4). This type of instrument has limited sensitivity (as most of the light does not fall on the detector) and requires the use of an external source of wavelength calibration (because there is no high-precision laser wavelength to reference). In contrast to the former, the working mechanism of the latter is based on the Michelson Interferometer experimental setup (Figs. 2.5 and 2.6), allowing a simultaneous collection of all the wavelengths. A Michelson interferometer consists of a source, a beam splitter (essentially a half-silvered mirror), a fixed mirror, and a mirror that moves forth and back at a constant velocity (being timed according to the very precise laser wavelength). A collimated beam of the light source

TABLE 2.2 IR and Raman frequencies of common inorganic compounds.

	Infrared spectroscopy	Raman spectroscopy
Functional group	Position and intensity	Position and intensity
NH_4^+	3100 s	3100 w
	1410 s	1410 w
CN–	2100 m	2080 s
CO_3^{2-}	1450 vs	1065 s
	880 m	
	710 w	
HCO_3^-	1650 m	1270 m
	1320 vs	1030 s
NO_3^-	1390 vs	1040 s
	830 m	
	720 w	
NO_2^-	1270 vs	1320 s
	820 w	
SO_4^{2-}	1130 vs	980 s
	620 m	
PO_4^{3-}	1030 vs	940 s
	570 m	
TiO_2	660 vs	–
	540 vs	

s, strong; *m*, medium; *w*, weak; *v*, very; –, zero.
Data referenced from P. Larkin, Infrared and Raman spectroscopy: principles and spectral interpretation, second ed., Elsevier, 2018.

striking the beam splitter will be separated into two beams with equal intensity. One beam is transmitted through the beam splitter to the moving mirror and the other reflected off the beam splitter to the fixed mirror. After being reflected at the two mirrors, the two beams return to the beam splitter where each beam is again half split and then rejoined with half of the light from the other interferometer arm. It makes up two output beams: one sent to the detector and the other lost to the source. It is obvious that the path difference between the two beams is variable (i.e., the optical retardation) as the moving mirror scans a defined distance. Hence, an interference pattern generated at the beam

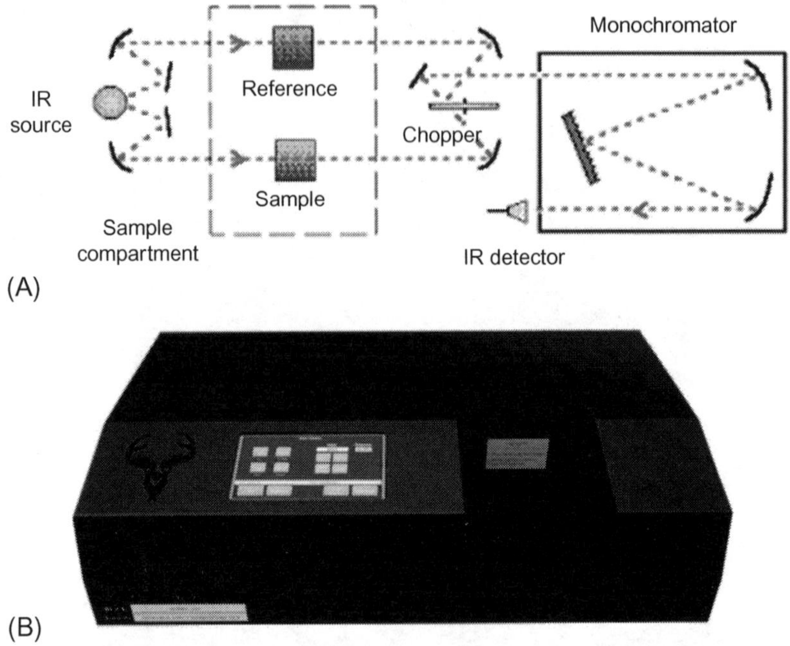

FIG. 2.4 Dispersive IR spectrometer: (A) a typical diagram and (B) a Buck Scientific Model 530 IR Spectrometer as example. *((A) Reproduced with permission from https://www.chemicool.com/definition/fourier_transform_infrared_spectrometer_ftir.html, Accessed 7 January 2020 and (B) Reproduced with permission from https://www.bucksci.com/products/buck-m530-quick-scan-infrared-spectrophotometer, Accessed 7 January 2020.)*

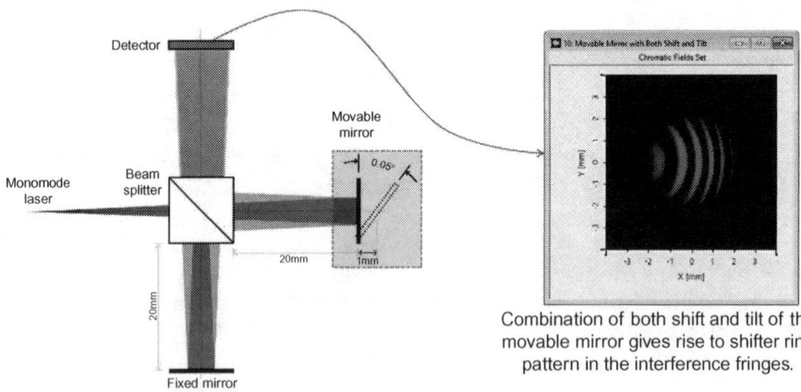

FIG. 2.5 Laser-based Michelson interferometer and interference fringe exploration. *(Reproduced with permission from https://www.lighttrans.com/use-cases/application-use-cases/laser-based-michelson-interferometer-and-interference-fringe-exploration.html, Accessed 25 August 2019.)*

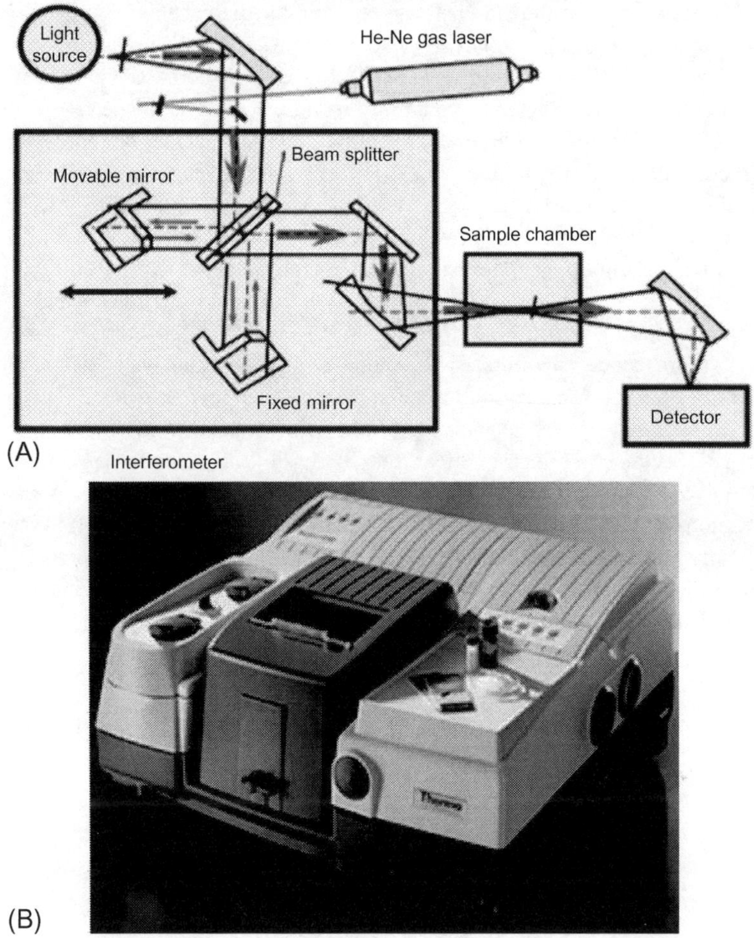

FIG. 2.6 FTIR spectrometer: (A) a schematic diagram and (B) a Nicolet 6700 FTIR spectrometer as example. *((A) Reproduced with permission from https://covalentmetrology.com/ftir/, Accessed 25 August 2019 and (B) Reproduced with permission from https://mmrc.caltech.edu/FTIR/FTIR.html, Accessed 25 August 2019.)*

splitter will be constructive or destructive if optical retardation values are an integral number of wavelengths (0, λ, 2λ, ….) or intervals of $\lambda/2$ ($\lambda/2$, $3\lambda/2$, …), respectively. If a monochromatic source is used, a cosine interferogram is produced for the interference signal that the detector receives. For a polychromatic source, the interferogram is the sum of all cosine waves generated by each wavelength and complete constructive interference (for all wavelengths) only occurs if the two mirrors are equidistant from the beam splitter. Once an interferogram is collected, it could be translated into a spectrum by applying the Fast Fourier Transform algorithm proposed by Cooley and Turkey in the 1960s [10] that converts the measured intensity versus mirror displacement signal into a plot of intensity versus frequency. Nowadays, Fourier transform instruments (in particular, Fourier Transform InfraRed (FTIR) spectrometers) have been widely used in research labs due to their being advantageous over dispersive instruments, i.e., the Multiplex or Felgett advantage (all the wavenumbers of light are observed at once), throughput or Jacquinot's advantage (higher signal-to-noise ratio and resolution) and Connes' advantage (excitation frequency accuracy and precision of better than 0.01 wavenumbers) [11]. It is noted that in an FTIR spectrometer, sampling happens just prior to the detector and its collimating optics; whereas in commercial FT-Raman spectrometers, a 1064 nm Nd:YAG (neodymium-doped yttrium aluminum garnet) laser is mostly used to greatly reduce fluorescence encountered for many compounds and very steep filters (notch or edge pass) are required to attenuate the laser signal from reaching the detector when letting the weak emitted light signal transmit [12].

In IR spectroscopic measurement, an IR transmitting material is quite often needed to aid sampling, e.g., NaCl or KBr windows are used for the majority of applications and ZnSe frequently used for aqueous solutions. In an effort to obtain acceptable quality FTIR spectra, the development of requisite sample-preparation skills is required. Especially, it is of vital importance to take into account Baiulescu's conclusion that "no analysis is better than the sample itself" [13].

G.E. Baiulescu
(1931-2009)

For IR transmission, the peaks of interest should have an absorbance between 0.3 and 0.9. Liquids and solutions could be easily measured by FTIR spectroscopy. Neat (low volatility) liquids can be prepared as a capillary film (uniformly thick without holes or voids) between two plates. Samples that have been dissolved in a volatile solvent can be prepared by casting a thin film (ca. 5-μm thickness, ideally amorphous to eliminate scattering and crystallinity effects) to allow solvent evaporation. Solid-powdered samples (an average particle size of at least 0.5 μm) could be prepared as (i) dispersing fine particles (30–50 mg) in nonvolatile liquid paraffin (Nujol) to form a paste being then sandwiched between two IR transmitting windows (typically KBr or NaCl) or (ii) mixing dried KBr powder and finely ground particles (5 mg) to form a clear disc when hydraulically pressed in a die under high pressure. Although solids are more regularly treated as KBr discs than as mulls because KBr shows no absorption over the entire IR transmission range and requires much less amount of sample, some water introduced from the sample grinding (KBr is very hygroscopic) can complicate spectral interpretation. Thus, Nujol mull sample preparation should be employed to confirm the presence of OH- or NH-type species or to analyze a particularly hygroscopic material. Fig. 2.7 displays unacceptable quality FTIR spectra obtained by either Nujol mulls or KBr discs due to a nonuniform sample film or distribution of sample, e.g., the Nujol mull preparation of starch offers a so-called false spectrum (the most intense band broaden and the weak bands strengthened).

More commonly, FTIR spectral information could be achieved by other sampling techniques such as attenuated total reflection (ATR) [14, 15] and diffuse reflectance infrared Fourier transform spectroscopy (DRIFTS) [16] (Fig. 2.8).

In the case of ATR, a beam of IR light is passed through the ATR crystal, reflecting at least once off the internal surface in contact with the sample; the angle of incidence determines a number of reflections. The radiation at the

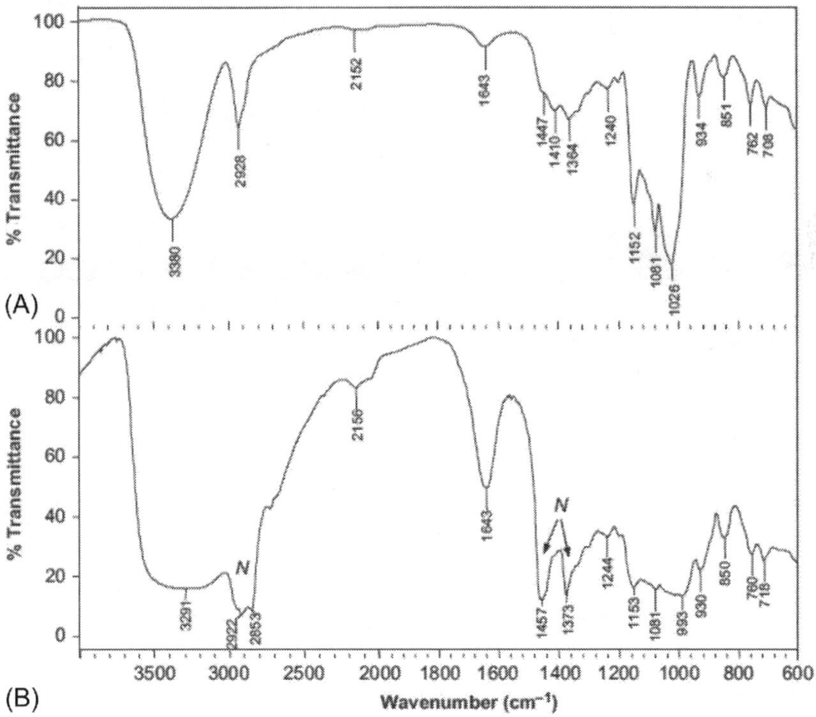

FIG. 2.7 The FT-IR spectra of starch prepared as (A) water cast film on a ZnSe plate and (B) Nujol mull. The N marks the Nujol bands. *(Reproduced P. Larkin, Infrared and Raman spectroscopy: principles and spectral interpretation, second ed., Elsevier, 2018.)*

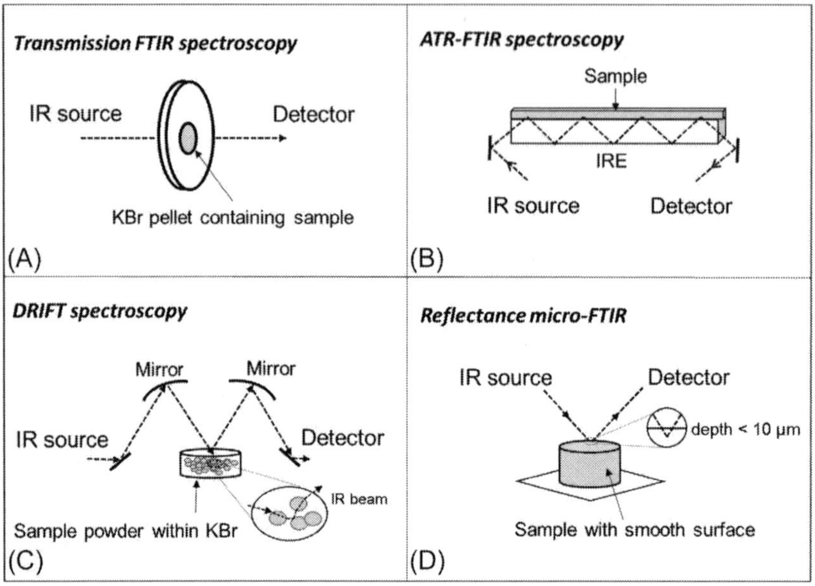

FIG. 2.8 Simplified schematics of common FTIR analysis modes, including: (A) transmission FTIR; (B) attenuated total reflectance (ATR)-FTIR. Note that the penetration depth is dependent on the physical characteristics of internal reflection element (IRE) material and the angle of incidence; (C) diffuse reflectance infrared Fourier transform (DRIFT) spectroscopy; (D) reflectance micro-FTIR. The penetration depth for reflectance micro-FTIR is usually less than 10 μm. *(Reproduced with permission from Y. Chen, C. Zou, M. Mastalerz, S. Hu, C. Gasaway and X. Tao, Applications of Micro-Fourier Transform Infrared Spectroscopy (FTIR) in the geological sciences—a review, Int. J. Mol. Sci. 16(12) (2015) 30223–30250.)*

reflection point probes the sample in the form of an evanescent wave with its penetration depth of ca. 0.5–2 μm and is then detected when existing the crystal. To make this evanescent effect work, the crystal must be of not only excellent IR transmitting properties but also higher refractive index than the sample being studied (i.e., optical materials typically used such as Ge, ZnSe, thallium halides, and diamond). The reflection will be attenuated at a frequency within an absorption band, while all light is reflected at frequencies far away from an absorption band. For most modern IR spectrometers, this contact sampling technique can be performed by mounting the ATR accessory in the spectrometer's sample compartment. It is suitable for characterizing for both liquids (pouring a shallow amount on the surface of the crystal) and solids (firmly clamped) that are too thick or strongly IR absorbing to be analyzed by more traditional transmission methods. On the contrary, DRIFTS is applied for analyzing powders and rough surface solids without prior preparation. The reflected and transmitted amounts of IR light irradiated on the bulk sample (being loosely packed in a cup) are dependent upon shape, compactness, refractive index, reflectivity, and absorption characteristics of the particles under study. In the DRIFTS accessory, collection optics (an ellipsoid or paraboloid mirror) are specifically designed to reject the specularly reflected radiation (directly reflecting off the surface with equal angles of incidence and reflectance) and collect as much as possible the diffuse reflected light (penetrating into the sample and subsequently scattering in all angles). It is indicated that dilution (ca. 5% relative to the powder matrix containing nonabsorbent substances such as KBr or KCl) is applicable to highly absorbent samples, and sample's particle size should be less than 50 μm to properly control the contribution of reflection from the surface. Because spectral distortions are generated by a constantly varying effective path length defined by the penetration depth of the beam into the sample, DRIFTS data do not bear a direct numerical relationship between peak intensity and concentration. To linearize diffuse reflectance data, usually in the Mid-IR region, the Kubelka-Munk function could be applied. In the near-IR region that contains very little of the specular component, nonetheless, no Kubelka-Munk conversion is necessary. Hence, the reflected energy in near-IR spectra is mainly useful for qualitative analysis and mostly given in % Reflectance (R) or log (1/R) units.

Dissimilar to IR spectroscopy, it is possible to record Raman spectra with a minimum of sampling handling and preparation (i.e., Raman spectra can be directly measured on the sample in a container in many times). For SERS measurements, it is necessary to examine how an analyte adsorbs or binds to the surface. Because the SERS enhancement can be significant, a very powerful laser beam may generate a signal overwhelming an instrument detector and/or potentially damage the substrate. Ordinarily, SERS substrates are prepared in the form of metal nanoparticles (20–100 nm in diameter, e.g., silver or gold colloids) suspended in solution or a flat surface with a metal layer deposited on top. A few microliters of the colloidal metal solution are then applied to the sample or mixed with the sample solution (50:50, for example). Once prepared, the sample is ready for Raman experiment or it can be placed on a microscope slide allowing to dry before analysis [16].

Vibrational microspectroscopic techniques have been rapidly emerging as effective tools for characterization of heterogeneous samples in pharmaceutical and biomedical sciences [17]. They involve coupling a microscope to an IR or Raman spectrometer to possibly allow structural visualization and chemical composition mapping (Figs. 2.9 and 2.10). Commercial FTIR microscopes can be suitable to a number of sampling techniques such as specular reflectance, diffuse reflectance, micro-ATR, grazing angle, and conventional transmission measurements. For transmission IR microscopy, the sample should be thin (5–10 μm), smooth and flat for minimal alteration of the optical path as well as large enough (ca. 25 μm) for minimizing diffraction of light. In a different way, reflection IR microscopy almost requires no sample preparation; for specular reflectance data, a Kramers-Kronig transformation is often used to provide more absorbance-like spectra. Being equipped with focal plane array detectors consisting of a matrix of 16×16 up to 128×128 detector elements, it is feasible to generate up to 16,000 pixels/spectra simultaneously enabling

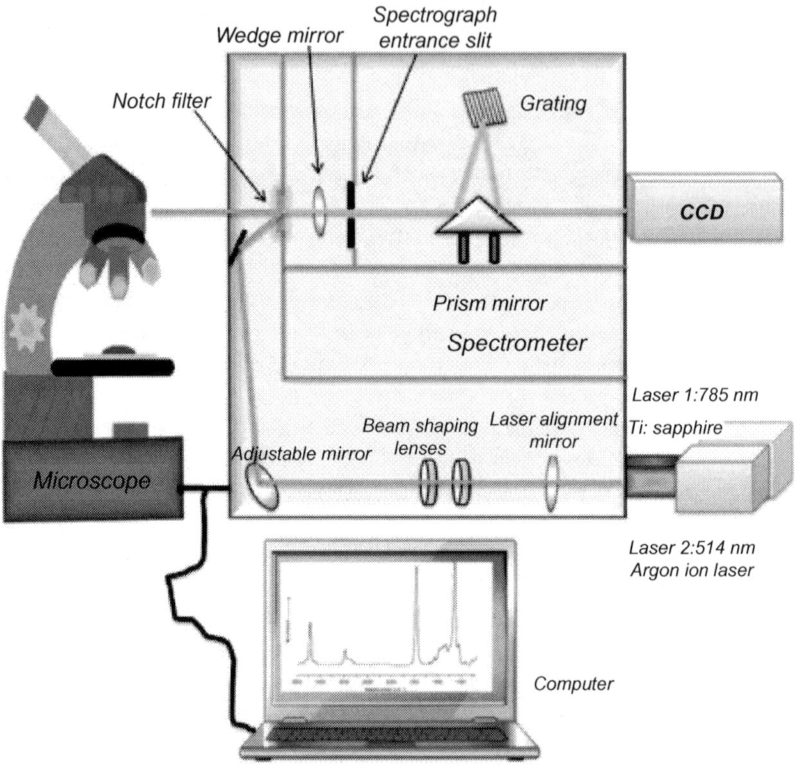

FIG. 2.9 Ray diagram of Raman microspectrometer. *(Reproduced with permission from S. Bhawana, G. Rekha, K. Srividya, Vinay BN, N. Upendra, N. Dipankar, M. Geetashree, S. Vani, S. Kumaravel, U. Siva, Application of vibrational microspectroscopy to biology and medicine, Curr. Sci. 102(2) (2012) 232–244.)*

FTIR microscopy to be a promising imaging technique. For Raman instrumentation, the choice of the laser line (e.g., 514.5-nm line of an argon ion laser or 532-nm line of a less-expensive frequency-doubled Nd:YAG laser) is eventually dependent on a good compromise between limited acquisition time and high spectral resolution. Typical Raman lateral spatial resolution is often quoted as being around 1 μm, while Raman depth resolution is possible in the order of 1–2 μm when incorporating a fully adjustable confocal pinhole aperture (i.e., a true confocal design).

It was proved that the flexible use of portable Raman, FTIR, and near-IR spectrometers could open up a broad range of on-site and in-the-field measurements [18]. For instance, IR measurements could be realized for liquids, powders, or solids with smooth surfaces by using the TruDefender Fourier transform (FT) handheld analyzer (Thermo Fisher Scientific Inc., United States) (Fig. 2.11). This FTIR spectrometer is workable in the spectral range of 4000–650 cm^{-1} (with 4 cm^{-1} spectral resolution) and designed for rapid, field-based analysis

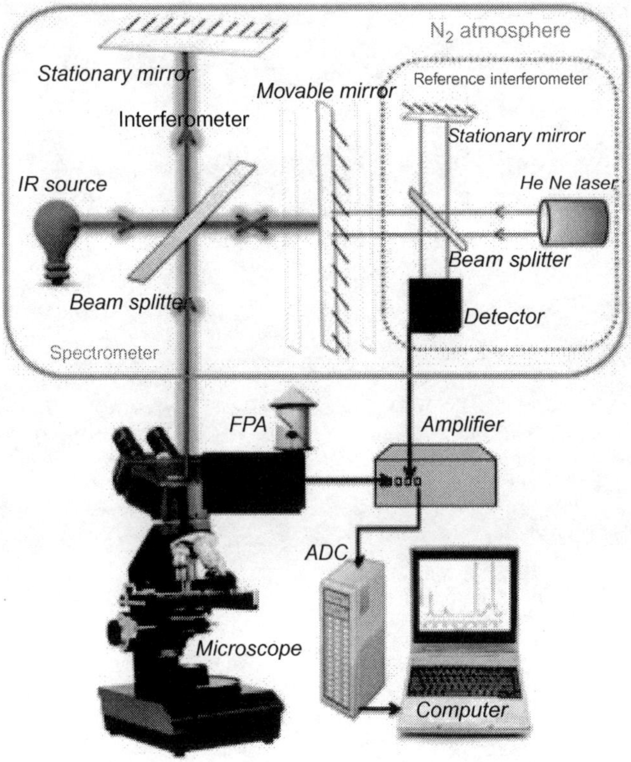

FIG. 2.10 Ray diagram of FTIR microspectrometer. *(Reproduced with permission from S. Bhawana, G. Rekha, K. Srividya, Vinay BN, N. Upendra, N. Dipankar, M. Geetashree, S. Vani, S. Kumaravel, U. Siva, Application of vibrational microspectroscopy to biology and medicine, Curr. Sci. 102(2) (2012) 232–244.)*

for its being small (19.8 × 11.2 × 5.3 cm), light (1.3 kg), rugged with more than 2 h of battery life, and requiring little maintenance. Unknown liquids and solids could be also in situ qualitatively identified by using the portable FirstDefender TruScan Raman spectrometer (Thermo Fisher Scientific Inc., United States) (Fig. 2.12). This Raman spectrometer operates with the 785-nm laser source, a maximum power of 300 mW, and a thermoelectrically cooled charge-coupled device (CCD) detector with 2048 pixels. Samples in plastic bags could be investigated (without interference from the packaging material) when being positioned into by the laser focus of this portable spectrometer.

It is noteworthy to state that the development in miniaturization of IR spectrometers has substantially benefited from advanced micro-technologies such as micro-electro-mechanical systems (MEMS) [19], micro-opto-electro-mechanical systems (MOEMS) [20], micro-mirror arrays, and linear variable filters (LVFs) [21]. It resulted in a sharp reduction in spectrometer's size and weight (100–200 g), while still having a good performance due to a highly precise

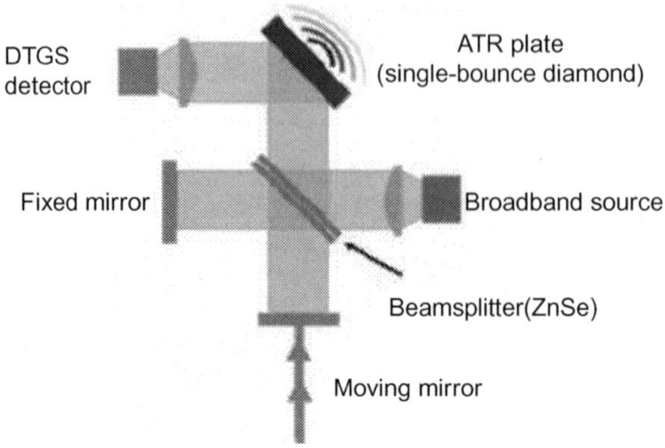

FIG. 2.11 Optical scheme of the TruDefender FT handheld analyzer. *(Reproduced with permission from D. Sorak, L. Herberholz, S. Iwascek, S. Altinpinar, F. Pfeifer, H.W. Siesler, New developments and applications of handheld Raman, mid-infrared, and near-Infrared spectrometers, Appl. Spectrosc. Rev. 47 (2012) 83–115).*

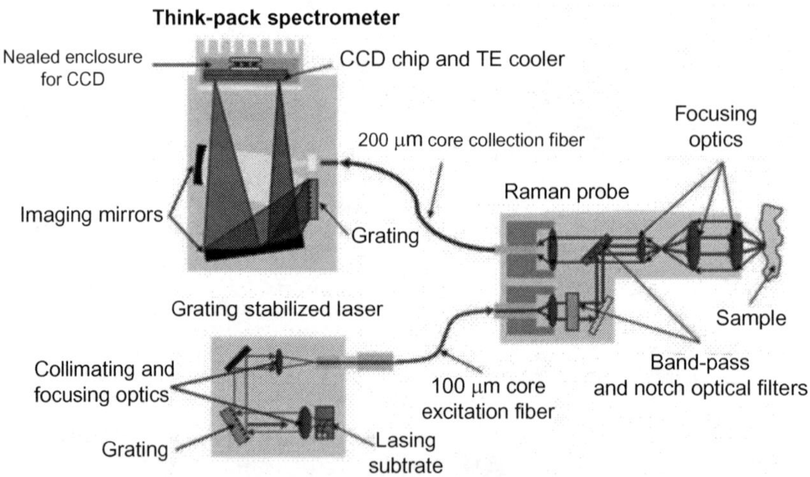

FIG. 2.12 Optical configuration of the portable FirstDefender TruScan Raman spectrometer. *(Reproduced with permission from D. Sorak, L. Herberholz, S. Iwascek, S. Altinpinar, F. Pfeifer, H.W. Siesler, New developments and applications of handheld Raman, mid-infrared, and near-infrared spectrometers, Appl. Spectrosc. Rev. 47 (2012) 83–115.)*

implementation of key elements in the final device. In comparison with Raman and Mid-IR spectrometers, this miniaturization has been much better driven for near-IR instrumentation [22] (e.g., Fig. 2.13). This could considerably reduce the cost for handheld NIR spectrometers and offer a broader dissemination of such instruments for daily applications by a community of nonexpert users.

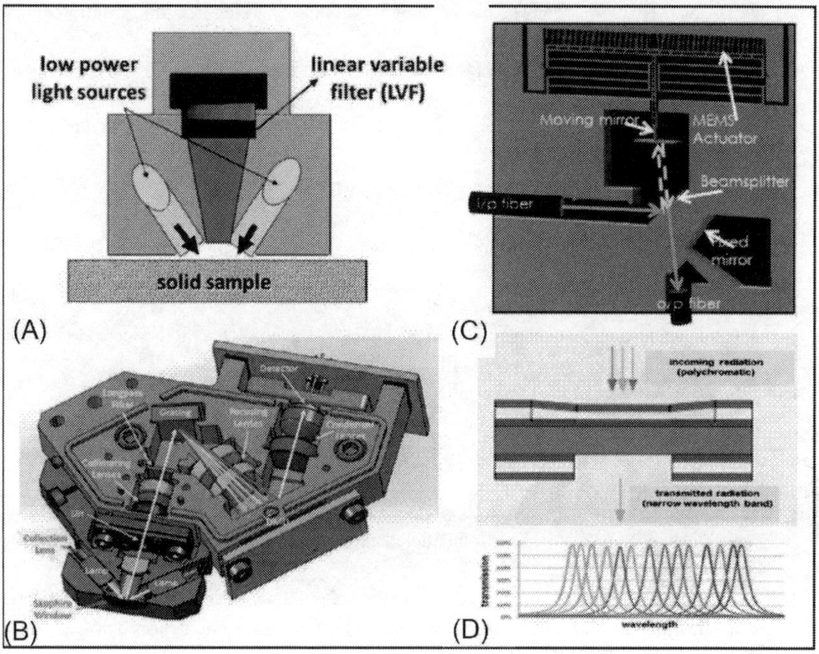

FIG. 2.13 Handheld NIR spectrometers based on different monochromator principles (A) VIAVI MicroNIR 1700, linear variable filter, (B) Texas Instruments DLP NIRscan Nano EVM, digital micromirror device (DMDTM), (C) Si-Ware Systems, MEMS-based Michelson interferometer, (D) Spectral Engines NIR spectrometer with tunable Fabry-Perot interferometer. *(Reproduced with permission from H. Yan, H.W. Siesler, Hand-held near-infrared spectrometers: state-of-the-art instrumentation and practical applications, NIR news, 29(7) (2018) 8–12.)*

References

[1] https://www.britannica.com/science/spectroscopy, (Accessed January 7, 2020).

[2] D.A. Skoog, D.M. West, F.J. Holler, S.R. Crouch, Introduction to spectrochemical methods—Part V: spectrochemical analysis, in: Fundamentals of Analytical Chemistry, ninth ed., Brooks/Cole, 2014, pp. 650–682. (Chapter 24).

[3] P.R. Griffiths, Introduction to vibrational spectroscopy. Introduction to the theory and practice of vibrational spectroscopy, in: J.M. Chalmers, P.R. Griffiths (Eds.), Handbook of Vibrational Spectroscopy, Volume 1: Theory and Instrumentation, Wiley, 2002.

[4] P. Larkin, Infrared and Raman Spectroscopy: Principles and Spectral Interpretation, 2nd ed., Elsevier, 2018.

[5] H. Baranska, An introduction to Raman scattering, in: H. Baranska, A. Labudzinska, J. Terpinski (Eds.), Laser Raman Spectrometry: Analytical Applications, Ellis Horwood, Chichester, 1987, pp. 9–31.

[6] B.B. Johnson, W.L. Peticolas, The resonant Raman effect, Annu. Rev. Phys. Chem. 27 (1976) 465–491.

[7] R. Pilot, R. Signorini, C. Durante, L. Orian, M. Bhamidipati, L. Fabris, A review on surface-enhanced Raman scattering, Biosensors 9 (2019) 57.

[8] E. Silberman, H.W. Morgan, The use of Group Theory in the interpretation of Infrared and Raman spectra, ORNL/TM-5666 1977.
[9] J.M. Chalmers, P.R. Griffiths, Sampling techniques for vibrational spectroscopy, in: Handbook of Vibrational Spectroscopy, vol. 2, Wiley, 2002.
[10] J.W. Cooley, J.W. Tukey, An algorithm for the machine calculation of complex Fourier series, Math. Comput. 19 (1965) 297–301.
[11] https://www.newport.com/n/introduction-to-ftir-spectroscopy, [(Accessed December 31, 2019)].
[12] B. Chase, A new generation of Raman instrumentation, Appl. Spectrosc. 48 (7) (1994) 14A–19A.
[13] G.E. Baiulescu, Moral ageing of analytical methods, Pure Appl. Chem. 52 (1980) 2525–2539.
[14] N.J. Harrick, Surface chemistry from spectral analysis of totally internally reflected radiation, J. Phys. Chem. 64 (9) (1960) 1110–1114.
[15] J. Fahrenfort, Attenuated total reflection: a new principle for the production of useful infra-red reflection spectra of organic compounds, Spectrochim. Acta 17 (1961) 698–709.
[16] https://assets.thermofisher.com/TFS-Assets/CAD/Product-Bulletins/D19663~.pdf, [(Accessed January 7, 2020)].
[17] M. Diem, Vibrational microspectroscopy (MSP), in: Modern Vibrational Spectroscopy and Micro-Spectroscopy: Theory, Wiley, Instrumentation and Biomedical Applications, 2015, pp. 235–250. (Chapter 11).
[18] D. Sorak, L. Herberholz, S. Iwascek, S. Altinpinar, F. Pfeifer, H.W. Siesler, New developments and applications of handheld Raman, mid-infrared, and near-infrared spectrometers, Appl. Spectrosc. Rev. 47 (2012) 83–115.
[19] L.P. Schuler, J.S. Milne, J.M. Dell, L. Faraone, MEMS-based microspectrometer technologies for NIR and MIR wavelengths, J. Phys. D. Appl. Phys. 42 (2009) 133001.
[20] A. Kenda, S. Lüttjohann, T. Sandner, M. Kraft, A. Tortschanoff, A. Simon, A compact and portable IR analyzer: progress of a MOEMS FT-IR system for mid-IR sensing, in: Proc. SPIE 8032, Next-Generation Spectroscopic Technologies IV, 80320O (12 May 2011), 2011, https://doi.org/10.1117/12.883841.
[21] N.A. O'Brien, C.A. Hulse, D.M. Friedrich, F.J. Van Milligen, M.K. von Gunten, F. Pfeifer, H.W. Siesler, Miniature near-infrared (NIR) spectrometer engine for handheld applications. in: Proc. SPIE 8374, Next-Generation Spectroscopic Technologies V, 837404 (17 May 2012), 2012, https://doi.org/10.1117/12.917983.
[22] H. Yan, H.W. Siesler, Hand-held near-infrared spectrometers: state-of-the-art instrumentation and practical applications, NIR news 29 (7) (2018) 8–12.

Part II

Biomedical analysis applications

Chapter 3

Body fluid analysis

Biological fluids, in general, represent the most important source of samples in the diagnosis and prognosis of a disease. This is because a physiological or pathological condition may be well indicated by a change in concentration and/or composition of a specific constituent (i.e., biomarker) in body fluids. At the present time, laboratory testing can be performed on any type of body fluids other than blood such as cerebrospinal fluid, drainage fluid, peritoneal fluid, feces, urine, amniotic fluid, esoteric fluids (e.g., sweat, tear, and saliva), gastric juice, mucus, and others. Surely, this helps doctors develop an appropriate therapeutic treatment as well as monitor the effectiveness of a particular therapy.

Taking into account the demand for noninvasive, rapid, and inexpensive diagnostic methods, vibrational spectroscopy (Raman and IR) was extensively investigated for body fluid analysis in the last two decades [1–7].

So far, a wide range of body fluids have been sampled for this type of investigation, such as serum [8, 9], blood [10, 11], tear [12], urine [13, 14], breast milk [15], and cerebrospinal fluids [16, 17].

In clinical biochemistry, FTIR-ATR was probably the first vibrational spectroscopic technique proposed as an alternative to chemical or enzymatic methods as early as 1980s for a multicomponent analysis of human heparinized plasma (i.e., total protein, total cholesterol, triglycerides, glucose, urea, and uric acid) [18]. The IR quantitative analysis of plasma or blood described here was reagent-free with small sample volumes of about 200 μL. It could somewhat replace routine chemical and biochemical analysis thanks to its reasonable precision and reproducibility with a possible large sample throughput in clinics.

Up to now, there has been a list of some compounds IR spectroscopically quantifiable in different clinical samples (as displayed in Fig. 3.1).

It is noted that although very specific fingerprint bands can be observed for the constituents of body fluids from approximately 3 to 20 μm, this mid-IR spectral range is not currently used in clinics. In contrast, IR measurements are possibly used for evaluation of any change in the composition and/or structure of biological samples (as typically illustrated in Fig. 3.2).

Basically, water (the principal constituent of body fluids) absorbs light in the IR region very well and has distinctive bands (H–O–H deformation at 1640 cm^{-1}, water association band at 2130 cm^{-1}, H–O elongation at 3360 cm^{-1}). It means that the Mid-IR spectrum of an aqueous biological sample is dominated by the intense absorption of water due to fundamental O–H stretching vibrations.

Vibrational Spectroscopy Applications in Biomedical, Pharmaceutical and Food Sciences
https://doi.org/10.1016/B978-0-12-818827-9.00004-4

© 2020 Elsevier Inc. All rights reserved.

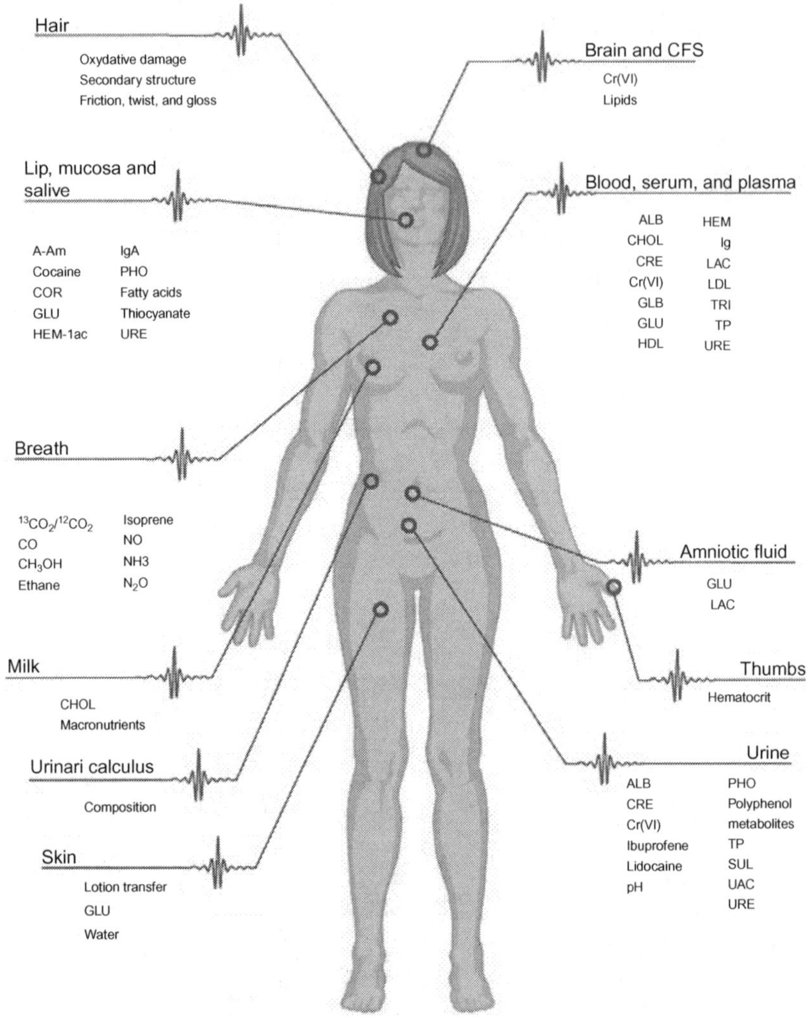

FIG. 3.1 Graphic summary of the parameters determined in clinical samples by IR spectroscopy-based methodologies. *(Reproduced with permission from D. Perez-Guaita, S. Garrigues, M. de la Guardia, Infrared-based quantification of clinical parameters, TrAC Trends Anal. Chem. 62 (2014) 93–105.)*

Fig. 3.3A and B shows the recovery of the absorption pattern for the dissolved species of a typical serum specimen by spectrally subtracting pure water from IR spectra [6]. This subtraction could be done if the absorbance of water at 1645 cm^{-1} was in the range of 1–1.5. For such measurement, an optical path-length of 6–10 mm was required, i.e., a small volume of the sample was sandwiched between removable barium fluoride or calcium fluoride windows being detached by a Teflon© ring spacer. The spectral interference of water could be

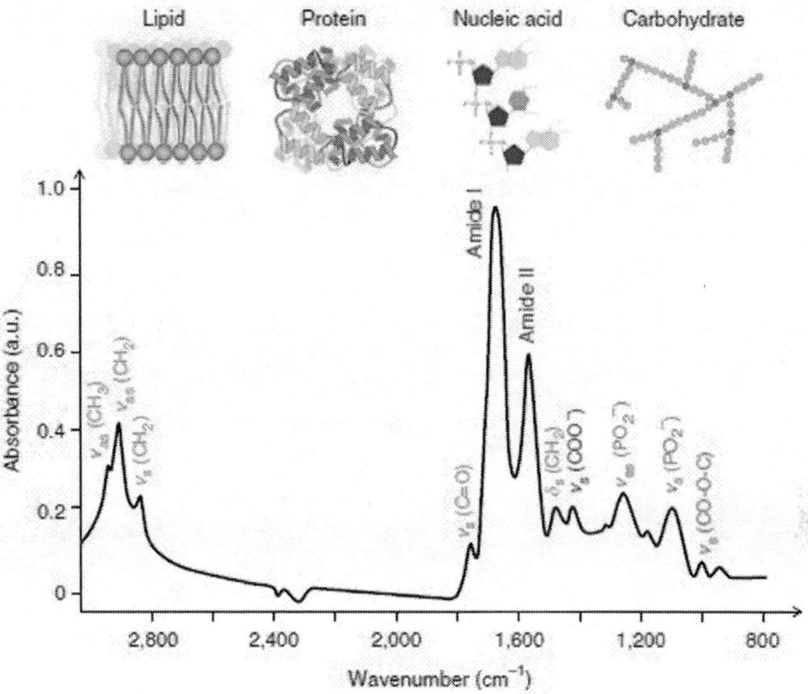

FIG. 3.2 IR spectra for different cellular components. *(Reproduced with permission from M.J. Baker, J. Trevisan, P. Bassan, R. Bhargava, H.J. Butler, K.M. Dorling, P.R. Fielden, S.W. Fogarty, N.J. Fullwood, K.A. Heys, C. Hughes, P. Lasch, P.L. Martin-Hirsch, B. Obinaju, G.D. Sockalingum, J. Sulé-Suso, R.J. Strong, M.J. Walsh, B.R. Wood, P. Gardner, F.L. Martin, Using Fourier transform IR spectroscopy to analyze biological materials, Nat. Protoc. 9(8) (2014) 1771–1791.)*

also removed by spreading about 5–50 μL of sample on an appropriate substrate and acquiring a transmission spectrum of the resultant film (e.g., Fig. 3.4). This approach could also eliminate water-solute interactions to provide an inherently better spectral resolution.

For body fluid samples, matrix compositions are very complex so that vibrational spectra usually consist of overlapping absorption bands of the main biomolecules and interfering substances. For this reason, the application of multivariate analysis is indispensable in order to process very high-dimensional data as schematically presented for FTIR analysis of complex biological systems in Fig. 3.5. There is no ambiguity on the point that this tactic has been favorably exploited in the modern application of vibrational spectroscopy for analyzing various types of body fluids in relation to commonly encountered diseases worldwide as well as forensic investigation (as described later).

Alzheimer's disease (AD) is an age-associated neurodegenerative disorder typified by amnesia and cognitive impairment due to the death of brain cells [19]. AD pathogenesis is widely thought to be driven by accumulated amyloid-β

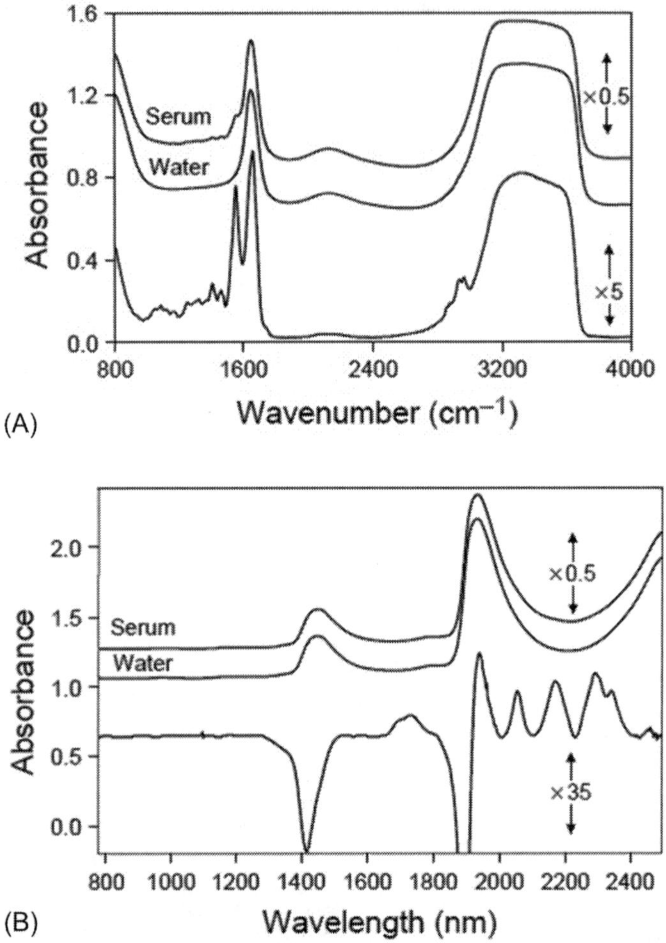

FIG. 3.3 MIR (A) and near-infrared (B) absorption spectra of serum and water, collected with an optical path length of 6 mm and the residual spectra with the spectrum of water subtracted from each *(solid lines). (Reproduced with permission from R.A. Shaw, H.H. Mantsch, Infrared spectroscopy in clinical and diagnostic analysis, in: R.A. Meyers (Ed.), Encyclopedia of Analytical Chemistry, John Wiley & Sons Ltd., Chichester, 2011, ISBN 0-471-97670-9.)*

peptide in extracellular senile plaques [20] and hyperphosphorylated tau protein in intracellular neurofibrillary tangles [21]. It is true that doctors cannot offer a cure for AD and there is no way to stop or slow its progression, but an early AD diagnosis can be beneficial. To diagnose AD, a combination of different tests must be performed (mental status, neuropsychological, brain-imaging as well as laboratory tests to rule out other disorders causing symptoms similar to AD). It is, however, a time-consuming procedure that may be shortened by the utilization of vibrational spectroscopic techniques.

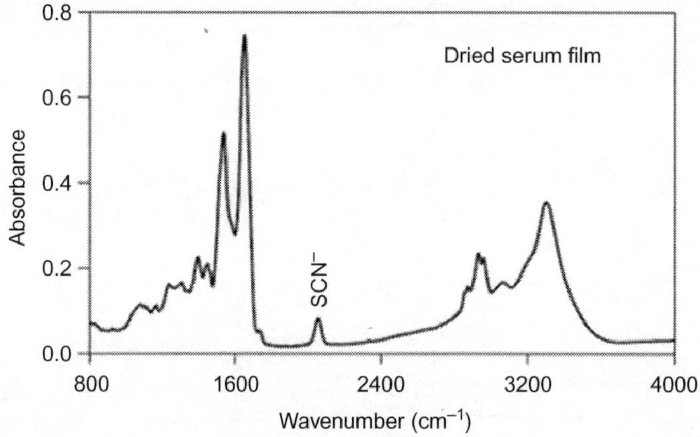

FIG. 3.4 Absorption (transmission) spectrum for a serum film dried onto a barium fluoride window. *(Reproduced with permission from R.A. Shaw, S. Kotowich, M. Leroux, H.H. Mantsch, Multianalyte serum analysis using mid-infrared spectroscopy, Ann. Clin. Biochem. 35 (1998) 624–632.)*

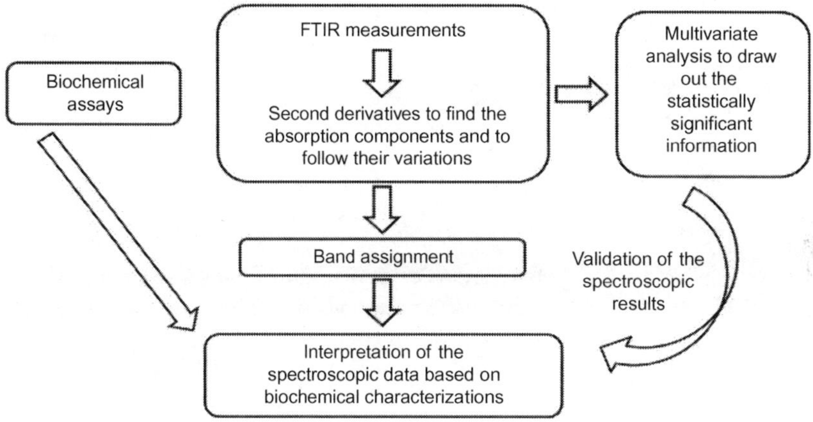

FIG. 3.5 Scheme of the FTIR approach to study complex biological system. *(Reproduced with permission from D. Ami, P. Mereghetti, S.M. Doglia, Multivariate analysis for Fourier transform infrared spectra of complex biological systems and processes, in: L.V. de Freitas, A.P. Barbosa Rodrigues de Freitas (Eds.), Multivariate Analysis in Management, Engineering and the Sciences, IntechOpen, 2013, https://doi.org/10.5772/53850.)*

In a pilot study, FTIR spectroscopy coupled with artificial neural network was confirmed to be a simple and cost-effective tool for distinguishing healthy from AD subjects (88.5% sensitivity and 80% specificity) as compared to ordinary assessments of cerebrospinal fluid (CSF) tau and β-amyloid$_{1-42}$ proteins (99% sensitivity and 86% specificity) [22]. Fig. 3.6 shows typical CSF FTIR spectra of a healthy control and an AD patient, with two spectral fingerprint

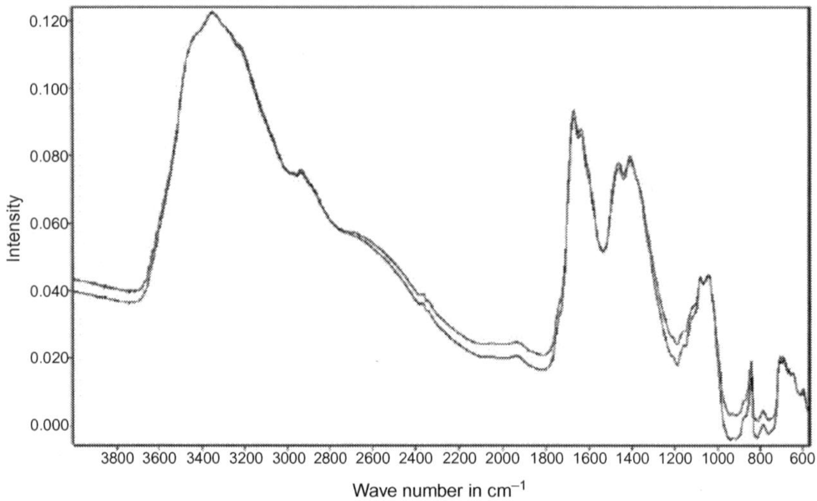

FIG. 3.6 Typical CSF FT-IR spectra of an AD patient and a healthy control. *(Reproduced with permission from M. Griebe, M. Dafferstshofer, M. Stroick, M. Syren, P. Ahmad-Nejad, M. Neumaier, J. Backhaus, M.G. Hennerici, M. Fatar, Infrared spectroscopy: a new diagnostic tool in Alzheimer disease, Neurosci. Lett. 420(1) (2007) 29–33.)*

regions identified (i.e., 3100–2700 cm^{-1} (C–H stretching vibrations associated with the lipids) and 1300–900 cm^{-1} (C–O, C–O–P, and P=O vibrations).

By using 2-dimensional IR spectroscopy, the main spectral features of peripheral mononuclear leukocytes, namely, β-sheets (1640–1625 cm^{-1}) [23] and carbonyl (1750–1700 cm^{-1}) bands, were revealed to have stronger intensity from AD patients as compared to their aged-matched healthy controls [24]. Carmona and coworkers performed receiver operating characteristic curve analysis (ROC), plotting (1 − specificity) versus sensitivity, by exploiting the entire data set of protein β-sheet percentages (estimated by employing a peak ratio second-derivative spectral treatment, Fig. 3.7). This IR strategy was shown to be promising for AD diagnostics (i.e., healthy controls were better differentiated from moderate and mild AD patients (90% sensitivity and 90.5% specificity) than from severe AD patients (82.1% sensitivity and 90.5% specificity for the curve involving three stages of AD). In a successive study, this research group described the changes in IR and Raman spectra in the course of AD with regard to blood plasma composition [25]. In addition to the frequency upshifting of the Raman band located near 744 cm^{-1} (owing to platelets), a stronger intensity of Aβ-peptide and globulin bands was seen with AD patients compared with age-matched healthy controls (Figs. 3.8 and 3.9). Using linear discrimination analysis and ROC curves, AD plasma could be diagnostically discriminated from healthy one with an accuracy of about 94%.

In another development, Kleiren and colleagues described an ATR-FTIR biosensor for specific detection and quantification of the various forms of

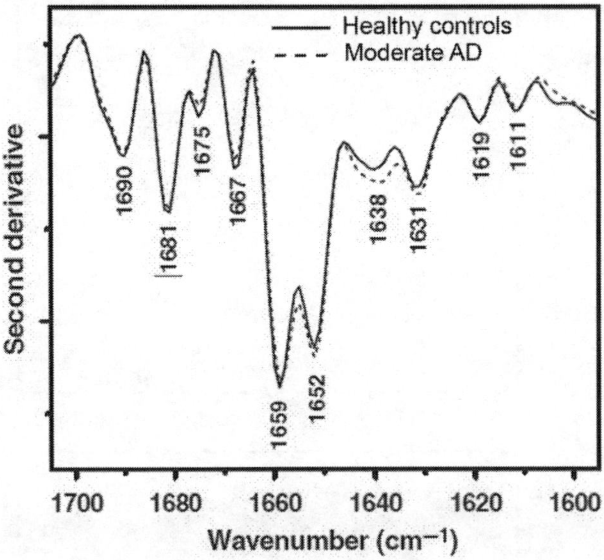

FIG. 3.7 Mean second-derivative spectra in the 1700–1600 cm^{-1} region of mononuclear leukocytes from healthy controls *(solid line)* and patients with moderate AD *(dashed line)*. *(Reproduced with permission from P. Carmona, M. Molina, M. Calero, F. Bermejo-Pareja, P. Martínez-Martín, I. Alvarez, A. Toledano, Infrared spectroscopic analysis of mononuclear leukocytes in peripheral blood from Alzheimer's disease patients, Anal. Bioanal. Chem. 402 (2012) 2015–2021.)*

amyloid β peptide (Aβ) if incubated in deuterated water [26] (as can be seen in Fig. 3.10). The principle of this detection was established on the "BIA-ATR" technique [27], i.e., Aβ was recognized by particular antibodies being already grafted on the surface of a chemically treated ATR element (a germanium crystal). This conformation-sensitive biosensor could offer improved perspectives in early-stage diagnosis of AD, for instance, the differentiation between the toxic fibrillar and oligomeric forms of Aβ(1–42) and Aβ(1–40).

Diabetes mellitus (DM, commonly known as diabetes) is a metabolic disease that affects millions of people worldwide [28]. It is characterized by the impairment of the body to produce enough insulin or to effectively use insulin for maintaining proper levels of sugar (glucose) in the blood. The vast majority of diabetic patients are classified into type I (autoimmune disease, accounts for 10% of people with diabetes) and type II (adult-onset or noninsulin-dependent diabetes). For controlling diabetes, blood glucose testing can be exercised at home by using a blood glucose meter, but this process may be painful (when pricking fingers) and inconvenient (anxiety if no education provided for interpreting and acting on test results).

The practicability of slightly invasive measurement of glucose concentration from aqueous solutions dried down under ambient conditions was determined by using an FTIR spectrometer equipped with a Golden Gate single reflection

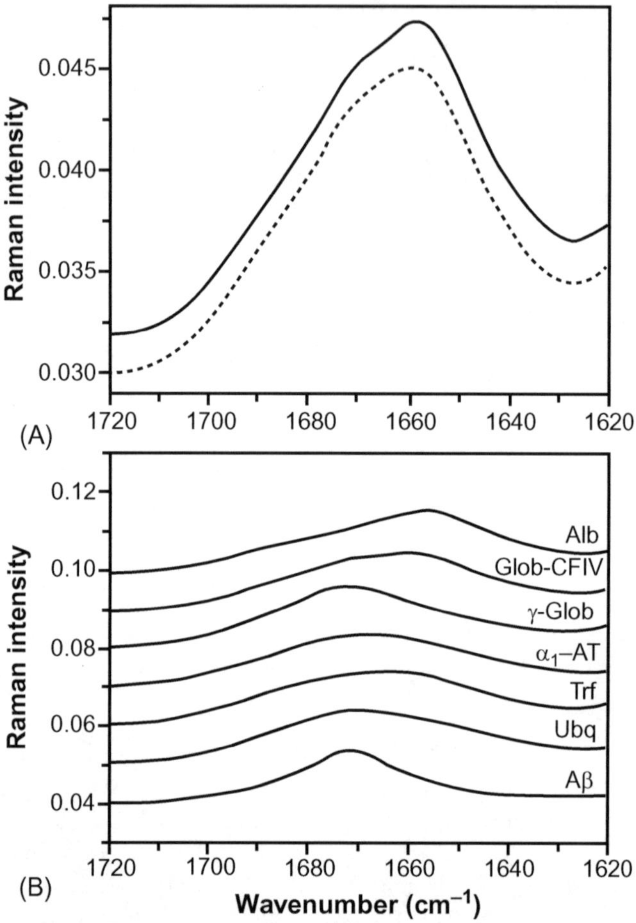

FIG. 3.8 The 1720 to 1620cm^{-1} region of (A) mean Raman spectra of blood plasma from healthy controls *(solid line)* and patients with moderate AD *(dashed line)* and (B) Raman spectra of albumin (Alb), globulins Cohn fraction IV (Glob-CFIV), γ-globulin (γ-Glob), α1-antitrypsin (α1-AT), transferrin (Trf), ubiquitin (Ubq), and $A\beta_{1-40}$ amyloid peptide (Aβ). *(Reproduced with permission from P. Carmona, M. Molina, E. López-Tobar, A. Toledano, Vibrational spectroscopic analysis of peripheral blood plasma of patients with Alzheimer's disease, Anal. Bioanal. Chem. 407 (2015) 7747–7756.)*

diamond ATR accessory and a DLATGS (deuterated L-alanine triglycine sulfate) pyroelectric detector [29]. The rationale for this approach was supported by the fact that protein concentration could be much diluted in biofluid samples (e.g., from microdialysis probes). Data showed that quantification could be achieved in the range 10–600 mg/dL for 7-nL sample volumes (Fig. 3.11); for 10 mg/dL sample, the lowest detectable amount of glucose was 0.7 ng with a signal-to-noise ratio of 10. It was worth investigating this nanoliter-sample ATR

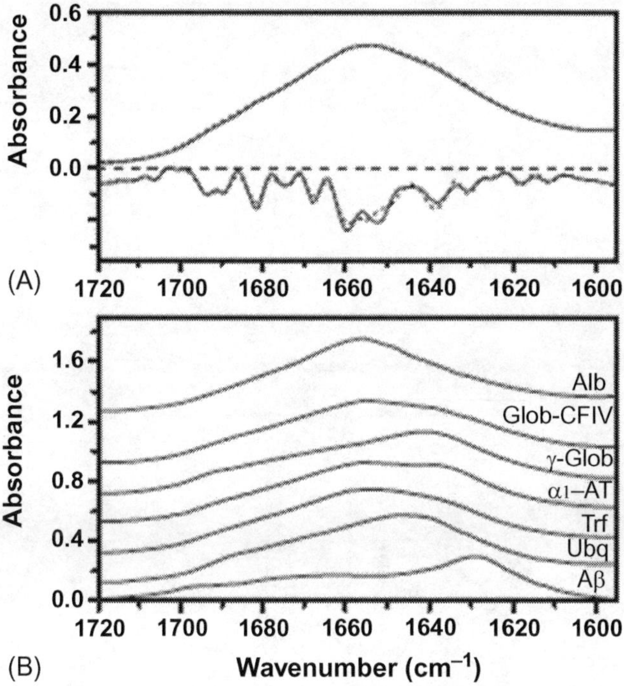

FIG. 3.9 The 1720- to 1595-cm^{-1} region of (A) mean infrared spectra of blood plasma from healthy controls *(solid line)* and patients with moderate AD *(dashed line)* (upper) and their second derivative spectra (lower) and (B) infrared spectra of albumin (Alb), globulins Cohn fraction IV (Glob-CFIV), γ-globulin (γ-Glob), α1-antitrypsin (α1-AT), transferring (Trf), ubiquitin (Ubq), and $A\beta_{1-40}$ amyloid peptide (Aβ). *(Reproduced with permission from P. Carmona, M. Molina, E. López-Tobar, A. Toledano, Vibrational spectroscopic analysis of peripheral blood plasma of patients with Alzheimer's disease, Anal. Bioanal. Chem. 407 (2015) 7747–7756.)*

technique for more complex biofluids such as interstitial fluid or serum samples, supposing that other fluid deposition techniques and spot area of sample be considered thoroughly.

For classification and diagnosis of type II diabetes, the combination of ATR-FTMIR and support vector machine (SVM) as classification algorithm could establish an effective method [30]. The spectra of serum samples from 55 healthy volunteers and 65 patients clinically confirmed were pretreated by multiple scattering correction, Savitzky-Golay smoothing, wavelet transform, and PCA. It was indicated that the accuracy of the diagnosis model was much improved when adopting the three algorithms (genetic algorithm, grid search method, and particle swarm optimization algorithm) to optimize SVM parameters. Even though the spectra of different pathologic groups were similar in shape (as comparatively presented in Fig. 3.12), their absorption intensities

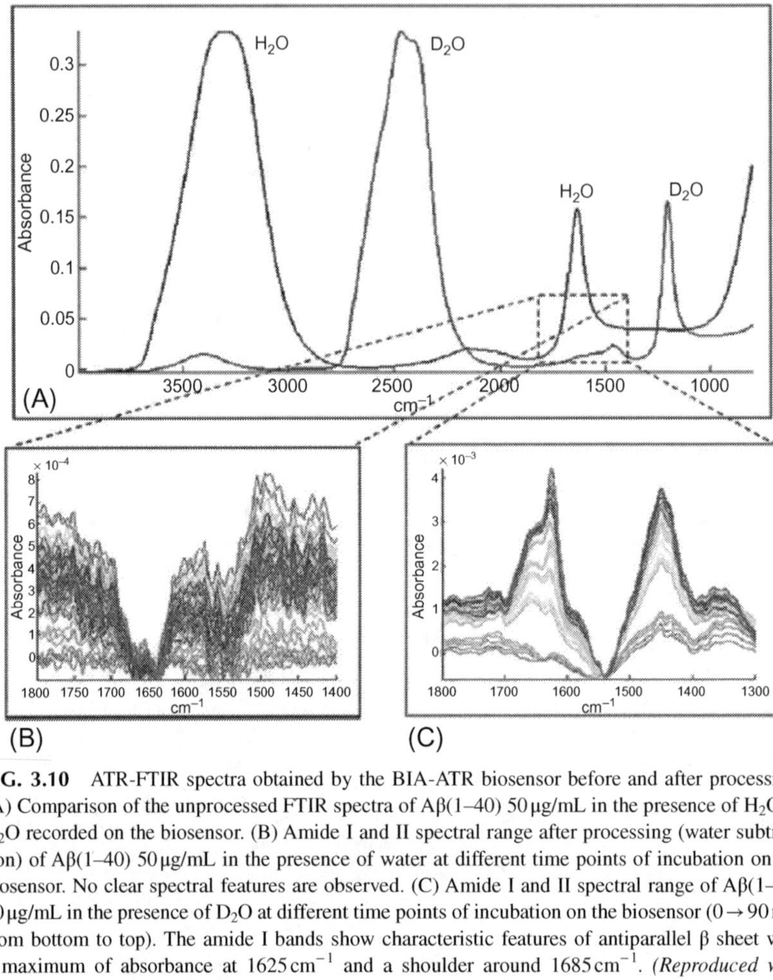

FIG. 3.10 ATR-FTIR spectra obtained by the BIA-ATR biosensor before and after processing. (A) Comparison of the unprocessed FTIR spectra of Aβ(1–40) 50 μg/mL in the presence of H_2O or D_2O recorded on the biosensor. (B) Amide I and II spectral range after processing (water subtraction) of Aβ(1–40) 50 μg/mL in the presence of water at different time points of incubation on the biosensor. No clear spectral features are observed. (C) Amide I and II spectral range of Aβ(1–40) 50 μg/mL in the presence of D_2O at different time points of incubation on the biosensor (0→90 min from bottom to top). The amide I bands show characteristic features of antiparallel β sheet with a maximum of absorbance at 1625 cm^{-1} and a shoulder around 1685 cm^{-1}. *(Reproduced with permission from E. Kleiren, J.M. Ruysschaert, E. Goormaghtigh, V. Raussens, Development of a quantitative and conformation-sensitive ATR-FTIR biosensor for Alzheimer's disease: the effect of deuteration on the detection of the Aβ peptide, Spectroscopy 24 (2010) 61–66.)*

varied due to a change in protein composition and conformation in the long-term environment of high blood sugar levels of the type II diabetes as compared with the normal.

It is suggested that glucose, α-amylase, and ghrelin appetite hormone could be specific salivary biomarkers for potential diagnosis of diabetes [31–33]. In order to monitor diabetes-specific alterations to saliva at the molecular and sub-molecular levels, IR spectroscopy was developed as a point-of-care diagnostic tool for diabetes [34]. In this research, specific IR spectral features embedded in human saliva were correlated with their corresponding ones in serum. Linear discriminant analysis was applied for identifying six spectral regions

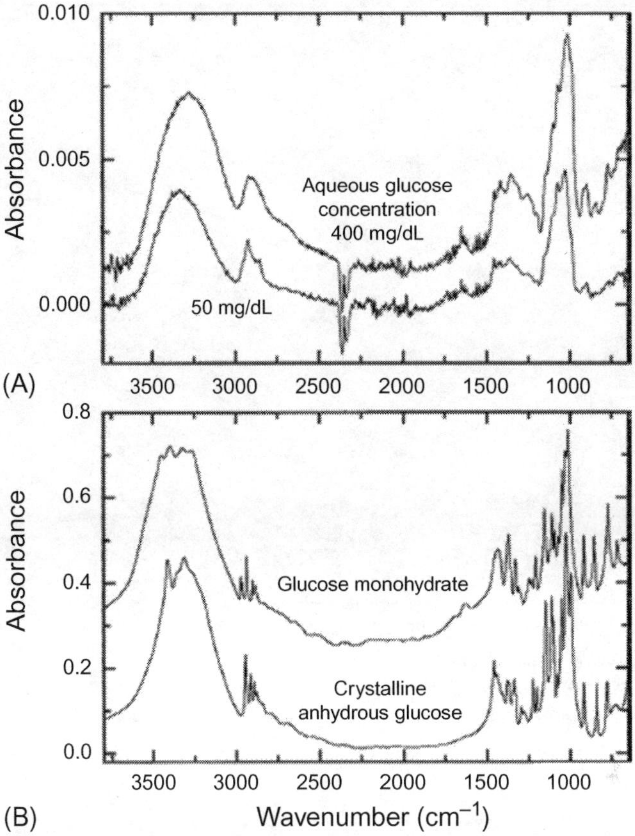

FIG. 3.11 (A) ATR spectra of two dry-film samples from 7 nL of aqueous solutions of different glucose concentrations in the midinfrared spectral range; (B) absorbance spectra of crystalline anhydrous glucose and glucose monohydrate prepared as KBr pellets obtained by transmission measurements. *(Reproduced with permission from E. Diessel, S. Willmann, P. Kamphaus, R. Kurte, U. Damm, H.M. Heise, Glucose quantification in dried-down nanoliter samples using mid-infrared attenuated total reflection spectroscopy, Appl. Spectrosc. 58(4) (2004) 443–450.)*

that best contribute to the demarcation of normal and diabetic groups (as shown in Fig. 3.13).

In another case, FTIR spectroscopy was able to report changes in salivary pattern of normal pregnant women aged 18–35 in each trimester [35]. Qualitative analysis was based on the type of spectral signatures that differentiate progesterone levels over the course of pregnancy (i.e., an increase in the secretion of progesterone was seen in the 1st trimester, which followed by a decrease in the next two consecutive trimesters), whereas quantification obtained by using the intensity ratio among absorption bands. This finding was profitably employed for comparing normal and diabetic pregnant women in each trimester (as illustrated in Fig. 3.14) [36].

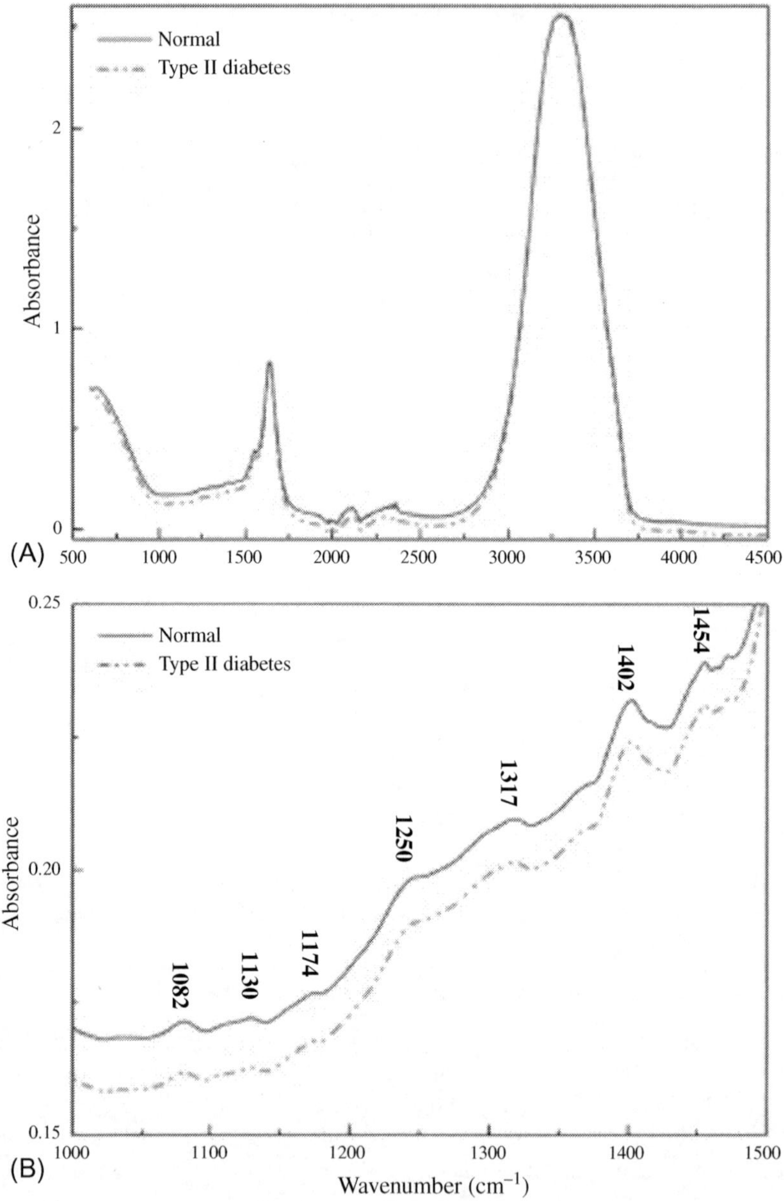

FIG. 3.12 Comparison of the average Fourier transform midinfrared attenuated total reflection spectrum of normal serum (*blue solid curve*, $n=55$) and type II diabetes (*red dotted curve*, $n=62$) serum samples: (A) wavenumber range from 4500 to 600 cm^{-1} and (B) wavenumber ranged from 1500 to 1000 cm^{-1}. *(Reproduced with permission from F. Tao, L. Yuanpeng, L. Fucui, H. Furong, Rapid diagnosis of type II diabetes using Fourier transform mid-infrared attenuated total reflection spectroscopy combined with support vector machine, Anal. Lett. 51(9) (2018) 1400–1416.)*

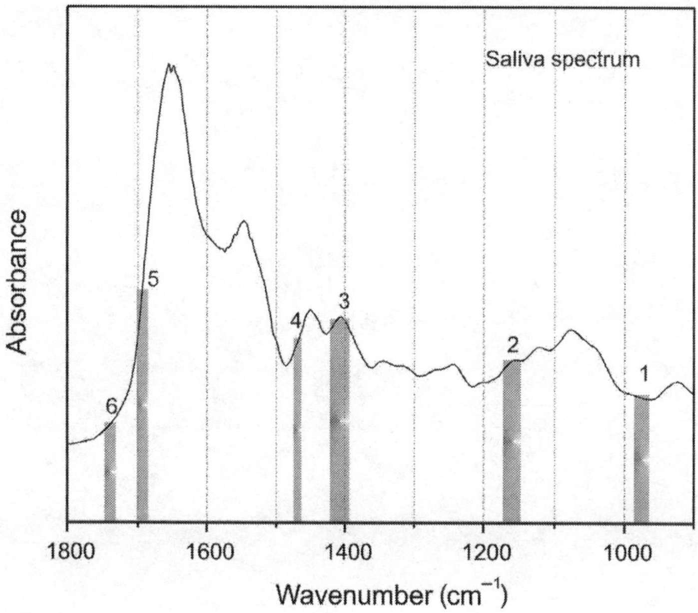

FIG. 3.13 Linear discriminant analysis of the normal and diabetic groups. The bars identify the six spectral regions selected by the optimal regional selection algorithm that best contribute to the differentiation of normal and diabetic groups by linear discriminant analysis. *(Reproduced with permission from D.A. Scott, D.E. Renaud, S. Krishnasamy, P. Meriç, N. Buduneli, Ş. Çetinkalp, K.Z. Liu, Diabetes-related molecular signatures in infrared spectra of human saliva, Diabetol. Metab. Syndr. 2 (2010) 48.)*

It is known that HIV (human immunodeficiency virus) causes not only a progressive, multifactorial breakage of the human immune system (sooner or later resulting to AIDS (acquired immunodeficiency syndrome)) but also metabolic complications (e.g., cardiovascular disease, hyperlipidemia, metabolic syndrome, and osteoporosis) due to the continued use of potent antiretroviral therapy [37].

Current HIV diagnosis is mainly relied on serological assays such as enzyme-linked immunosorbent assay (ELISA) and polymerase chain reaction (PCR). Interestingly, Sakudo and colleagues showed that NIR spectroscopy may yield a fast, reagent-free diagnostic method for HIV-1 infection [38]. For estimation of HIV-1 concentration, chemometric analysis (PLS regression and leave-one-out cross-validation) was applied to NIR spectra in the region 600–1000 nm obtained from plasma samples of healthy donors and preserologically HIV-1-infected individuals. NIR spectroscopic model for HIV-1 produced a good correlation with those acquired by HIV-1 p24 ELISA (the reference method). The alteration in constituents of HIV-1-infected plasma was also revealed by coupling Vis-NIR spectroscopic data with principal component analysis (PCA) and soft independent modeling of class analogy (SIMCA) (e.g., Fig. 3.15) [39].

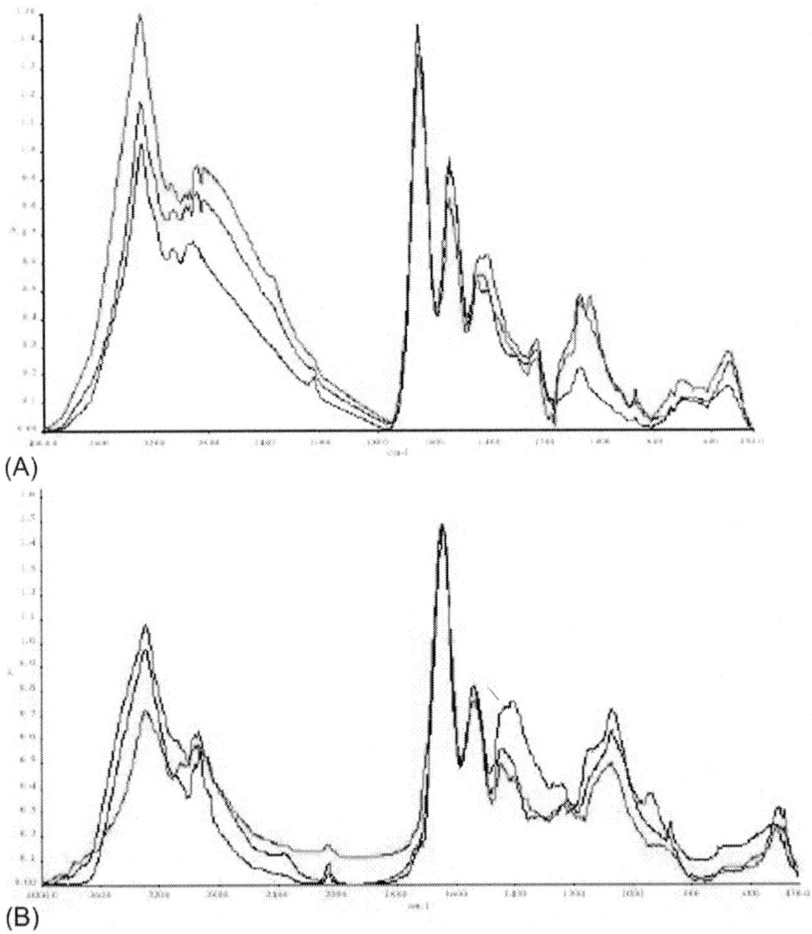

FIG. 3.14 Comparison of (A) normal and (B) diabetic pregnant women in each trimester. *(Reproduced with permission from R. Raziya Sultana, S.N. Zafarullah, N. Hephzibah Kirubamani, Utility of FTIR spectroscopic analysis of saliva of diabetic pregnant women in each trimester, Indian J. Sci. Technol. 4(8) (2011) 967–970.)*

The approach of mid ATR-FTIR spectroscopy-based metabolomics was determined to be reliable, simple, and predictive in distinguishing sera of HIV-infected-treatment-naïve (HIV^{pos} ART^{neg}) and HIV-infected treatment-experienced (HIV^{pos} ART^{pos}) subjects from those of uninfected control subjects [40]. Significant differences were displayed in the spectral regions linked to molecules affected by HIV/ART interference (as indicated by conventional biomedical analysis) when comparing HIV^{pos} ART^{neg} and HIV^{pos} ART^{pos} subjects with uninfected controls (Fig. 3.16). By applying multivariate pattern recognition techniques (PLS-DA and OPLS-DA), results obtained from this study were

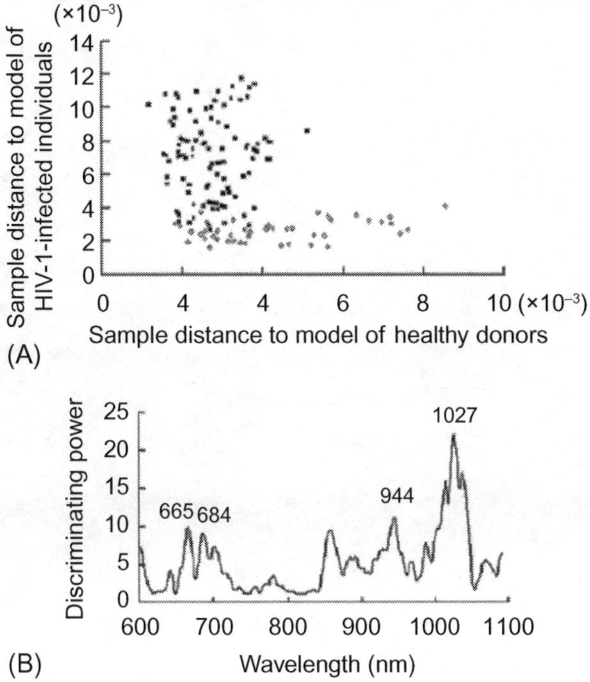

FIG. 3.15 Soft modeling of class analogy (SIMCA) of visible and near-infrared (Vis-NIR) spectra from the plasma of HIV-1-infected and healthy individuals. Three consecutive Vis-NIR spectra of plasma from 35 HIV-1-infected and 15 healthy individuals were processed by mean centering, smoothing, and standard normal variate and subjected to a SIMCA analysis. (A) A Coomans plot of the SIMCA model of HIV-1-infected individuals *(closed squares)* and healthy donors *(open diamonds)* is shown. (B) The discriminating power of the SIMCA model shows several important peaks differentiating the plasma of HIV-1-infected individuals from that of healthy donors. Samples from two individuals were hemolysed and excluded as outliers. *(Reproduced with permission from M.K. Bahmani, A. Khosravi, R. Miri, R. Yukieiwabu, K. Shiikuta, A. Sakudo, A spectroscopic characterization of human immunodeficiency virus type-1-infected plasma by principal component analysis and soft independent modeling of class analogy of visible and near-infrared spectra, Mol. Med. Rep. 2 (2009) 805–809.)*

in line with those reached by using more sensitive metabolomic methodologies (NMR and MS) [41, 42].

Cancer is a condition where cells grow and reproduce uncontrollably in a given part of the body. The cancerous cells can sometimes spread to other areas (i.e., metastasis) causing an invasion and destruction of neighboring healthy tissues and organs. As reported by the World Health Organization, it was the second principal cause of death globally (ca. 9.6 million deaths in 2018, a death ratio of one-sixth due to cancer) [43]. For most types of cancer, biopsy (i.e., examination of a piece of tissue or a sample of cells from the body) is the gold standard diagnosis that doctors recommend when finding something suspicious during a physical exam or other tests [44]. Nonetheless, it is not a perfect

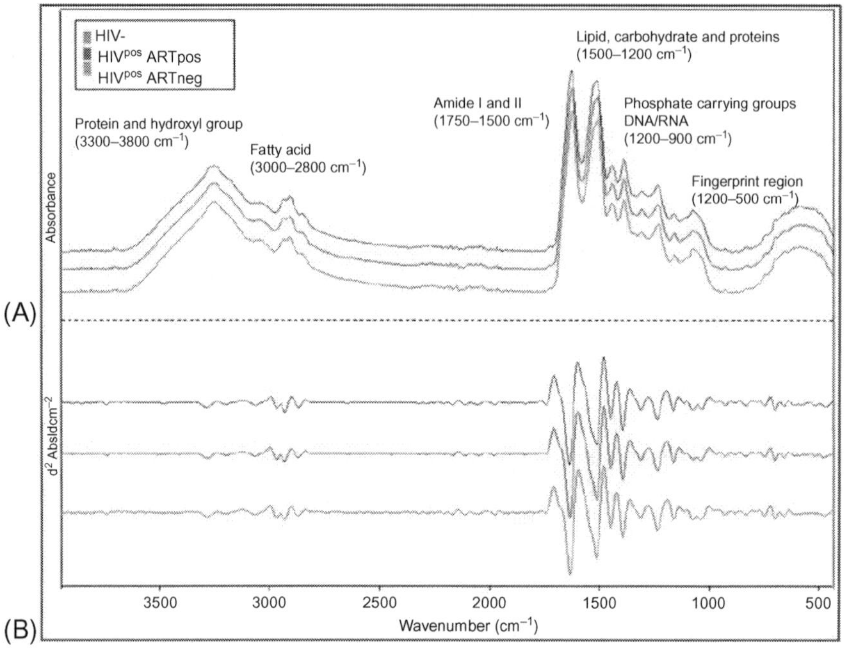

FIG. 3.16 Representative averaged ATR-FTIR spectra of serum obtained from (A) HIV uninfected controls *(blue)*, HIV^{pos} ART^{pos} *(red)*, and HIV^{pos} ART^{neg} *(pink)*. (B) Second-derivative spectra of HIV-negative controls, HIV^{pos} ART^{pos} and HIV^{pos} ART^{neg}. *(Reproduced with permission from L. Sitole, F. Steffens, T.P.J. Krüger, D. Meyer, Mid-ATR-FTIR spectroscopic profiling of HIV/AIDS sera for novel systems diagnostics in global health, OMICS 18(8) (2014) 514–523.)*

procedure because its interpretation may be subjected to intra- and interobserver variation [45] and sampling error [46]. It is noteworthy that there may be some molecules secreted by a tumor into the body fluids or a specific response of the body to cancerous existence (cancer biomarkers). It means that as a screening test, the utilization of Raman and IR spectroscopic techniques may be considerably helpful for suggesting the presence of cancer.

It was demonstrated in a pilot study that the coupling of ATR-FTIR spectroscopy with a proposed classification machine was able to obtain accurate class prediction (normal vs. cancer) by interrogating peripheral blood samples from endometrial cancer, ovarian cancer, and controls (healthy as well as those with benign gynecological conditions) [47]. Classification results (based on the accommodation of an arbitrary number of feature extraction and classification algorithms) of four datasets from endometrial/ovarian × serum/plasma were up to 96.7% and 81.7% for ovarian cancer and endometrial cancer (e.g., Fig. 3.17), respectively.

In other study, ATR-FTIR spectroscopy also manifested its high levels of accuracy (≥95%) in analyzing urine samples for detection of endometrial and

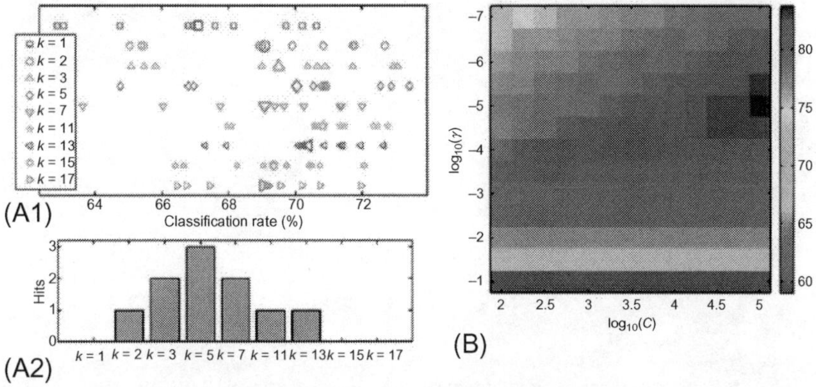

FIG. 3.17 Example results from classifier tuning. (A1) Finding the optimal k (number of nearest neighbors) for the k-NN classifier (ovarian plasma data). Each iteration of a 10-fold crossvalidation finds one classification rate for each k. The average for each k is represented by a bigger marker. (A2) Histogram showing the number of times that each k was selected. Note that although $k=11$ has the highest average in (A1), the most frequent choice was $k=5$. (B) Finding the optimal (C, g) for the SVM classifier (ovarian plasma data). This image map shows only the average classification rate of a 10-fold crossvalidation for each (C, g) pair. As with k-NN, the chosen parameters for each crossvalidation iteration may differ. *(Reproduced with permission from K. Gajjar, J. Trevisan, G. Owens, P.J. Keating, N.J. Wood, H.F. Strinfellow, P.L. Martin-Hirsch, F.L. Martin, Fourier-transform infrared spectroscopy coupled with a classification machine for the analysis of blood plasma or serum: a novel diagnostic approach for ovarian cancer, Analyst 138 (2013) 3917–3926.)*

ovarian cancers [48]. Fig. 3.18 displays the changes in IR absorbance for six spectral peaks responsible for differentiation between healthy and ovarian cancer patients.

For detection of breast cancer (the most common female cancer), IR measurement of only dried 1-μL serum samples was recommended by Backhaus et al. [49] (Figs. 3.19 and 3.20). The classification of 196 patients, in this investigation, was done by cluster analysis (sensitivity: 98% and specificity: 95%) and artificial neural networks (sensitivity: 92% and specificity: 100%). On testing the potential for interference from other diseases, the assignment of breast cancer patients to the correct class achieved only 79%. This observation was explained by a very low number of patients in this group (86 out of 3119), but it was confirmed that carcinoma in situ with a very low extension (2 mm) could be positively diagnosed.

In a context other, silver nanoparticles (Ag NP)-based surface-enhanced Raman spectroscopy (SERS) was explored for noninvasive detection of nasopharyngeal cancer by biochemical analysis of blood plasma samples [50]. Specific biomolecular differences were found to be a reduction in the percentage of saccharide and amino acid contents and an expansion in the relative amounts of nucleic acid, collagen, phenylalanine, and phospholipids in the blood plasma of patients pathologically affirmed nasopharyngeal carcinomas (WHO types I, II, and III) as compared to healthy subjects (Fig. 3.21). Linear discriminant analysis—principal component analysis—based spectral classification could

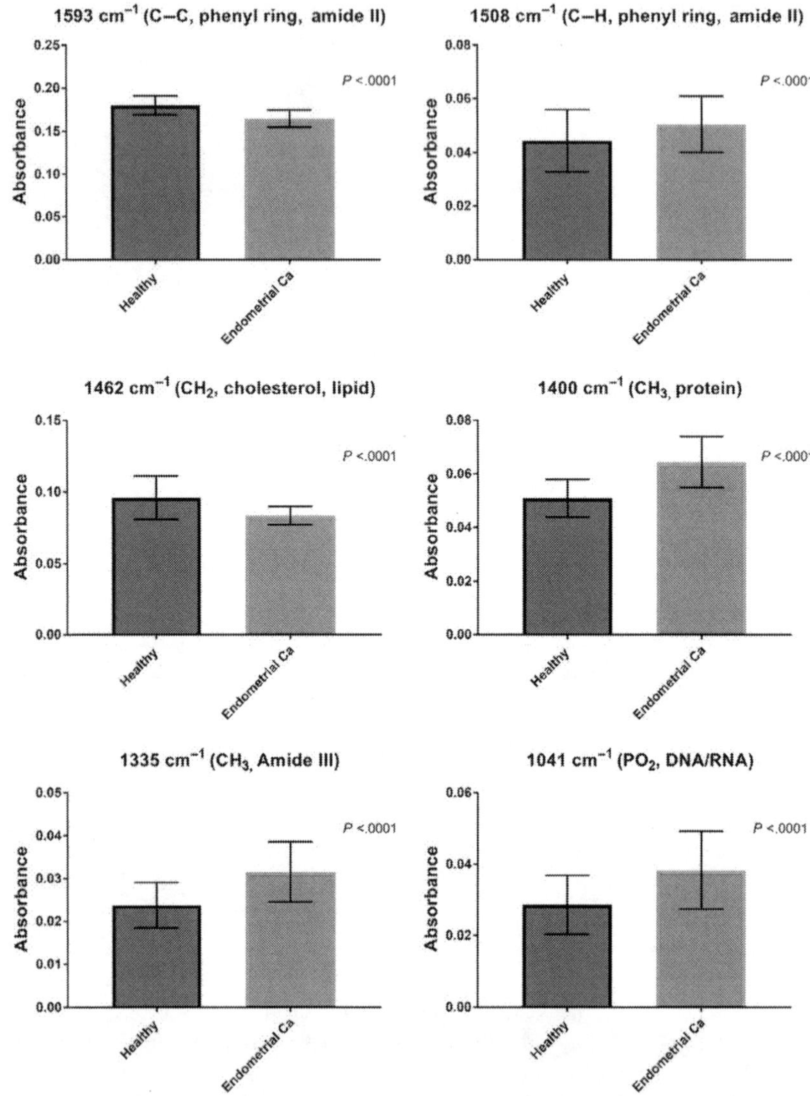

FIG. 3.18 Analysis of the top six discriminatory peaks between healthy controls and endometrial cancer patients. *(Reproduced with permission from M. Paraskevaidi, C.L.M. Morais, K.M.G. Lima, K.M. Ashton, H.F. Stringfellow, P.L. Martin-Hirsch, F.L. Martin, Potential of mid-infrared spectroscopy as a non-invasive diagnostic test in urine for endometrial or ovarian cancer, Analyst 143(13) (2018) 3156–3163.)*

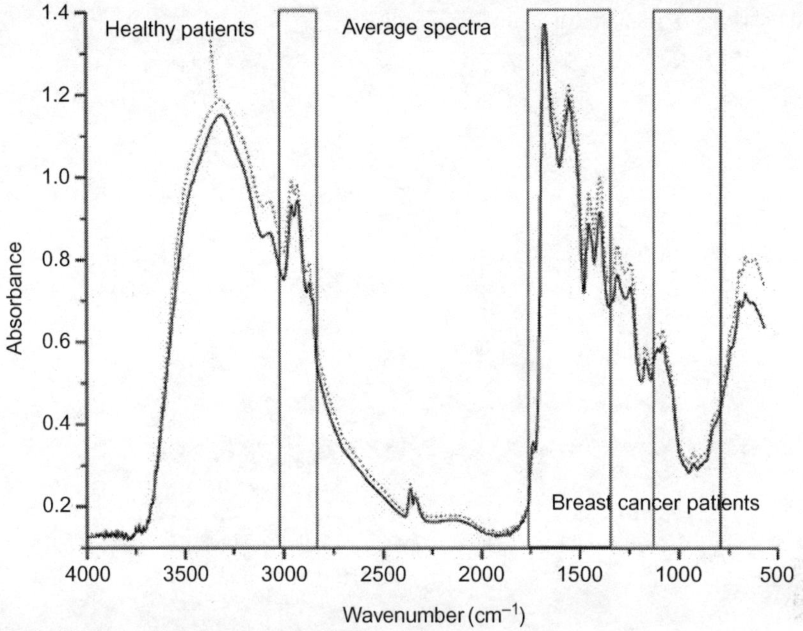

FIG. 3.19 Overview spectra in transflection, spectral region 4000–500 cm^{-1}. *(Reproduced with permission from J. Backhaus, R. Mueller, N. Formanski, N. Szlama, H.G. Meerpohl, M. Eidt, P. Bugert, Diagnosis of breast cancer with infrared spectroscopy from serum samples, Vib. Spectrosc. 52 (2010) 173–177.)*

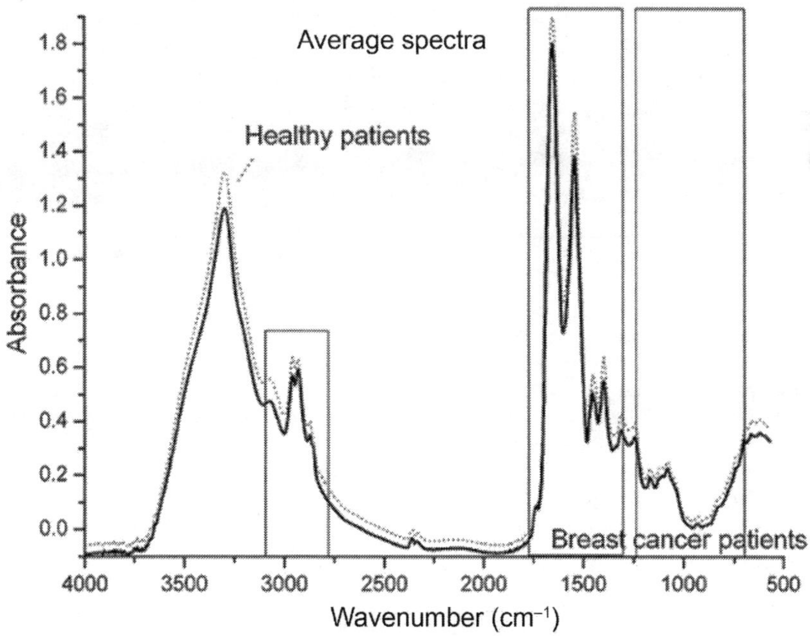

FIG. 3.20 Overview spectra in transmission, spectral region: 4000–500 cm^{-1}. *(Reproduced with permission from J. Backhaus, R. Mueller, N. Formanski, N. Szlama, H.G. Meerpohl, M. Eidt, P. Bugert, Diagnosis of breast cancer with infrared spectroscopy from serum samples, Vib. Spectrosc. 52 (2010) 173–177.)*

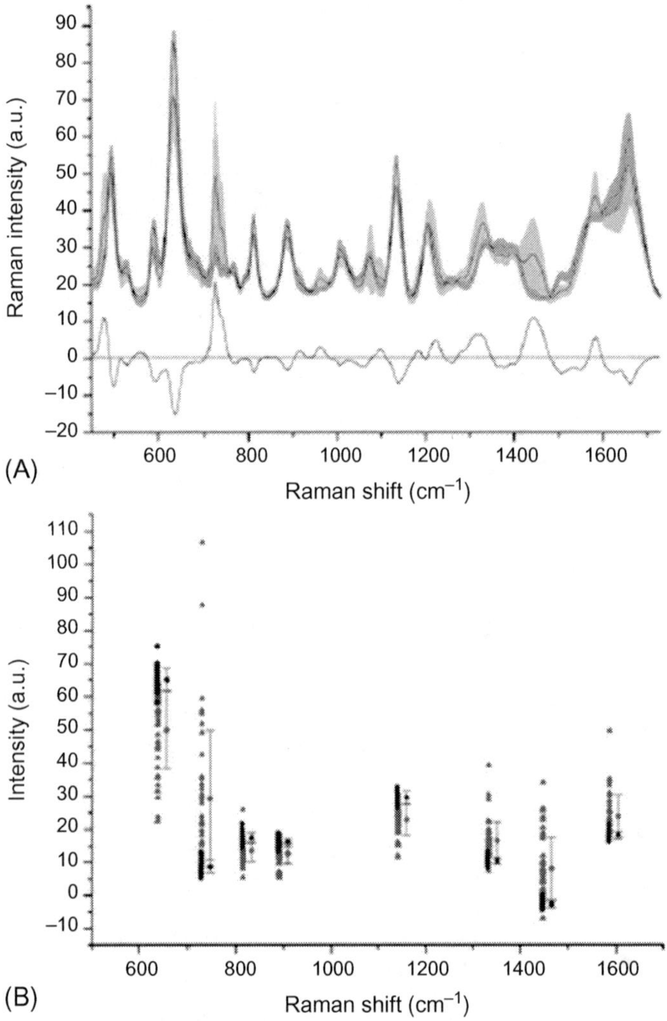

FIG. 3.21 (A) Comparison of the mean spectrum for the normal blood plasma (*black line*, $n=33$) versus that of the nasopharyngeal cancer (*red line*, $n=43$). The *shaded areas* represent the standard deviations of the means. Also shown at the bottom is the difference spectrum. (B) Plot showing the intensity value of the selected peaks with the most distinguishable differences between normal human plasma *(black)* and nasopharyngeal cancer plasma *(red)*. The corresponding mean intensities and standard deviations are also displayed on the side. *(Reproduced with permission from S.Y. Feng, J.J. Pan, Y.A. Wu, D. Lin, Y.P. Chen, G.Q. Xi, J.Q. Lin, R. Chen, Study on gastric cancer blood plasma based on surface-enhanced Raman spectroscopy combined with multivariate analysis, Sci. China Life Sci. 54 (2011) 828–834.)*

help this cancer diagnostic method be highly sensitive (90.7%) and specific (100%). Using the same methodological approach, SERS measurements in combination with multivariate analysis were additionally able to effectively classify gastric cancer from normal samples with diagnostic values (sensitivity: 79.5% and specificity: 91%) [51].

In a different study, serum FTIR spectroscopy was investigated to distinguish gastric cancer patients from healthy persons [52]. Absorbance bands in IR spectra showed that the distinguishment could be efficiently relied on a lower value of the H2959/H2931 ratio (representing CH_3/CH_2, the lipids supposed to have shorter chains and/or more branched chains in healthy persons' serum) and a lower value of the RNA/DNA ratio (likely due to an increase in DNA content in gastric cancer patients' serum), as can be seen in Fig. 3.22.

Lung cancer is considered to be the most common cancer in men worldwide. Despite significant advancements in the management of oncological late stage, its survival remains poor because most patients (ca. 75%) experience advanced disease (stage III/IV) at the time of diagnosis [53].

To develop a high-throughput and cost-effective diagnosis of lung cancer, FTIR spectroscopy was evaluated for identifying biochemical changes in sputum [54]. According to principal components analysis, a panel of 5 prominent significant IR wavenumbers at 964, 1024, 1411, 1577, and 1656 cm^{-1} (involved with putative changes in nucleic acid, glycogen, and protein levels in tumors, as seen in Fig. 3.23) was able to spectrally separate cancer cases (more centrally localized tumors) from normal cases into two apparent groups (group 1: 100%

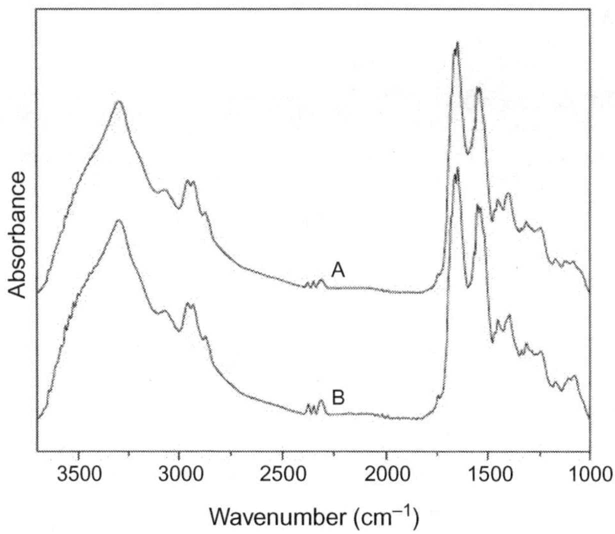

FIG. 3.22 Average IR spectra of gastric cancer patients' serum (a) and healthy persons' serum (b). *(Reproduced with permission from D. Sheng, Y. Wu, X. Wang, D. Huang, X. Chen, X. Liu, Comparison of serum from gastric cancer patients and from healthy persons using FTIR spectroscopy, Spectrochim. Acta A: Mol. Biomol. Spectrosc. 116 (2013) 365–369.)*

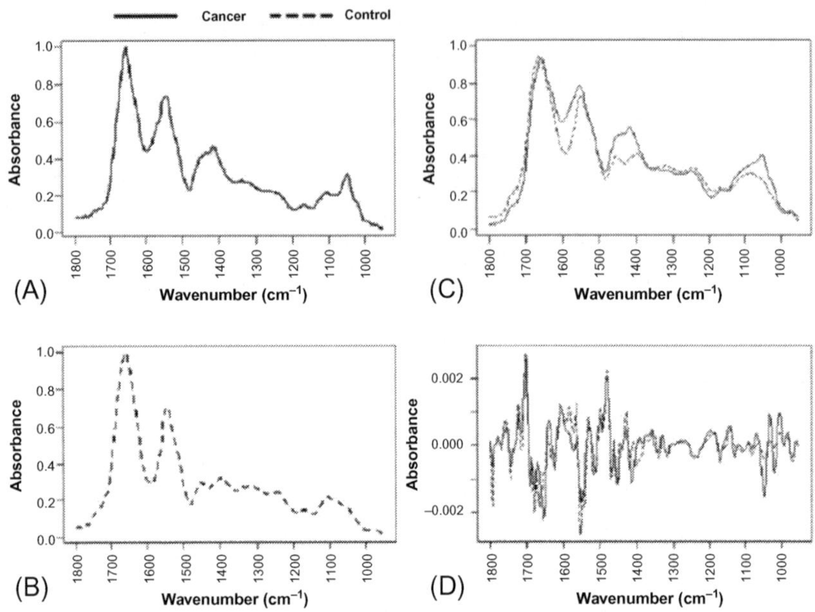

FIG. 3.23 Raw example FTIR spectra for cancer and normal sputum. Raw example FTIR spectra between wavenumbers 950 and 1800 cm^{-1} for (A) cancer sputum and (B) normal sputum. (C) Median raw spectra for cancer and normal sputa. (D) Second-derivative spectra for cancer and normal sputa. *(Reproduced with permission from P.D. Lewis, K.E. Lewis, R. Ghosal, S. Bayliss, A.J. Lloyd, J. Wills, R. Godfrey, P. Kloer, L.A.J. Mur, Evaluation of FTIR spectroscopy as a diagnostic tool for lung cancer using sputum, BMC Cancer 10 (2010) 640.)*

cancer cases; group 2: 92% normal cases). Notably, these wavenumbers could also identify lung cancer patients who had been diagnostically identified with breast cancer. Based on this study, FTIR spectroscopy was also suggested to detect peripheral lung tumors by using sputum as well as patients having a presumed high risk for lung cancer. It is because this technique was successful in 48% cases of bronchoscopically invisible lung cancers.

In another FTIR spectroscopic study, serum components were found to be markedly different between lung cancer and normal subjects, i.e., nucleic acid content (indicated by an increase of the absorbance of the $\nu_s PO_2^-$ functional group of nucleic acids and a shift of the peak to higher wave number in cancer patients); glycogen content (lowered in cancer patients); spectral characteristics of C–O groups of protein (moved to higher wave number in cancer patients) [55].

Chronic infection with hepatitis C virus is a main reason of liver-associated morbidity and mortality, and typically predisposes to progressive hepatic fibrosis [55]. Evaluating hepatic fibrosis is therefore of utmost importance in the choice of treatment options. In an effort to classify serum samples from chronic hepatitis C (CHC) patients in conformity with the degree of hepatic fibrosis, the discriminant potential of FTIR spectroscopy coupled with support vector machine was inspected [56]. Fig. 3.24 presents the mean IR spectra and their corresponding first derivative spectra for two classes of hepatic fibrosis:

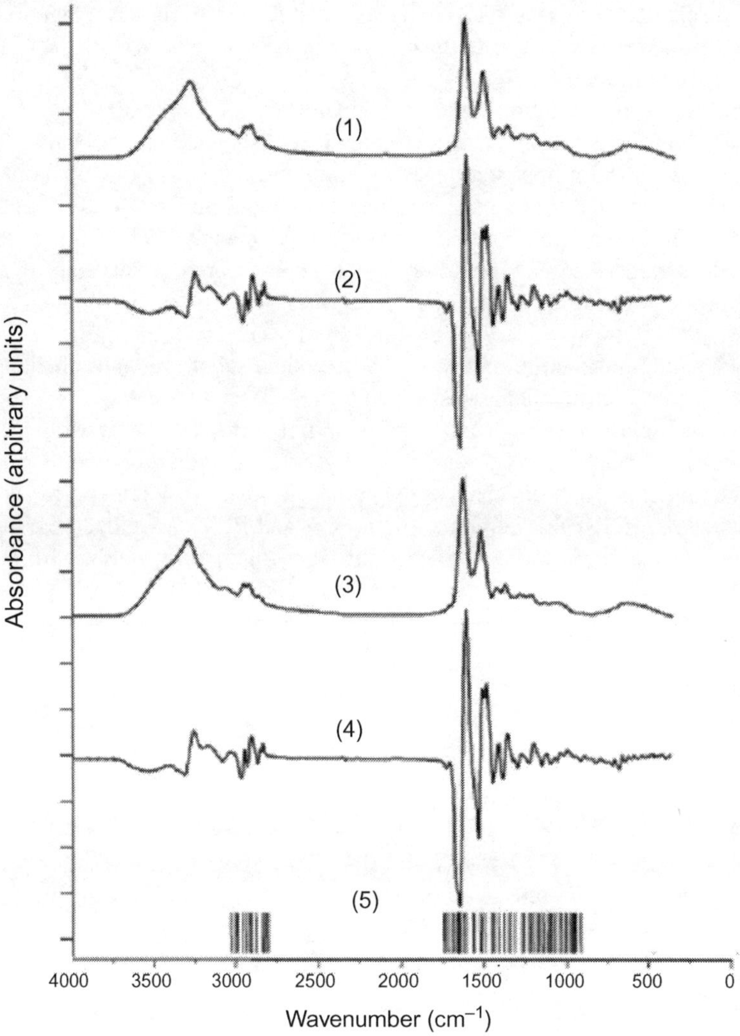

FIG. 3.24 Mean infrared absorbance spectra from serum samples of patients without hepatic fibrosis (trace 1) and patients with extensive hepatic fibrosis (trace 3). Traces 2 and 4 are first derivative of mean spectra from groups with no hepatic fibrosis and extensive hepatic fibrosis, respectively. All spectra were vector-normalized. Trace 5 shows the discriminant spectral wavelengths used in the SVM classification model after optimization. *(Reproduced with permission from G. Sebastiani, K. Gkouvatsos, K. Pantopoulos, Chronic hepatitis C and liver fibrosis, World J. Gastroenterol. 20(32) (2014) 10033–11053.)*

F0 (no hepatic fibrosis) and F3–F4 (extensive hepatic fibrosis). When applying to all spectra, the classifier developed was proved to be sensitive (90.1%) and specific (100%).

In forensic investigations, one of the major tasks is to identify body fluids at trace levels discovered at a crime scene. It was shown that confocal Raman microscopy at 785-nm excitation wavelength could confirm identification of body fluids (vaginal fluid, semen, saliva, sweat, and blood) usually found in dry traces for forensic purpose in a nondestructive manner [57]. Although each body fluid was tested only once throughout this research, preliminary results were promising (i.e., dry traces of canine and human semen exhibited very remarkably distinguishable Raman signatures) (as can be seen in Fig. 3.25).

Estimating postmortem interval (aka time since death) remains challenged in the human forensic medicine since conventional methods reckon on subjective estimation of body signs alone in the initial period (normally within 24-h postmortem). Much efforts were devoted by Zhang's research group to the interpretation of postmortem changes in animal specimens by FTIR spectroscopy. For example, ATR-FTIR combined with PCA and PLS was utilized for interrogation of rabbit plasma harvested at different time points within 48-h post-

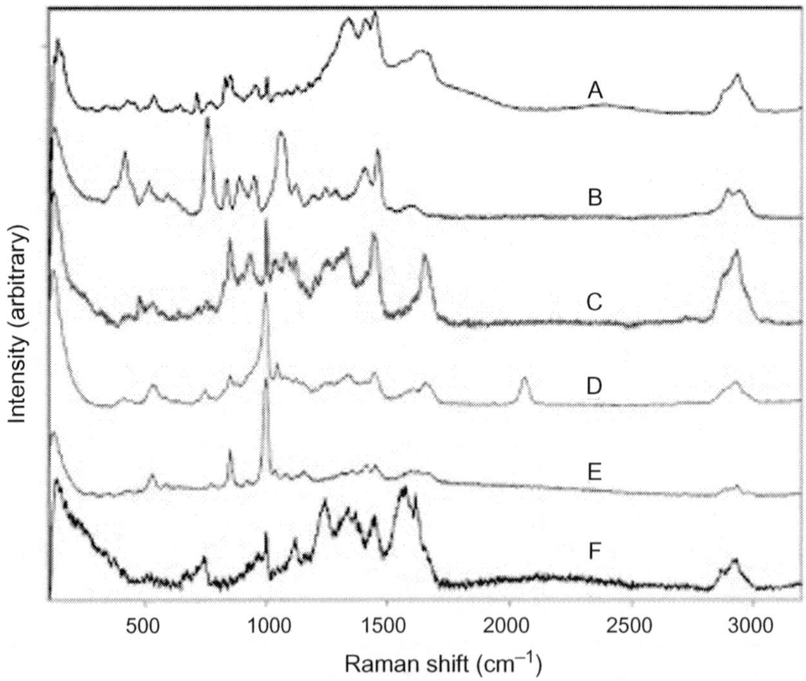

FIG. 3.25 Raman spectra of (a) human semen, (b) canine semen, (c) vaginal fluid, (d) saliva, (e) sweat, (f) and blood, with a 785-nm excitation. *(Reproduced with permission from K. Virkler, I.K. Lednev, Raman spectroscopy offers great potential for the nondestructive confirmatory identification of body fluids, Forensic Sci. Int. 181 (2008) e1–e5.)*

mortem [58]; the application of ATR-FTIR coupled with two-dimensional correlation analysis and chemometrics (PLS and nu-support vector machine) was performed for collecting exhaustive biochemical information from rabbit pericardial fluid within 48-h postmortem at 6-h intervals (Figs. 3.26 and 3.27) [59]. Overall, encouraging data obtained from this type of research could provide experimental and theoretical basis for future determination of postmortem interval with human biofluids as forensic evidence.

Identification is not always straightforward in crime scene investigation in that many body fluids stains are invisible and existent in very small quantities or mixtures. It is especially crucial for discriminating semen stain associated with sexual abuse cases. With regard to this aim, near-infrared hyperspectral imaging (NIR-HSI) in combination with simple chemometric techniques (PCA and CLS) demonstrated the potential of visualizing stains of semen, vaginal fluid, and urine on fabrics [60]. It was proved that NIR-HSI was able to locate and discriminate regions containing semen from those containing vaginal fluid along the same stain (as shown in Fig. 3.28). Even though the identification was not optimal and NIR-HSI-CLS could not be introduced as a technique for confirmation, this methodology was suggested as a valuable supplementary tool to render visible a stain containing traces of semen positively confirmed (Fig. 3.29).

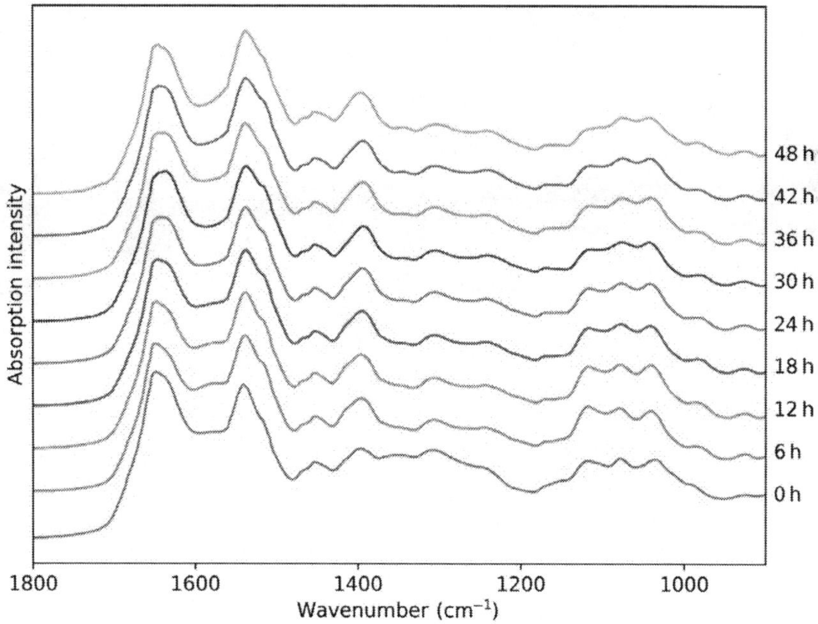

FIG. 3.26 A comparison of average spectra with SNV normalization among PMI groups from 0 to 48h postmortem. *(Reproduced with permission from J. Zhang, B. Li, Q. Wang, X. Wei, W. Feng, Y. Chen, P. Huang, Z. Wang, Application of Fourier transform infrared spectroscopy with chemometrics on postmortem interval estimation based on pericardial fluids, Sci. Rep. 7 (2017) 18013.)*

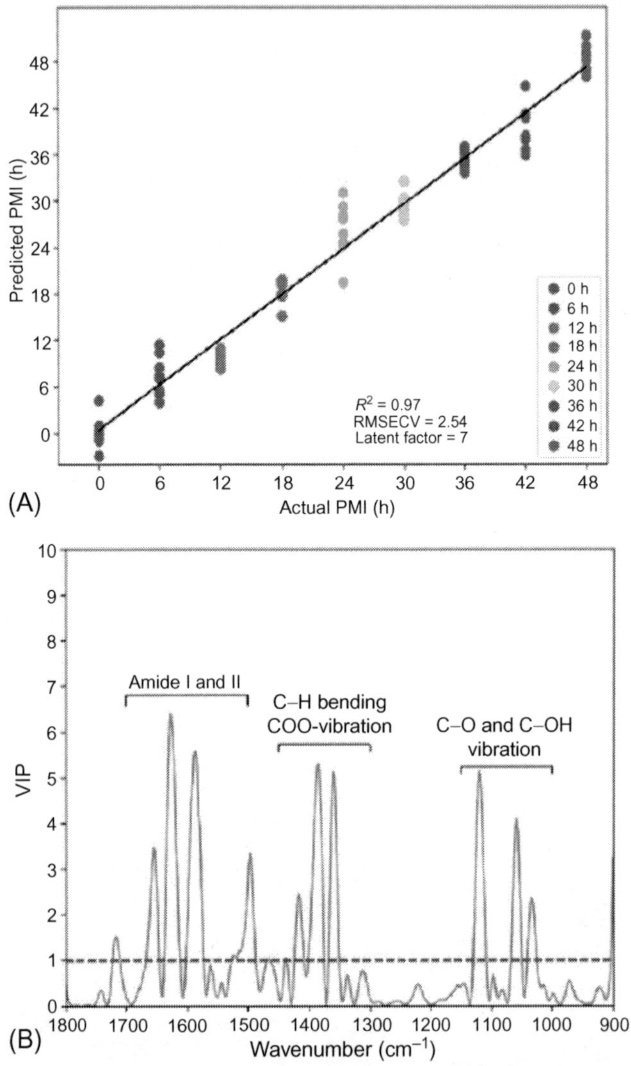

FIG. 3.27 The crossvalidation results of the PLS model using spectral variables within 1800–900 cm^{-1}. (A) The regression plot between the predicted and actual PMI. The *black line* represents the reference line where the predicted PMI scores are closer to it, the higher fitting of goodness will be. (B) The plot of VIP scores displays the contribution of the spectral variables to the distinction in the PLS model. The variables with VIP scores above 1.0 (marked by a *red dot line*) are considered most significant, and their assignments are symbolized. *(Reproduced with permission from J. Zhang, B. Li, Q. Wang, X. Wei, W. Feng, Y. Chen, P. Huang, Z. Wang, Application of Fourier transform infrared spectroscopy with chemometrics on postmortem interval estimation based on pericardial fluids, Sci. Rep. 7 (2017) 18013.)*

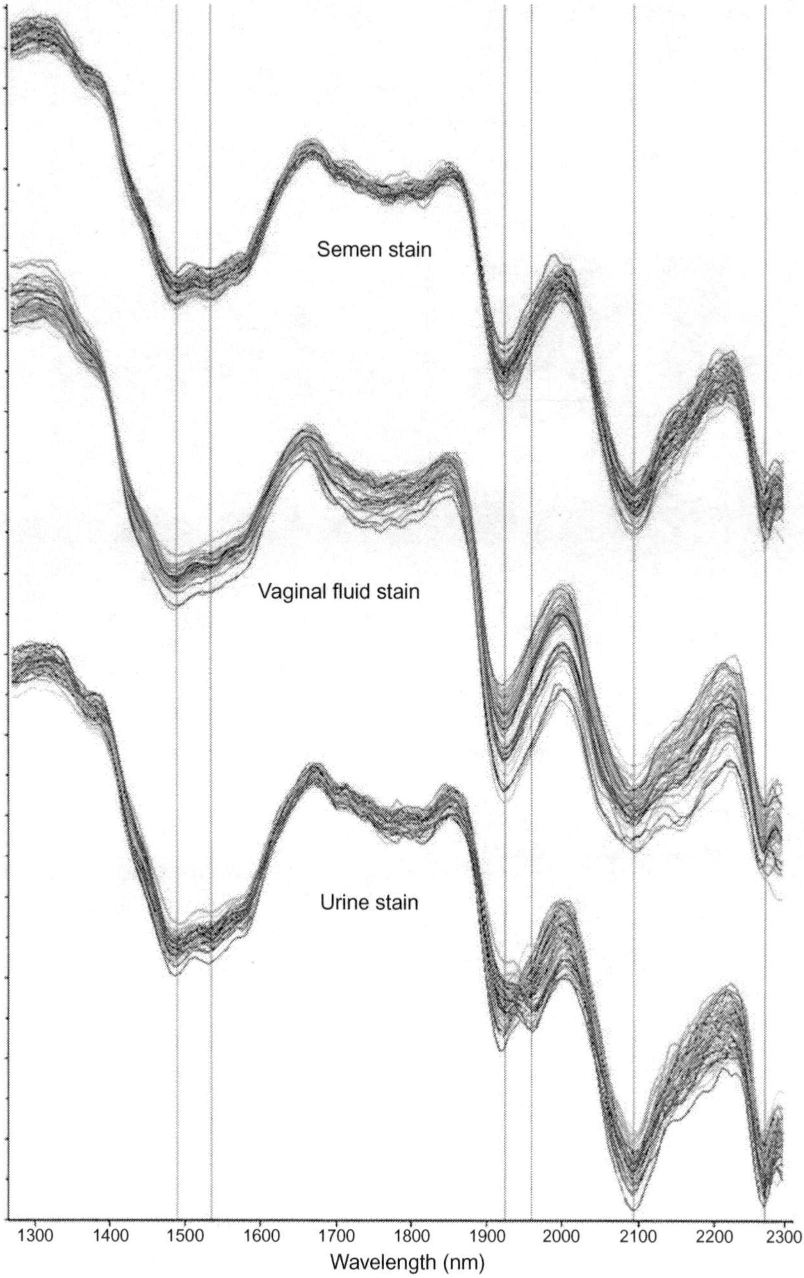

FIG. 3.28 NIR spectra from stains of semen, vaginal fluid, and urine on cotton fabrics from 1270 to 2300 nm. *(Reproduced with permission from F. Zapata, F.E. Ortega-Ojeda, C. García-Ruiz, Revealing the location of semen, vaginal fluid and urine in stained evidence through near infrared chemical imaging, Talanta 166 (2017) 292–299.)*

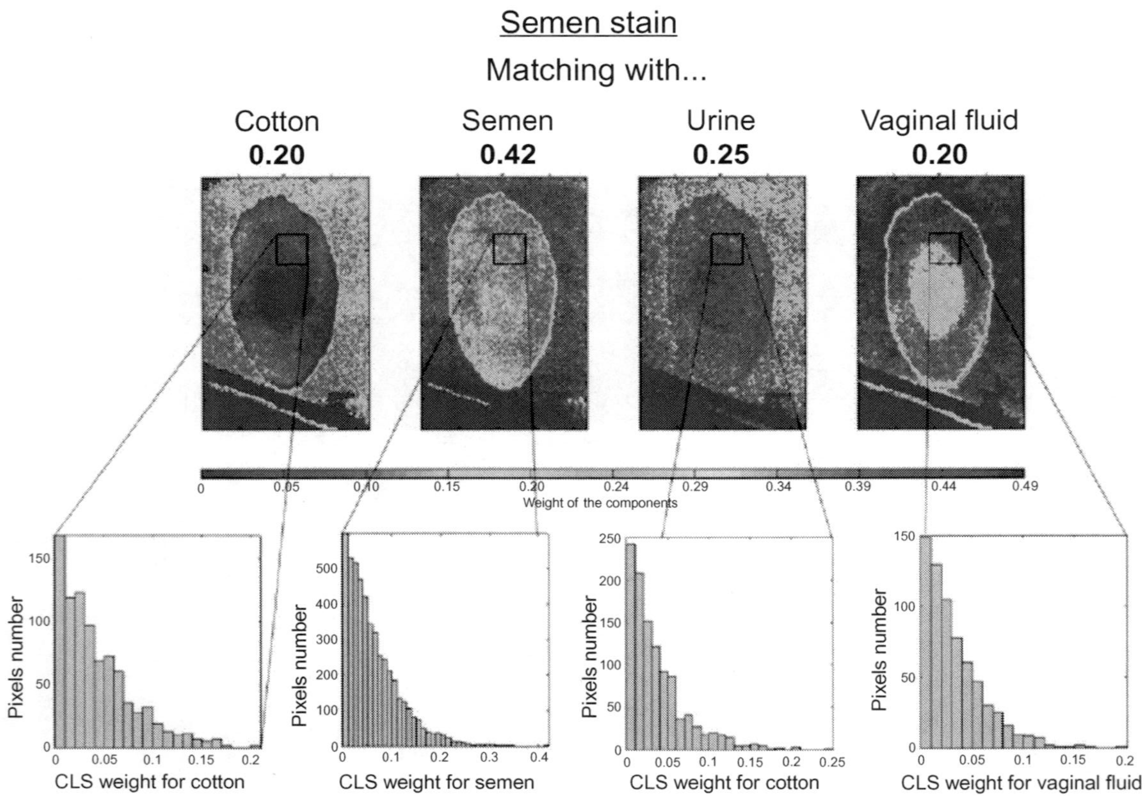

FIG. 3.29 CLS classification model applied to a semen stain on cotton fabric. The CLS colored maps for each class (cotton, semen, urine, and vaginal fluid) are displayed. The maximum CLS weight values obtained for each class within the selected stained region is indicated above every color map. The histogram containing the number of pixels within the selected region and their corresponding CLS weight to each class are provided below every color map. *(Reproduced with permission from F. Zapata, F.E. Ortega-Ojeda, C. García-Ruiz, Revealing the location of semen, vaginal fluid and urine in stained evidence through near infrared chemical imaging, Talanta 166 (2017) 292–299.)*

References

[1] A.A. Bunaciu, Ş. Fleschin, V.D. Hoang, H.Y. Aboul-Enein, Vibrational spectroscopy in body fluids analysis, Crit. Rev. Anal. Chem. 47 (1) (2017) 67–75.

[2] M.J. Baker, J. Trevisan, P. Bassan, R. Bhargava, H.J. Butler, K.M. Dorling, P.R. Fielden, S.W. Fogarty, N.J. Fullwood, K.A. Heys, C. Hughes, P. Lasch, P.L. Martin-Hirsch, B. Obinaju, G.D. Sockalingum, J. Sulé-Suso, R.J. Strong, M.J. Walsh, B.R. Wood, P. Gardner, F.L. Martin, Using Fourier transform IR spectroscopy to analyze biological materials, Nat. Protoc. 9 (8) (2014) 1771–1791.

[3] R.A. Shaw, H.H. Mantsch, Infrared spectroscopy in clinical and diagnostic analysis, in: R.A. Meyers (Ed.), Encyclopedia of Analytical Chemistry, John Wiley & Sons Ltd., Chichester, ISBN: 0-471-97670-9, 2011.

[4] D. Perez-Guaita, S. Garrigues, M. de la Guardia, Infrared-based quantification of clinical parameters, TrAC Trends Anal. Chem. 62 (2014) 93–105.

[5] A.A. Bunaciu, S. Fleschin, H.Y. Aboul-Enein, Biomedical investigations using Fourier transform-infrared microspectroscopy, Crit. Rev. Anal. Chem. 10 (2014) 132–139.

[6] R.A. Shaw, S. Kotowich, M. Leroux, H.H. Mantsch, Multianalyte serum analysis using mid-infrared spectroscopy, Ann. Clin. Biochem. 35 (1998) 624–632.

[7] C.M. Orphanou, L. Walton-Williams, H. Mountain, J. Cassella, The detection and discrimination of human body fluids using ATR FT-IR spectroscopy, Forensic Sci. Int. 252 (2015) e10–e16.

[8] J. Depciuch, M. Parlinska-Wojtan, Comparing dried and liquid blood serum samples of depressed patients: an analysis by Raman and infrared spectroscopy methods, J. Pharm. Biomed. Anal. 150 (2018) 80–86.

[9] D. Perez-Guaita, J. Ventura-Gayate, C. Perez-Rambla, M. Sancho-Andreu, S. Garrigues, M. de la Guardia, Evaluation of infrared spectroscopy as a screening tool for serum analysis: impact of the nature of samples included in the calibration set, Microchem. J. 106 (2013) 202–211.

[10] A.J. Berger, T.W. Koo, I. Itzkan, G. Horowitz, M.S. Feld, Multicomponent blood analysis by near-infrared Raman spectroscopy, Appl. Optics 38 (13) (1999) 2916–2926.

[11] Y. Li, R. Chen, L. Liu, S. Feng, B. Huang, Human blood analysis by IR and Raman spectroscopy. in: B. Chance, M. Chen, A.E.T. Chiou, Q. Luo (Eds.), Optics in Health Care and Biomedical Optics: Diagnostics and Treatment II, Proc. of SPIEvol. 5630, SPIE, Bellingham, WA, 2005, https://doi.org/10.1117/12.570540. 1605-7422/04/$15.

[12] D. Borchman, G.N. Foulks, M.C. Yappert, D. Tang, D.V. Ho, Spectroscopic evaluation of human tears lipids, Chem. Phys. Lipids 147 (2) (2007) 87–102.

[13] K. Oliver, A. Vilasi, A. Maréchal, S.H. Moochhala, R.J. Unwin, P.R. Rich, Infrared vibrational spectroscopy: a rapid and novel diagnostic and monitoring tool for cystinuria. Sci. Rep. 6 (2016) 34737, https://doi.org/10.1038/srep34737.

[14] C. Chenn, L. Yang, J. Zhao, Y. Yuan, C. Chen, J. Tang, H. Yang, Z. Yan, H. Wang, X. Lv, Urine Raman spectroscopy for rapid and inexpensive diagnosis of chronic renal failure (CRF) using multiple classification algorithms, Optik 203 (2020) 164043.

[15] G. Fusch, N. Rochow, A. Choi, S. Fusch, S. Poeschl, A.O. Ubah, S.Y. Lee, P. Raja, C. Fusch, Rapid measurement of macronutrients in breast milk: how reliable are infrared milk analyzers? Clin. Nutr. 34 (2015) 465–476.

[16] M. Shojima, E. Watanabe, Y. Mayanagi, Cerebral blood oxygenation after cerebrospinal fluid removal in hydrocephalus measured by near infrared spectroscopy, Surg. Neurol. 62 (4) (2004) 312–318.

[17] A. Sevinc, D. Yonar, F. Severcan, Investigation of neurodegenerative diseases from body fluid samples using Fourier transform infrared spectroscopy, Biomed. Spectrosc. Imaging 4 (4) (2015) 341–357.

[18] G. Janatsch, J.D. Kruse-Jarres, R. Marbach, H.M. Heise, Multivariate calibration for assays in clinical chemistry using attenuated total reflection infrared spectra of human blood plasma, Anal. Chem. 61 (1989) 2016–2023.

[19] D.J. Selkoe, Alzheimer's disease: genes, proteins, and therapy, Physiol. Rev. 81 (2) (2001) 741–766.

[20] J. Kang, H.G. Lemaire, A. Unterbeck, J.M. Salbaum, C.L. Masters, K.H. Grzeschik, G. Multhaup, K. Beyreuther, B. Müller-Hill, The precursor of Alzheimer's disease amyloid A4 protein resembles a cell-surface receptor, Nature 325 (1987) 733–736.

[21] A. Delacourte, A. Défossez, Biochemical characterization of an immune serum which specifically marks neurons in neurofibrillary degeneration in Alzheimer's disease, C. R. Acad. Sci. III 303 (1986) 439–444.

[22] M. Griebe, M. Dafferstshofer, M. Stroick, M. Syren, P. Ahmad-Nejad, M. Neumaier, J. Backhaus, M.G. Hennerici, M. Fatar, Infrared spectroscopy: a new diagnostic tool in Alzheimer disease, Neurosci. Lett. 420 (1) (2007) 29–33.

[23] J.T. Pelton, L.R. McLean, Spectroscopic methods for analysis of protein secondary structure, Anal. Biochem. 277 (2) (2000) 167–176.

[24] P. Carmona, M. Molina, M. Calero, F. Bermejo-Pareja, P. Martínez-Martín, I. Alvarez, A. Toledano, Infrared spectroscopic analysis of mononuclear leukocytes in peripheral blood from Alzheimer's disease patients, Anal. Bioanal. Chem. 402 (2012) 2015–2021.

[25] P. Carmona, M. Molina, E. López-Tobar, A. Toledano, Vibrational spectroscopic analysis of peripheral blood plasma of patients with Alzheimer's disease, Anal. Bioanal. Chem. 407 (2015) 7747–7756.

[26] E. Kleiren, J.M. Ruysschaert, E. Goormaghtigh, V. Raussens, Development of a quantitative and conformation-sensitive ATR-FTIR biosensor for Alzheimer's disease: the effect of deuteration on the detection of the Aβ peptide, Spectroscopy 24 (2010) 61–66.

[27] M. Voue, E. Goormaghtigh, F. Homble, J. Marchand-Brynaert, J. Conti, S. Devouge, J. De Coninck, Biochemical interaction analysis on ATR devices: a wet chemistry approach for surface functionalization, Langmuir 23 (2007) 949–955.

[28] https://www.healthline.com/health/diabetes, (Accessible on 25 December 2019).

[29] E. Diessel, S. Willmann, P. Kamphaus, R. Kurte, U. Damm, H.M. Heise, Glucose quantification in dried-down nanoliter samples using mid-infrared attenuated total reflection spectroscopy, Appl. Spectrosc. 58 (4) (2004) 443–450.

[30] F. Tao, L. Yuanpeng, L. Fucui, H. Furong, Rapid diagnosis of type II diabetes using Fourier transform mid-infrared attenuated total reflection spectroscopy combined with support vector machine, Anal. Lett. 51 (9) (2018) 1400–1416.

[31] S. Aydin, A comparison of ghrelin, glucose, alpha-amylase and protein levels in saliva from diabetics, J. Biochem. Mol. Biol. 40 (1) (2007) 29–35.

[32] M.A. Belazi, A. Galli-Tsinopoulou, D. Drakoulakos, A. Fleva, P.H. Papanayiotou, Salivary alterations in insulin-dependent diabetes mellitus, Int. J. Paediatr. Dent. 8 (1) (1998) 29–33.

[33] A. Borg Andersson, D. Birkhed, K. Berntorp, F. Lindgarde, L. Matsson, Glucose concentration in parotid saliva after glucose/food intake in individuals with glucose intolerance and diabetes mellitus, Eur. J. Oral Sci. 106 (5) (1998) 931–937.

[34] D.A. Scott, D.E. Renaud, S. Krishnasamy, P. Meriç, N. Buduneli, Ş. Çetinkalp, K.Z. Liu, Diabetes-related molecular signatures in infrared spectra of human saliva, Diabetol. Metab. Syndr. 2 (2010) 48.

[35] R. Raziya Sultana, S.N. Zafarullah, N. Hephzibah Kirubamani, Saliva signature of normal pregnant women in each trimester, Indian J. Sci. Technol. 4 (5) (2011) 481–486.

[36] R. Raziya Sultana, S.N. Zafarullah, N. Hephzibah Kirubamani, Utility of FTIR spectroscopic analysis of saliva of diabetic pregnant women in each trimester, Indian J. Sci. Technol. 4 (8) (2011) 967–970.

[37] https://www.mdedge.com/jcomjournal/article/153805/infectious-diseases/metabolic-complications-hiv-infection, (Accessible on 25 December 2019).

[38] A. Sakudo, R. Tsenkova, T. Onozuka, K. Morita, S. Li, J. Warachit, Y. Iwabu, G. Li, T. Onodera, K. Ikuta, A novel diagnostic method for human immunodeficiency virus type-1 in plasma by near-infrared spectroscopy, Microbiol. Immunol. 49 (2005) 695–701.

[39] M.K. Bahmani, A. Khosravi, R. Miri, R. Yukieiwabu, K. Shiikuta, A. Sakudo, A spectroscopic characterization of human immunodeficiency virus type-1-infected plasma by principal component analysis and soft independent modeling of class analogy of visible and near-infrared spectra, Mol. Med. Rep. 2 (2009) 805–809.

[40] L. Sitole, F. Steffens, T.P.J. Krüger, D. Meyer, Mid-ATR-FTIR spectroscopic profiling of HIV/AIDS sera for novel systems diagnostics in global health, OMICS 18 (8) (2014) 514–523.

[41] R. Hewer, J. Vorster, F.E. Steffens, D. Meyer, Applying biofluid 1H NMR-based metabonomic techniques to distinguish between HIV-1 positive/AIDS patients on antiretroviral treatment and HIV-1 negative individuals, J. Pharm. Biomed. Anal. 41 (2006) 1442–1446.

[42] C. Philippeos, F.E. Steffens, D. Meyer, Comparative 1H NMR-based metabonomic analysis of HIV-1 sera, J. Biomol. NMR 44 (2009) 127–137.

[43] World Health Organization, http://www.who.int/mediacentre/factsheets/fs297/eng, (Accessed 8 March 2018).

[44] https://www.cancer.net/navigating-cancer-care/diagnosing-cancer/tests-and-procedures/biopsy, (Accessible on 31 December 2019).

[45] The French METAVIR Cooperative Study Group, Intraobserver and interobserver variations in liver biopsy interpretation in patients with chronic hepatitis C, Hepatology 20 (1994) 15–20.

[46] P. Bedossa, D. Dargere, V. Paradis, Sampling variability of liver fibrosis in chronic hepatitis C, Hepatology 38 (2003) 1449–1457.

[47] K. Gajjar, J. Trevisan, G. Owens, P.J. Keating, N.J. Wood, H.F. Strinfellow, P.L. Martin-Hirsch, F.L. Martin, Fourier-transform infrared spectroscopy coupled with a classification machine for the analysis of blood plasma or serum: a novel diagnostic approach for ovarian cancer, Analyst 138 (2013) 3917–3926.

[48] M. Paraskevaidi, C.L.M. Morais, K.M.G. Lima, K.M. Ashton, H.F. Stringfellow, P.L. Martin-Hirsch, F.L. Martin, Potential of mid-infrared spectroscopy as a non-invasive diagnostic test in urine for endometrial or ovarian cancer, Analyst 143 (13) (2018) 3156–3163.

[49] J. Backhaus, R. Mueller, N. Formanski, N. Szlama, H.G. Meerpohl, M. Eidt, P. Bugert, Diagnosis of breast cancer with infrared spectroscopy from serum samples, Vib. Spectrosc. 52 (2010) 173–177.

[50] S. Feng, R. Chen, J. Lin, J. Pan, G. Chen, Y. Li, M. Cheng, Z. Huang, J. Chen, H. Zeng, Nasopharyngeal cancer detection based on blood plasma surface-enhanced Raman spectroscopy and multivariate analysis, Biosens. Bioelectron. 25 (2010) 2414–2419.

[51] S.Y. Feng, J.J. Pan, Y.A. Wu, D. Lin, Y.P. Chen, G.Q. Xi, J.Q. Lin, R. Chen, Study on gastric cancer blood plasma based on surface-enhanced Raman spectroscopy combined with multivariate analysis, Sci. China Life Sci. 54 (2011) 828–834.

[52] D. Sheng, Y. Wu, X. Wang, D. Huang, X. Chen, X. Liu, Comparison of serum from gastric cancer patients and from healthy persons using FTIR spectroscopy, Spectrochim. Acta A: Mol. Biomol. Spectrosc. 116 (2013) 365–369.

[53] S.B. Knight, P.A. Crosbie, H. Balata, J. Chudziak, T. Hussell, C. Dive, Progress and prospectcs of early detection in lung cancer, Open Biol. 7 (9) (2017) 170070.

[54] P.D. Lewis, K.E. Lewis, R. Ghosal, S. Bayliss, A.J. Lloyd, J. Wills, R. Godfrey, P. Kloer, L.A.J. Mur, Evaluation of FTIR spectroscopy as a diagnostic tool for lung cancer using sputum, BMC Cancer 10 (2010) 640.

[55] L. Zhao, D. Han, X. Sun, Study on cancer serum components by Fourier transform infrared spectroscopy, in: 3rd International Conference on Material, Mechanical and Manufacturing Engineering (IC3ME), 2015.

[56] G. Sebastiani, K. Gkouvatsos, K. Pantopoulos, Chronic hepatitis C and liver fibrosis, World J. Gastroenterol. 20 (32) (2014) 10033–11053.

[57] E. Scaglia, G.D. Sockalingum, J. Schmitt, C. Gobinet, N. Schneider, M. Manfait, G. Thiéfin, Noninvasive assessment of hepatic fibrosis in patients with chronic hepatitis C using serum Fourier transform infrared spectroscopy, Anal. Bioanal. Chem. 401 (2011) 2919–2925.

[58] J. Zhang, B. Li, Q. Wang, C. Li, Y. Zhang, H. Lin, Z. Wang, Characterization of postmortem biochemical changes in rabbit plasma using ATR-FTIR combined with chemometrics: a preliminary study, Spectrochim. Acta A: Mol. Biomol. Spectrosc. 173 (2017) 733–739.

[59] J. Zhang, B. Li, Q. Wang, X. Wei, W. Feng, Y. Chen, P. Huang, Z. Wang, Application of Fourier transform infrared spectroscopy with chemometrics on postmortem interval estimation based on pericardial fluids, Sci. Rep. 7 (2017) 18013.

[60] F. Zapata, F.E. Ortega-Ojeda, C. García-Ruiz, Revealing the location of semen, vaginal fluid and urine in stained evidence through near infrared chemical imaging, Talanta 166 (2017) 292–299.

Chapter 4

Tissues analysis

During the last decades, biomedical analysis has experienced a rapid expansion of the number of papers published on IR and Raman spectroscopic techniques [1–6]. Vibrational spectroscopic methods were proved to be able to replace histologic analysis that is prohibitively time-consuming for humans (about 2–3 weeks). This is due to the fact that they help better understand the biochemical composition, molecular structure, and molecular interactions in various types of sample, such as blood, bones, cells, and tissues. The pathogenesis of a given disease can be traced down to a single chemical substance (usually a protein), with either abnormal in structure or present in reduced amounts. Such changes should be reflected in the vibrational spectra (e.g., absorption bands characteristically associated with proteins, amides, O–P–O stretching of DNA phosphate backbone or C=O stretching of phospholipids, disulfide groups, etc.) and may be specific enough to be used as phenotic markers of the disease.

Actually, vibrational spectroscopy was applied for biomedical analysis a long time after the discovery of IR and Raman spectroscopy. By 1950, Elliot and Ambrose [7] used IR spectroscopy to study protein conformation. Blout and Mellors [8] and Woernley [9] were conducting IR experiments on human and animal tissues. Unfortunately, they were unable to identify discrete spectroscopic signatures from various tissues most likely due to the limitation of relatively unsophisticated instrumentation, little spectroscopic knowledge of biological molecules, and sample treatment highly introducing artifacts (fixatives and homogenized).

Around the same time, the coupling of a reflecting microscope to an IR spectrometer done by Thompson's group [10] was, in particular, useful for studying tissues with very spatial resolution. The first IR microscope was built, in 1947, by C.R. Burch [11] and manufactured by Perkin-Elmer, an American global corporation being the first to commercialized this type of microscope and further develop it with technological breakthroughs, as shown in Fig. 4.1.

Nevertheless, the interpretation of IR spectra of biological tissues was not very easy at that time, as clearly stated by Barer [12], as follows:

> *...the infra-red absorption spectrum gives what is essentially information concerning the presence or absence of certain specific chemical groups such as OH, CH, NH, CO, etc. In a complex molecule such as a protein, many such groups will be present and it may be wondered whether it is possible to distinguish different proteins by their infra-red spectra. This subject is still in its infancy.*

Vibrational Spectroscopy Applications in Biomedical, Pharmaceutical and Food Sciences
https://doi.org/10.1016/B978-0-12-818827-9.00005-6

© 2020 Elsevier Inc. All rights reserved.

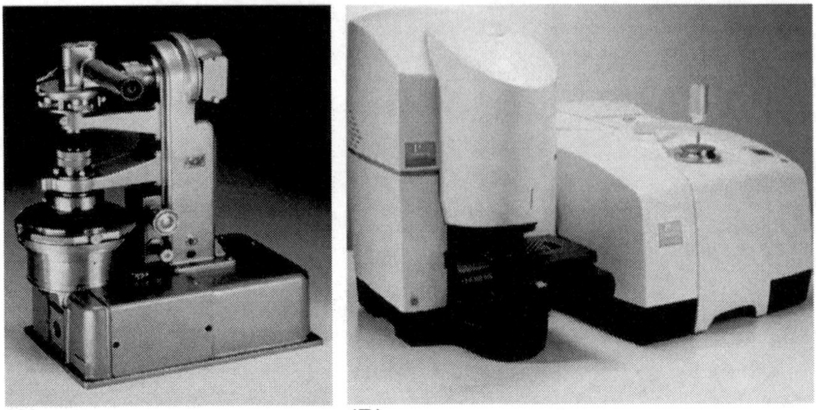

(A) (B)

FIG. 4.1 PerkinElmer microscopes: (A) the first and (B) the last model. *((A) Reproduced with permission from N. Sheppard, The historical development of experimental techniques in vibrational spectroscopy, in: J.M. Chalmers, P.R. Griffits (Eds.), Handbook of Vibrational Spectroscopy, John Wiley & Sons Ltd., 2002. (B) Reproduced with permission from https://www.perkinelmer.com/product/spotlight-400-std-frontier-mir-tgs-na-l1860116, accessed 18 October 2019.)*

Shermann [13] proved the utility of IR spectroscopy in tissue research, by studying the beef tendons purchased from a local butcher, as presented in Fig. 4.2.

Hermann [14] showed that FMIR technique (frustrated multiple infrared reflectance) had noteworthy advantages over transmission techniques to identify micrograms amounts of pesticides, potentially suggesting that lowering the limit of sensitivity in the IR region could help detect several changes with tissues.

Much later than IR spectroscopy, the first laser Raman spectrum of a protein was recorded in 1970 [15]. The main reason for this late application of Raman spectroscopy as a valuable analytical tool in biomedical scenarios was attributable to the weak nature of spontaneous Raman scattering (i.e., only one Raman scattered photon from 10^6 to 10^8 excitation photons).

It would be plausible to state that the new era of biological IR spectroscopy started in the 1980s, which has been accelerated by a number of key factors such as advances in instrumentation, computers, chemical and molecular biological methods [16].

In 1995, Mantsch and Jackson [17] clearly emphasized that it is possible to develop novel diagnostic methods by using IR spectroscopy to sensitively scrutinize biochemical events in human tissues underlying transformation from normal to disease state. This emphasis was comprehensible when (i) feasibly measuring an IR snapshot of the overall biochemical composition of each individual cell and (ii) fairly rapidly acquiring an IR microspectrum (in about 500 ms). With the aid of chemometrics [2, 18], the analysis of a hyperspectral dataset could detect small, but recurring differences, even in the existence of larger random variance.

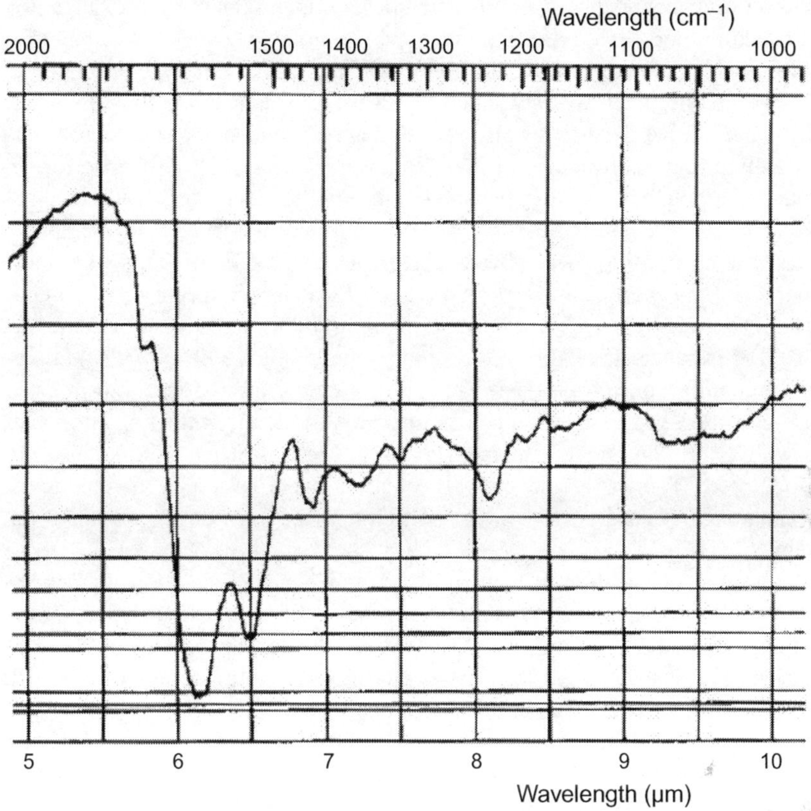

FIG. 4.2 ATR spectrum of beef tendon. *(Reproduced with permission from B. Sherman, Infrared spectroscopy by attenuated total reflection, Appl. Spectrosc. 18 (1964) 7–9.)*

In this way, spectral analysis was evolved and matured for cells, spectral cytopathology (SCP), and tissues (spectral histopathology, SHP) [19]. These two diagnostic methodologies were shown to be more sensitive than standard cytopathology and histopathology in terms of detecting a disease objectively, earlier, on a smaller scale, and requiring no use of stains and contrast-enhancing agents [20].

According to Max Diem, the information obtained from MIR spectral interpretation may revolutionize pathological performance in the 21st century [5, 21]. But, we must also recognize the important early contribution of Mantsch's research group on molecular spectroscopy in biodiagnostics [17].

Up-to-date data show that the discrimination between normal, dysplastic, and malignant tissues has been effectively done by IR and Raman techniques [2, 5, 6, 18, 20–25].

In the literature, bone was presumably the first tissue vibrational spectroscopically investigated. This composite tissue constitutes part of the vertebrate skeleton and consists of cells embedded in an abundant hard intercellular material. Its chemical composition includes a mineral phase (hydroxyapatite), an organic matrix [ca. type I collagen (90%), noncollagenous proteins (5%), lipids (2%)], and water [26]. To confirm X-ray diffraction results related to hydroxyapatite, bone was subjected to IR analysis as early as the 1950s [27, 28]. Quantitative IR analysis also revealed age dependency of amorphous and crystalline mineral fractions of bone, i.e., during early stages of bone formation, the amount of calcium phosphate amorphous decreases, while changes in crystallite size and perfection of crystallinity are observed for bone apatite [29]. The composition and physicochemical status of the mineral phase and matrix of bone were additionally probed by IR analysis (spectroscopy, microspectroscopy, and microspectroscopic imaging) of normal and diseased tissues [30]. For instance, an FT-IR spectrophotometer coupled with an optical microscope could produce high-quality spectra for calcifying tissues at 20-μm spatial resolution in about 1–2 min scanning time to identify mineralization sites as well as protein/mineral interactions [31]. Spectral maps could be also obtained by FTIR microscopic analysis of individual nonosteoporotic human osteonal bone from centers to peripheries in four orthogonal directions (in 10-μm steps) [32] (Fig. 4.3).

Terminal sterilization of bone allograft by Cobalt-60 radiation sources has been utilized for many years. Gamma rays may split polypeptide chain, causing collagen degradation in the bone matrix in a dose-dependent manner that consequently induces changes in the dynamic-mechanical properties of bone. To monitor such irradiation-induced alterations in biological systems, mid-FTIR spectroscopic analysis was proved to be useful for molecular characterization of DNA, proteins, and lipids, as synthetic or isolated macromolecules, membranes, cells, and tissues [33].

In a study, ATR-FTIR spectroscopic data were submitted to hierarchical cluster analysis (HCA) for evaluation of the level of similarity between spectral structures obtained from bone fragments nonirradiated and irradiated with different doses of ionizing radiation [34]. Fig. 4.4 shows the fingerprint spectral region, which gives information of vibrational modes related to organic and inorganic content of bone. This study indicated that the classification accuracy obtained with HCA increases with radiation dose due to biochemical effects provided by higher doses.

In human body, tendons are the fibrous connective tissues that transfer mechanical tension experienced by muscle contraction to the bones. Articular cartilage is the smooth, white tissue avascular and aneural that covers the surface of bones where they articulate or come together to form joints. Articular cartilage is mainly composed of water, type II collagen, proteoglycans, and chondrocytes [35].

Using Fourier transform infrared imaging (FTIRI), a three-dimensional cartilage or tendon block was investigated for amide anisotropies at different surfaces (Fig. 4.5) [36]. For tendon, a clear IR anisotropy (possibly due to a zigzag planar waveform exhibited by collagen fibers) was seen for both amide

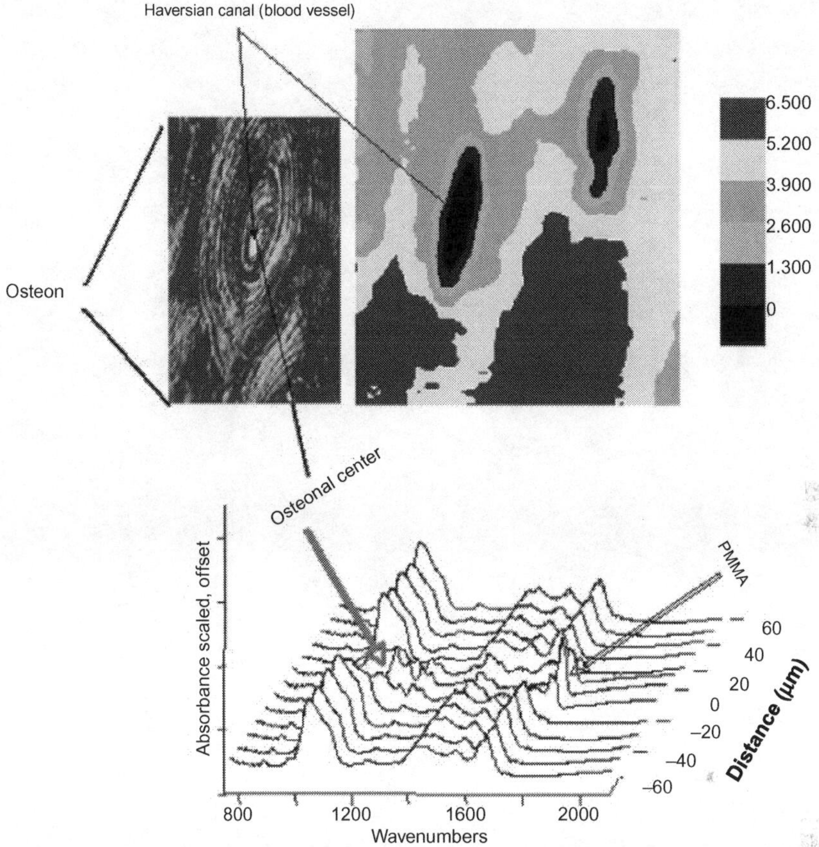

FIG. 4.3 Osteonal bone. An FTIR image illustrating the variation in mineral: matrix ratio along with a photomicrograph showing the concentric rings around the blood vessel in the center of the osteon, and individual spectra taken 20 μm apart along the osteon. *(Reproduced with permission from A. Boskey, R. Mendelshon, Infrared analysis of bone in health and disease, J. Biomed. Opt. 10(3) (2005) 031102.)*

I and amide II bands. For articular cartilage, IR anisotropy was similarly seen for parallel sections in the superficial zone and regular sections, which was also comparable to that for tendon's regular and parallel sections. The parallel sections from cartilage's radial zone, however, had a distinct amide I anisotropy and a nearly isotropic amide II absorption.

Articular cartilage can be anatomically described as a specific zonal structure composed of collagen network and tissue components. To detect depth dependency of structure and composition in intact patellae articular cartilage, FTIR microspectroscopy coupled with fuzzy c-means algorithm (i.e., unsupervised clustering analysis) was studied with rabbit and bovine samples [37]. Although the typical layered structure of articular cartilage was revealed by both clustering

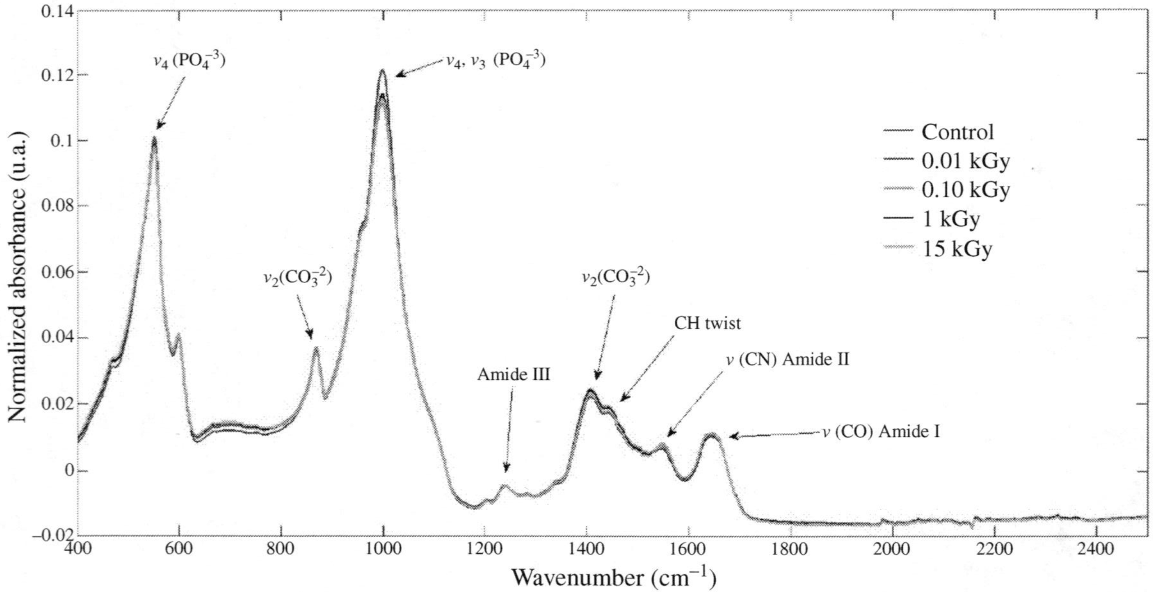

FIG. 4.4 Averaged spectra for nonirradiated bone (Control) and irradiated bone with different doses of ionizing radiation. *(Reproduced with permission from D.A. Dias, M.N. Veloso, P.A. Augusto de Castro, C.A. Lima, D.M. Zezell, Biochemical evaluation of bone submitted to ionizing radiation using ATR-FTIR spectroscopy associated to cluster analysis, in: 2015 International Nuclear Atlantic Conference, INAC 2015, São Paulo, Brazil, 2015.)*

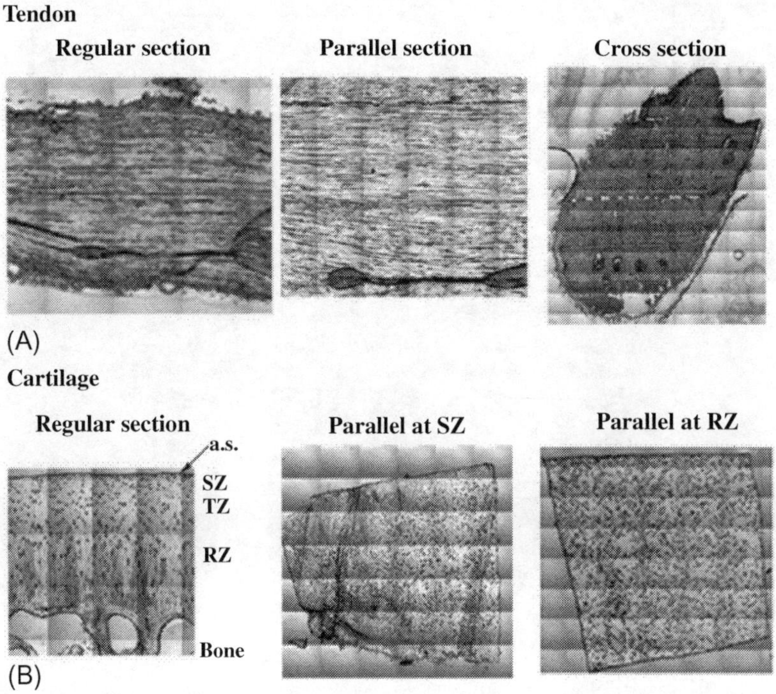

FIG. 4.5 FTIR visible images for tendon (A) and cartilage (B) (*a.s.*, articular surface). *(Reproduced with permission from N. Ramakrishnan, Y. Xia, A. Bidthanapally, Fourier-transform infrared anisotropy in cross and parallel sections of tendon and articular cartilage, J. Orthop. Surg. Res. 3 (2008) 48.)*

and polarized light microscopic images for both species (Fig. 4.6), some differences were also observed at the same spectral locations for clusters. The explanation for this fact was thought to be a significant decrease in estimated proteoglycan/collagen ratio from superficial to middle or deep zones. This finding suggests the potential use of FTIR microspectroscopy for discriminating between intact and degenerated or repaired cartilage tissues. Besides, Fourier transform infrared imaging spectroscopy (FT-IRIS) was applied for macromolecular characterization of orientation and composition in thin tissue sections such as cartilage, bone, and tendon [38]. FT-IRIS nonpolarized and polarized data were recorded in transflectance mode when using low emissivity (low-e) slides (i.e., glass microscope slides coated with a layer of silver-doped tin oxide to minimize the amount of UV and IR light passing through glass without compromising transmission of visible light and high reflection in the mid-IR region) and in transmittance mode (IR beam passes through sample without reflection) when using salt windows (such as BaF_2). Although the absorbance and peak position (in some cases) differed between transmittance and transflectance modes, FT-IRIS analysis qualitatively showed that the orientation of collagen

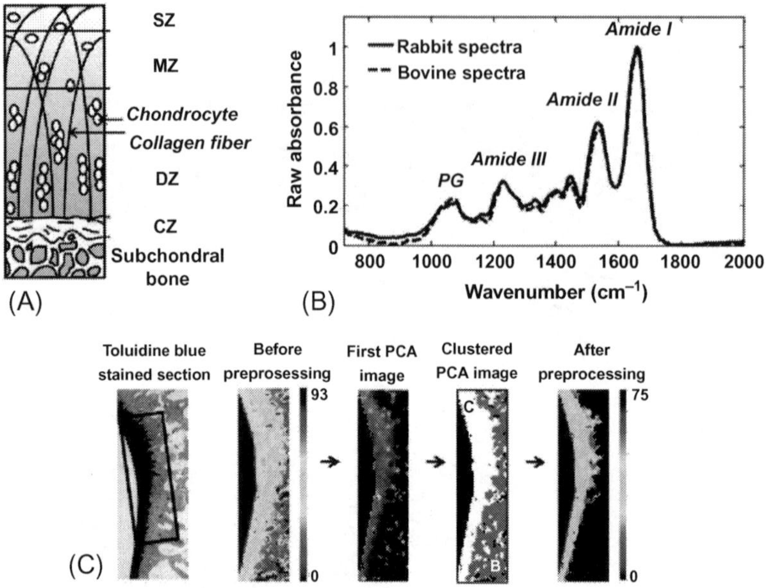

FIG. 4.6 (A) A schematic representation of the articular cartilage structure. The superficial zone (SZ), middle zone (MZ), deep zone (DZ), and calcified zone (CZ) of the tissue are indicated; (B) a typical FTIR microspectroscopy spectrum of rabbit and bovine articular cartilage. Spectral peaks of interest are indicated (amide I, amide II, amide III, and proteoglycan). Spectra are scaled so that the maximum value of the amide I peak (1585–1720 cm^{-1}) equals one; (C) preprocessing steps. Bone spectra and outliers were removed from the images using principal component analysis (PCA). The first, most discriminative, PCA image was built based on the proteoglycan (968–1140 cm^{-1}) region of infrared spectra and clustered using the fuzzy c-means algorithm. Pixels assigned to bone and outliers were removed. Integrated collagen absorbance images show the sample before and after preprocessing. *(Reproduced with permission from A. Hanifi, C. McGoverin, Y.T. Ou, F. Safadi, R. Spencer, N. Pleshko, Differences in infrared spectroscopic data of connective tissues in transflectance and transmittance modes, Anal. Chim. Acta 779 (2013) 41–49.)*

fibrils in articular cartilage was similar based on polarized data (by measuring absorbance ratio of amide I/amide II) for the two modalities.

Osteoarthritis, the most common joint disorder and a major cause of disability with socioeconomic burden, is characterized by progressive deterioration of the articular cartilage or the entire joint. The age-related increase in stiffness of the collagen network in human articular cartilage by accumulating advanced glycation end products crosslinks may putatively present a molecular mechanism, by which age is a predisposing factor for osteoarthritis development [39]. Currently, HPLC is the standard method for determination of crosslink concentrations in tissues. Instead of using this destructive method, crosslink concentrations could be feasibly analyzed by applying competitive adaptive reweighted sampling—PLS regression to measured average FTIR microspectra of standard unstained histological articular cartilage sections (Fig. 4.7) [40].

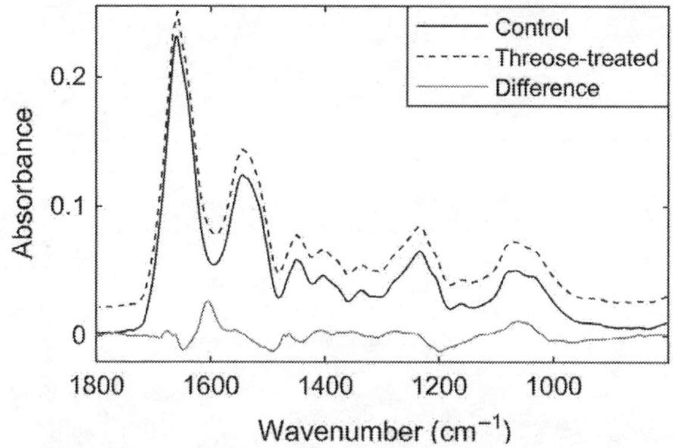

FIG. 4.7 Mean absorbance spectra of control samples *(solid black line)* and threose-treated samples *(dashed black line)*. For better visualization, an offset was added while the difference spectrum (a *solid gray line*) was multiplied by a factor of five. The difference spectrum presents negative peaks in the amide I (1700–1590 cm^{-1}) and the amide II (1590–1450 cm^{-1}) regions, and a positive peak in the carbohydrate region (1000–1100 cm^{-1}). *(Reproduced with permission from L. Rieppo, H.T. Kokkonen, K.A.M. Kulmala, V. Kovanen, M.J. Lammi, J. Töyräs, S. Saarakkala, Infrared microspectroscopic determination of collagen cross-links in articular cartilage, J. Biomed. Opt. 22(3) (2017) 035007.)*

Nonenzymatic glycation is the major posttranslational adjustment of long-life proteins in pathological processes such as aging, atherosclerosis, and diabetes. Thus, the development of in vitro nonenzymatic glycation of fibrillar collagen gels is valuable as it serves as a means of adjusting collagen mechanical properties when investigating cell behavior in response to biochemical changes in the extracellular matrix. For processing of type I collagen gels by nonenzymatic glycation, the documentation of the binding of ribose to collagen was realized by using FTIR spectroscopy [41]. In a biophotonic complementary way, Raman and FTIR microspectroscopic techniques could yield specific and characteristic spectral features as a robust marker for semi-quantitative recognition of nonenzymatic glycation of type I collagen (such as unchanged triple-helical structure, hydroxyproline and proline attributions, carbohydrate band) [42].

In medicine and dentistry, enamel and dentine are known to be the most important mineralized tissues successfully studied by vibrational spectroscopy [43–45]. Apart from energy-dispersive spectroscopy and microenergy-dispersive X-ray fluorescence spectrometry [46], the chemical content of dentine and enamel could be effectively determined by IR and Raman analysis. For example, the composition of dentine and enamel [protein/mineral ratio (2931/430 cm^{-1}), carbonate/phosphate ratio (1070/960 cm^{-1})] was chemically tested by Raman spectroscopic analysis before and after simulated oral cancer radiotherapy [47]; ATR-FTIR and Raman/fluorescence spectra were analyzed for carbonate : mineral ratio and phosphate group concentration within the

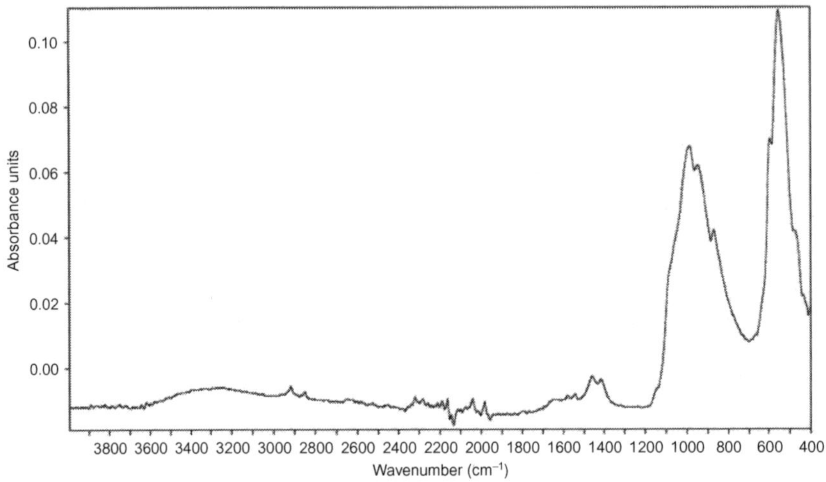

FIG. 4.8 A typical ATR spectrum of dental enamel. *(Reproduced with permission from C. de Carvalho Almança Lopes, P.H. Justino Oliveira Limirio, V. Resende Novais, P. Dechichi, Fourier transform infrared spectroscopy (FTIR) application chemical characterization of enamel, dentin and bone, Appl. Spectrosc. Rev. 53(9) (2018) 747–769.)*

hydroxyapatite molecule respectively, when investigating the effect of acidic and neutral 30% hydrogen peroxide on human tooth enamel [48]; Raman spectra were recorded to calculate inorganic, organic collagen and inorganic/organic content ratio of intracoronary dentin when being bleached with 35% hydrogen peroxide either activated or not by a 970-nm diode laser [49]; FTIR was performed for surface characterization of regular, demineralized, and deproteinized dentin in terms of functional groups such as phosphate, carbonate, C–N, and N–H bands [50]. A typical ATR infrared spectrum of dental enamel is presented in Fig. 4.8.

According to the medical definition [51], cancer is a broad term characterized by an uncontrolled cell division, which leads to abnormal growth of tissues and, in some cases, to metastasis. In the prevention of cancer, early detection is the most important factor. It has been proved since the 2000s that vibrational spectroscopic techniques (Raman and IR) in conjunction with modern chemometrics can extract and analyze biochemical signatures for nondestructive, rapid, and clinically relevant diagnostic assessment of various types of cancer as described below [52].

Prostate adenocarcinoma (CaP) is the most popular cancer in males. A complicated interaction between intraprostatic hormonal androgens, exogenous procarcinogens and intracellular metabolites could be etiologically linked to CaP. The prevalence of latent CaP is shown to be in all age groups evenly across different populations [53]. It means that for different ethnic groups, the same initiation process of CaP may occur with different molecular mechanisms [54]. In a study on segregating human prostate tissues for adenocarcinoma [55], benign

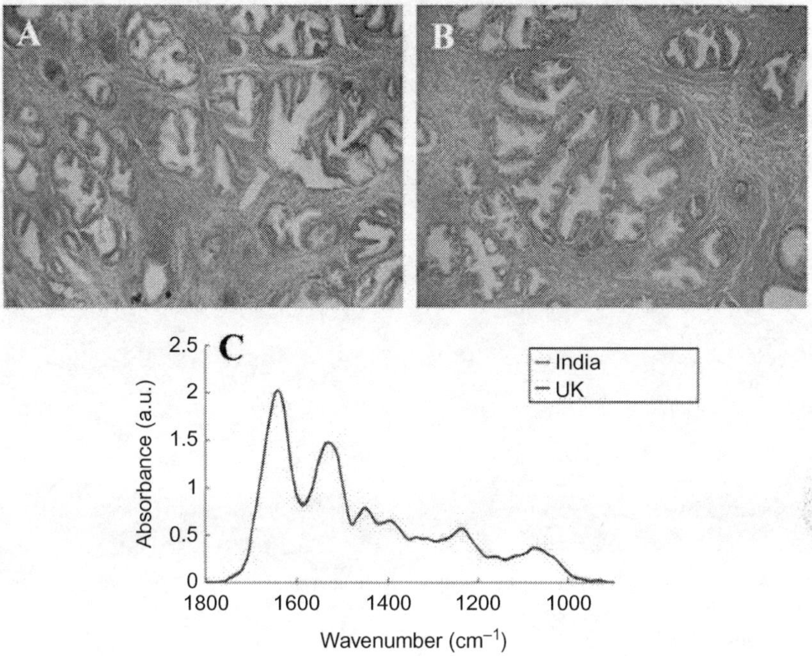

FIG. 4.9 ATR-FTIR spectroscopy analysis of India vs. UK prostate tissue using PCA-LDA. (A) Representative H&E image of India TZ prostate tissue. (B) Representative H&E image of UK TZ prostate tissue. (C) Average spectra of India vs. UK prostate tissue. *(Reproduced with permission from I.I. Patel, J. Trevisan, P.B. Singh, C.M. Nicholson, R.K. Gopala Krishnan, S.S. Matanhelia, F.L. Martin, Segregation of human prostate tissues classified high-risk (UK) versus low-risk (India) for adenocarcinoma using Fourier-transform infrared or Raman microspectroscopy coupled with discriminant analysis, Anal. Bioanal. Chem. 401 (2011) 969–982.)*

samples categorized as high-risk (UK) versus low-risk (India) demographic cohorts were examined by employing attenuated total reflection FTIR spectroscopy, FTIR and Raman microspectroscopies coupled with discriminant analysis to establish spectrally derived risk associations to CaP. Based on the finding of this study (Figs. 4.9–4.11), secondary protein structure variations were identified to be the main molecular biomarkers liable for susceptibility to clinically invasive CaP. Together with DNA alterations exclusively located in the epithelial cell layers of glandular elements, these biochemical changes may reveal vital clues about the etiology of CaP and its progression.

Colorectal carcinoma is the third most frequently diagnosed cancer and the fourth dominant cause of death worldwide. Colorectal lesions (usually presented as polyps, i.e., a well-circumscribed tissue mass protruding into the lumen of the colon) can be classified as nonneoplastic, adenomatous (adenoma), and adenocarcinoma. Adenomatous polyps are known to be premalignant lesions preceding colorectal cancer by 10–15 years [56]. For diagnosis of

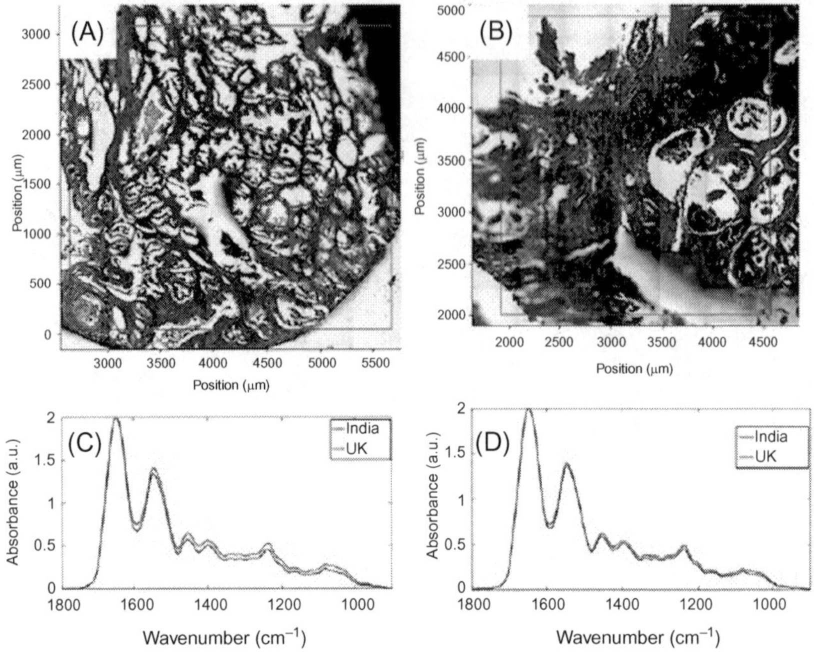

FIG. 4.10 FTIR microspectroscopy analysis of India *(red)* vs. UK *(blue)* prostate tissue architecture using PCA-LDA. (A) Representative bright field image of glandular area of prostate tissue with spectral acquisitions from the glandular epithelium *(red crosses)*. (B) Representative bright field image of glandular area of prostate tissue with spectral acquisitions from stroma *(red crosses)*. (C) Average spectra of India vs. UK glandular epithelium. (D) Average spectra of India vs. UK stroma. *(Reproduced with permission from I.I. Patel, J. Trevisan, P.B. Singh, C.M. Nicholson, R.K. Gopala Krishnan, S.S. Matanhelia, F.L. Martin, Segregation of human prostate tissues classified high-risk (UK) versus low-risk (India) for adenocarcinoma using Fourier-transform infrared or Raman microspectroscopy coupled with discriminant analysis, Anal. Bioanal. Chem. 401 (2011) 969–982.)*

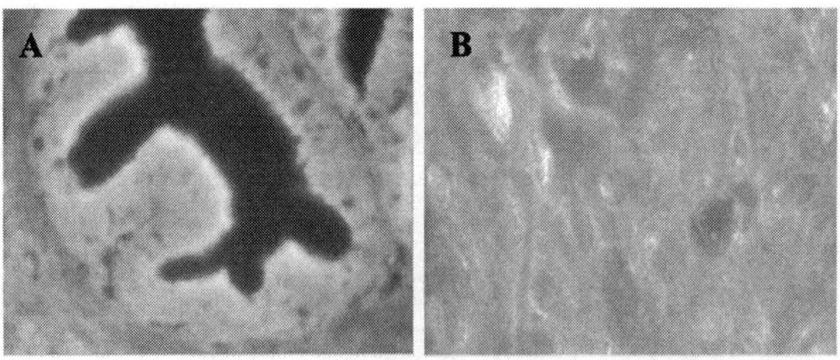

FIG. 4.11 Raman microspectroscopy analysis of India vs. UK prostate tissue architecture using PCA-LDA. (A) Representative dark field image of a single gland from prostate tissue. (B) Representative dark field image of stroma from prostate tissue. *(Reproduced with permission from I.I. Patel, J. Trevisan, P.B. Singh, C.M. Nicholson, R.K. Gopala Krishnan, S.S. Matanhelia, F.L. Martin, Segregation of human prostate tissues classified high-risk (UK) versus low-risk (India) for adenocarcinoma using Fourier-transform infrared or Raman microspectroscopy coupled with discriminant analysis, Anal. Bioanal. Chem. 401 (2011) 969–982.)*

human colorectal adenocarcinoma, the overall accuracy of FTIR imaging data classified by artificial neural network could reach 95% [57]. This is because each tissue structure in the colon wall presents its own distinct signature identified IR spectrally such as the lamina propria mucosae, lamina muscularis mucosae, crypts and lumen filled with mucus. An excellent agreement was reported for immunohistochemical, histopathological, and spectral histopathological results for colon cancer tissue sections [58]. In Figs. 4.12–4.14, the identification of alterations in human colecteral tissues with adenoma and adenocarcinoma by FTIR imaging data was judged to be highly sensitive and specific.

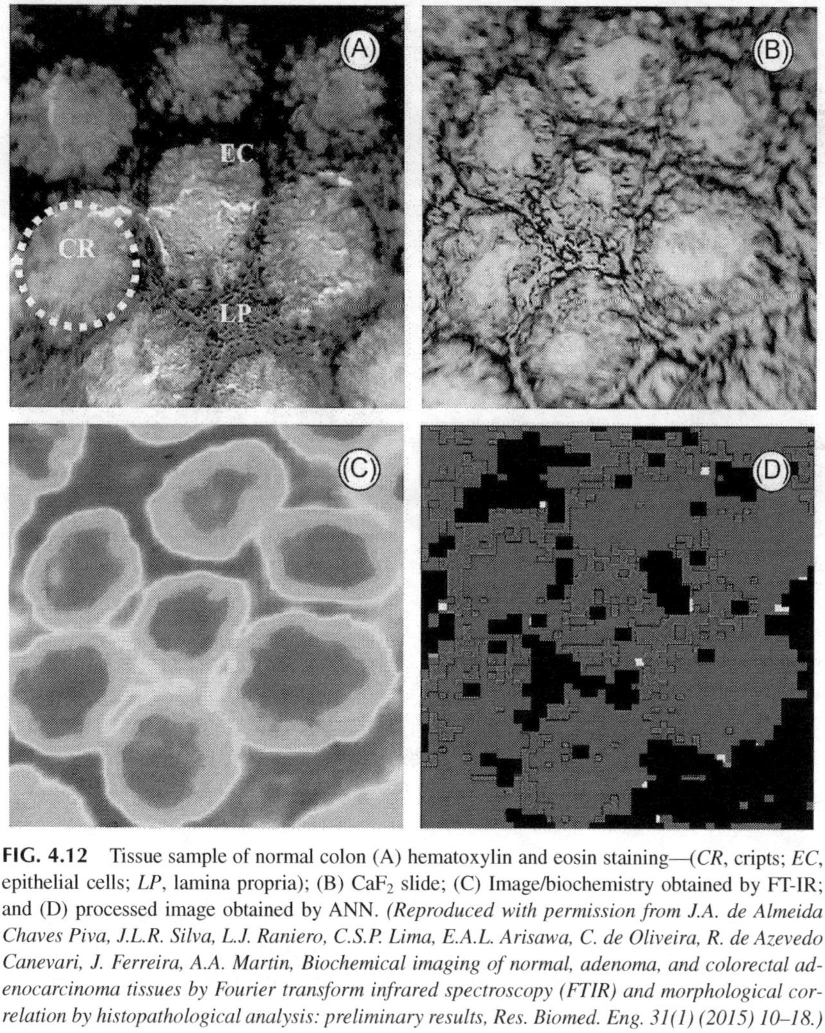

FIG. 4.12 Tissue sample of normal colon (A) hematoxylin and eosin staining—(*CR*, cripts; *EC*, epithelial cells; *LP*, lamina propria); (B) CaF_2 slide; (C) Image/biochemistry obtained by FT-IR; and (D) processed image obtained by ANN. *(Reproduced with permission from J.A. de Almeida Chaves Piva, J.L.R. Silva, L.J. Raniero, C.S.P. Lima, E.A.L. Arisawa, C. de Oliveira, R. de Azevedo Canevari, J. Ferreira, A.A. Martin, Biochemical imaging of normal, adenoma, and colorectal adenocarcinoma tissues by Fourier transform infrared spectroscopy (FTIR) and morphological correlation by histopathological analysis: preliminary results, Res. Biomed. Eng. 31(1) (2015) 10–18.)*

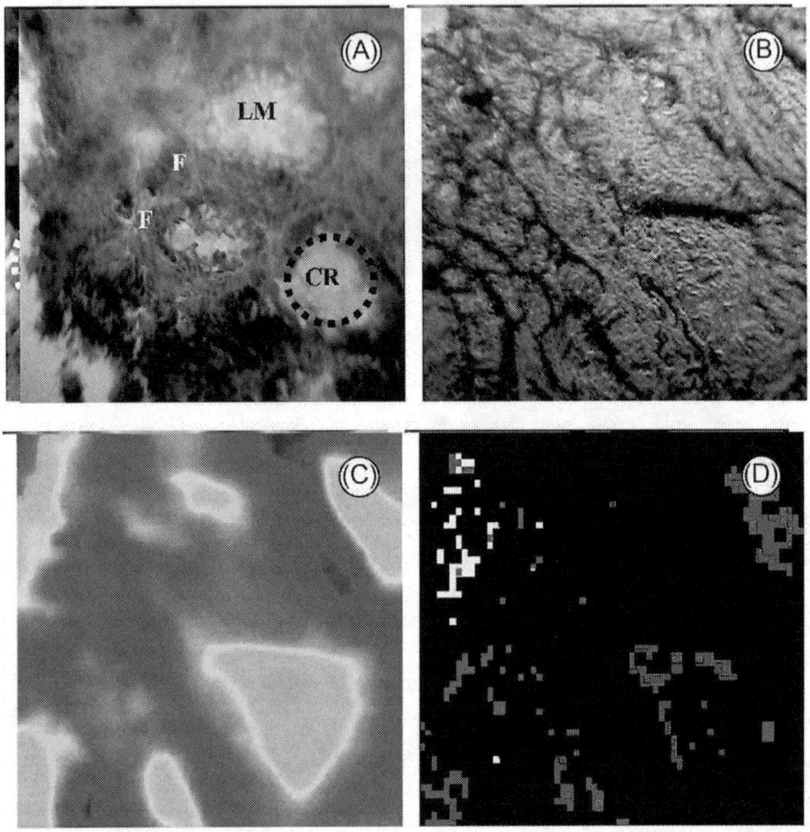

FIG. 4.13 Samples of polyp (severe dysplasia) (A) hematoxylin and eosin staining (*LM*, lumen; *CR*, cripts; *F*, fusion of glandular structures); (B) CaF_2 slide; (C) image/biochemistry obtained by FT-IR; and (D) processed image obtained by ANN. *(Reproduced with permission from J.A. de Almeida Chaves Piva, J.L.R. Silva, L.J. Raniero, C.S.P. Lima, E.A.L. Arisawa, C. de Oliveira, R. de Azevedo Canevari, J. Ferreira, A.A. Martin, Biochemical imaging of normal, adenoma, and colorectal adenocarcinoma tissues by Fourier transform infrared spectroscopy (FTIR) and morphological correlation by histopathological analysis: preliminary results, Res. Biomed. Eng. 31(1) (2015) 10–18.)*

For characterization of thyroid tissue, FTIR spectroscopic technique was also proved to be able to significantly discriminate benign nodules from healthy tissue in the area of the B-band between healthy tissue (1452.90 cm^{-1} biologically associated with lipids and proteins) and goiter (1069.80 cm^{-1} corresponding to DNA) as well as in the width of the C band (corresponding to DNA-RNA, DNA) between normal thyroid tissue and carcinoma [59].

In women, there are five major kinds of cancer that affect the reproductive organs, i.e., cervical, ovarian, uterine, vaginal, and vulvar. It was shown that the employment of formalin-fixed tissues was feasible in optical pathology of ovarian cancer [60]. Raman and FTIR spectral characteristics were similar for

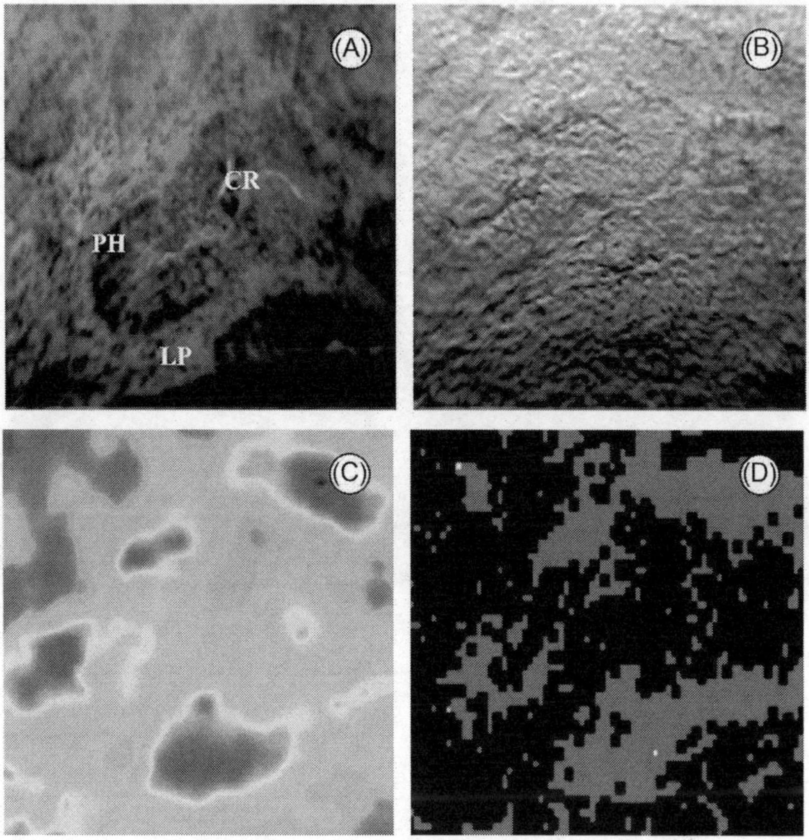

FIG. 4.14 Samples of adenocarcinoma. (A) hematoxylin and eosin staining (*CR*, cripts; *LP*, lamina propria; *PH*, cellular and nuclear pleomorphism, and hyperchromatic nuclear); (B) CaF_2 slide; (C) Image/biochemistry obtained by FT-IR; and (D) Processed image obtained by ANN. *(Reproduced with permission from J.A. de Almeida Chaves Piva, J.L.R. Silva, L.J. Raniero, C.S.P. Lima, E.A.L. Arisawa, C. de Oliveira, R. de Azevedo Canevari, J. Ferreira, A.A. Martin, Biochemical imaging of normal, adenoma, and colorectal adenocarcinoma tissues by Fourier transform infrared spectroscopy (FTIR) and morphological correlation by histopathological analysis: preliminary results, Res. Biomed. Eng. 31(1) (2015) 10–18.)*

normal and benign tissues (containing higher protein contents), but very different from those of malignant tissues (containing higher levels of lipids and DNA), as shown in Figs. 4.15 and 4.16. Malignant tissues could be well delineated from normal and benign ones by hierarchical cluster analysis of first-derivative Raman spectra (700–1700 cm^{-1}) and second-derivative FTIR spectra (1540–1680 and 1720–1780 cm^{-1}). Moreover, nonoverlapping subclusters were also observed for benign and normal tissues at a lower heterogeneity level.

For cervical cancer screening, it is highly possible that premalignant lesions of the cervix are neglected because many cells are potentially related to a high-grade squamous intraepithelial lesion (HSIL) lineage, although

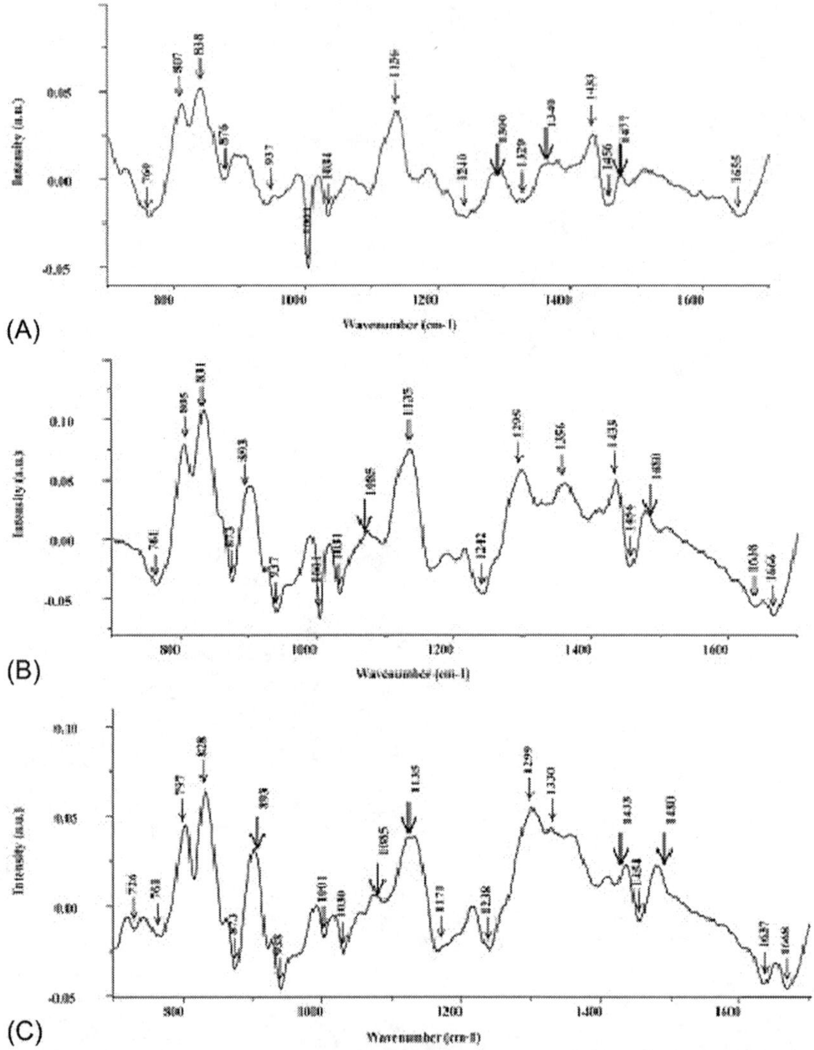

FIG. 4.15 Difference micro-Raman spectra of ovarian tissues: (A) benign-normal, (B) malignant-normal, and (C) malignant-benign. *(Reproduced with permission from C. Murali Krishna, G.D. Sockalingum, A.R. Bhat, L. Venteo, P. Kushtagi, M. Pluot, M. Manfait, FTIR and Raman microspectroscopy of normal, benign, and malignant formalin-fixed ovarian tissues, Anal. Bioanal. Chem. 387 (2007) 1649–1656.)*

cytologically considered as normal. Raman spectroscopy, however, appeared to be sensitive enough to detect HSIL regardless of the cell type being measured on the unstained slides, i.e., morphologically normal appearing cells or a mixed population of superficial and intermediate cells (Fig. 4.17).

For cervical cancer diagnosis, normal areas in the cancerous cervix were mostly served as control for in vivo Raman diagnosis. This spectral acquirement is impossible for cervical cancer detected at advanced stages. Provided

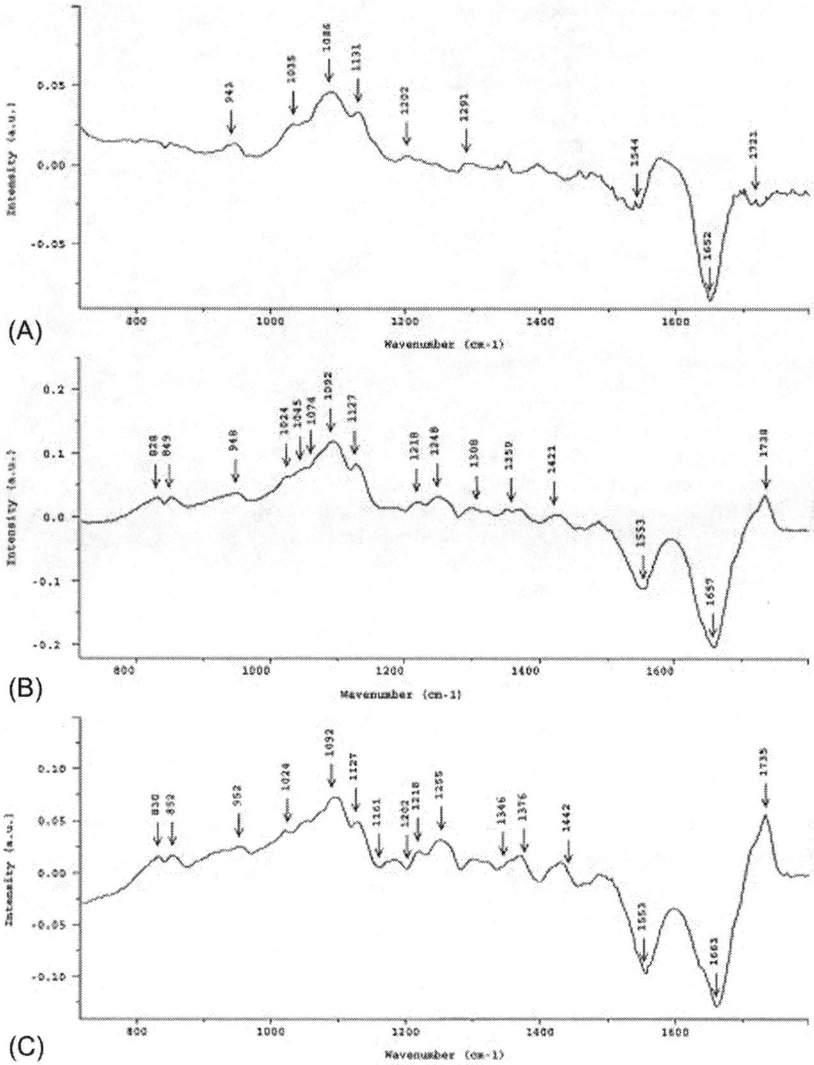

FIG. 4.16 Difference FTIR spectra of ovarian tissues: (A) benign-normal, (B) malignant-normal, and (C) malignant-benign. *(Reproduced with permission from C. Murali Krishna, G.D. Sockalingum, A.R. Bhat, L. Venteo, P. Kushtagi, M. Pluot, M. Manfait, FTIR and Raman microspectroscopy of normal, benign, and malignant formalin-fixed ovarian tissues, Anal. Bioanal. Chem. 387 (2007) 1649–1656.)*

that vagina and ectocervix are biochemically similar, the utility of the vagina was proposed as an internal control for classification of normal (characterized by collagenous proteins) and cancerous (abundant in DNA and noncollagenous proteins) conditions in the Indian population [61].

Breast carcinoma is believed to be most common cancer in women, of which the highest incidence rate was reported in the developed world (ca. 50%) [62].

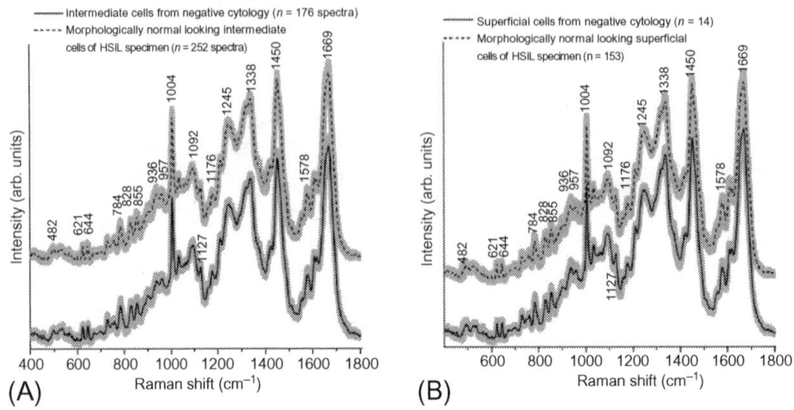

FIG. 4.17 (A) Mean Raman spectra ±1 standard deviation (SD) acquired from the intermediate cells of negative cytology ThinPrep specimens ($n=18$) and morphologically normal appearing intermediate cells of high-grade squamous intraepithelial lesion (HSIL) ThinPrep specimens ($n=17$). (B) Mean Raman spectra ±1 SD acquired from the superficial cells of negative cytology specimens ($n=18$) and morphologically normal appearing superficial cells of HSIL cytology ThinPrep specimens ($n=17$). *(Reproduced with permission from S. Duraipandian, D. Traynor, P. Kearney, C. Martin, J.J. O'Leary, F.M. Lyng, Raman spectroscopic detection of high-grade cervical cytology: using morphologically normal appearing cells, Sci. Rep. 8 (2018) 15048.)*

The diagnosis of breast cancer is usually relied on various types of technique such as histopathological, physical (e.g., ultrasonography, elastography, mammography, positron emission tomography, magnetic resonance), biological and optical (e.g., fluorescence tomography, photoacoustic imaging). Unluckily, none of these techniques can generate unique or concrete answers. Thus, comparative measurements were arranged to find out whether spectral differences existed between normal noncancerous breast tissue, breast cancer tissues before and after chemotherapy, and normal breast tissues in the vicinity of the tumor [63]. It was stated that the regions specific to carotenoids and lipids in Raman spectra as well as characteristic for IR signals of sugars and proteins could be exploited as excellent biomarkers in the diagnosis of breast cancer and monitoring of its chemotherapy effectiveness (e.g., Fig. 4.18).

According to the latest global cancer data published in 2018, lung cancer is accountable for the biggest number of deaths (ca. 1.8 million people died, 18.4% of the total) due to its insufficient prognosis [64]. It was shown that high-quality spectra of lung cancer tissues (malignant, adjacent to cancer, normal) could be favorably obtained in a few seconds by using ATR-FTIR [65] and confocal Raman microscopes. Spectral characteristics differed significantly between normal and malignant tissues (e.g., the ratios of 1245-cm^{-1}/1571-cm^{-1} Raman intensity and 1453-cm^{-1}/1645-cm^{-1} IR intensity were the most statistically different). In a pilot study, canonical discriminant analysis was performed on ATR-FTIR spectral data to discriminate malignant and nonmalignant lung tissues collected from 30 patients undergoing pulmonary lobectomy [66]. Fig. 4.19

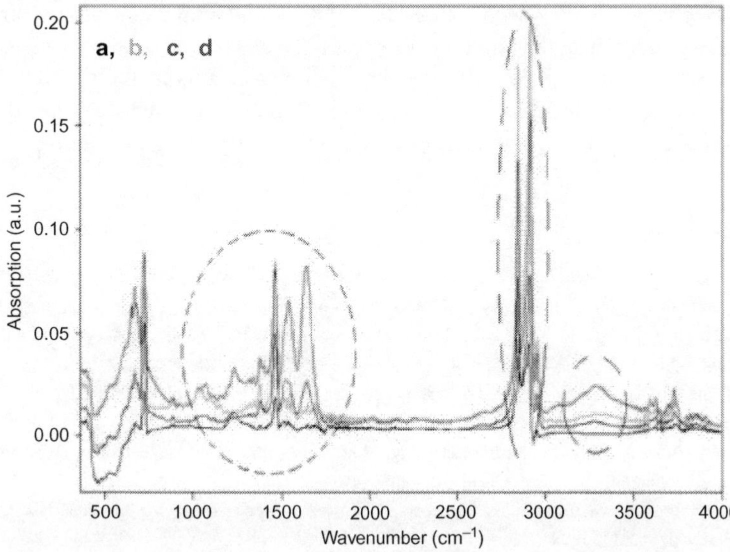

FIG. 4.18 Selection of the spectrum of absorbance versus wave number for tissue materials obtained from a female 54-year-old patient of triple-negative breast cancer: breast cancer tissues before chemotherapy (A), normal breast tissues received around cancerous breast region (B), breast cancer tissues after chemotherapy (C), and sample noncancerous normal breast tissue (D). *(Reproduced with permission from S. Rehman, Z. Movasaghi, J.A. Darr, I.U. Rehman, Fourier transform infrared spectroscopic analysis of breast cancer tissues; identifying differences between normal breast, invasive ductal carcinoma, and ductal carcinoma in situ of the breast, Appl. Spectrosc. Rev. 45 (2010) 355–368.)*

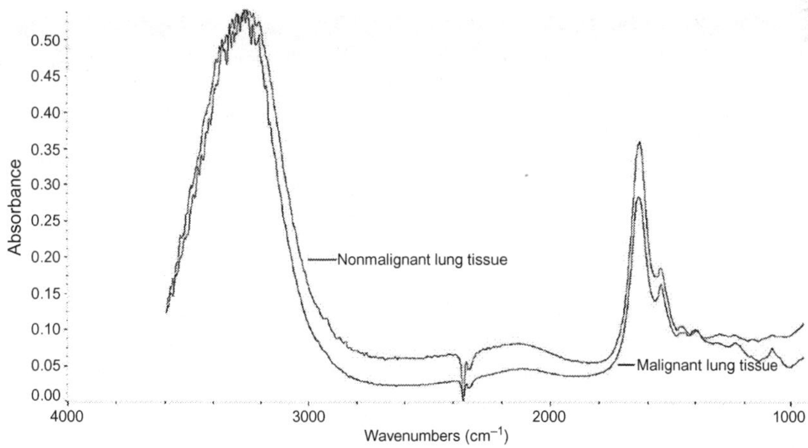

FIG. 4.19 ATR-FTIR spectra of nonmaglinant and malignant lung tumors. *(Reproduced with permission from X. Sun, Y. Xu, J. Wu, Y. Zhang, K. Sun, Detection of lung cancer tissue by attenuated total reflection-Fourier transform infrared spectroscopy—a pilot study of 60 samples, J. Surg. Res. 179 (2013) 33–38.)*

displays characteristic spectra of nonmalignant and malignant tissues from the same patient. The specificity and sensitivity of the discriminants were all 96.7%. When being joined with adaptable optic fiber, this method could offer in vivo intraoperative detection and differentiation, helping surgeons fast diagnose, decide the extent of dissection, and avoid needless surgical trauma.

References

[1] G. Hertzburg, Molecular Spectra and Molecular Structure, Volume II: Infrared and Raman Spectra of Polyatomic Molecules, Krieger Publishing Company, 1945.
[2] P. Lasch, J. Kneipp, Biomedical Vibrational Spectroscopy, Wiley-Interscience, Hoboken, NJ, 2008, pp. 121–147.
[3] T. Theophanides (Ed.), Infrared Spectroscopy—Life and Biomedical Sciences, Intech, Croatia, 2012.
[4] E.W. Ciurczak, B. Igne, Pharmaceutical and Medical Applications of Near-Infrared Spectroscopy, second ed., CRC Press Taylor & Francis, 2015.
[5] M. Diem, P.R. Griffiths, J.M. Chalmers, Vibrational Spectroscopy for Medical Diagnosis, John Wiley and Sons, Chichester, UK, 2008.
[6] R. Salzer, H.W. Siesler, Infrared and Raman Spectroscopic Imaging, Wiley-VCH Verlag, Weinheim, Germany, 2008.
[7] A. Elliot, E.J. Ambrose, Structure of synthetic polypeptides, Nature 165 (1950) 921–922.
[8] E.K. Blout, R.C. Mellors, Infrared spectra of tissues, Science 110 (2849) (1949) 137–138.
[9] D.L. Woernley, Infrared absorption curves for normal and neoplastic tissues and related biological substances, Cancer Res. 12 (7) (1952) 516–523.
[10] R. Barer, A.R.H. Cole, H.W. Thompson, Infra-red spectroscopy with the reflecting microscope in physics, chemistry and biology, Nature 163 (4136) (1949) 198–201.
[11] C.R. Burch, Reflecting microscope, Nature 152 (1943) 748–749.
[12] R. Barer, Aspects of ultra-violet and infra-red microspectrography with the Burch reflecting microscope, Discuss. Faraday Soc. 9 (1950) 369–378.
[13] B. Sherman, Infrared spectroscopy by attenuated total reflection, Appl. Spectrosc. 18 (1964) 7–9.
[14] T.S. Hermann, Identification of trace amounts of organophosphorus pesticides by frustrated multiple internal reflectance spectroscopy, Appl. Spectrosc. 19 (1965) 10–14.
[15] R.C. Lord, N.T. Yu, Laser-excited Raman spectroscopy of biomolecules: I. Native lysozyme and its constituent amino acids, J. Mol. Biol. 50 (2) (1970) 509–524.
[16] H. Shin, M.K. Markey, A machine learning perspective on the development of clinical decision support systems utilizing mass spectra of blood serum, J. Biomed. Inform. 39 (2) (2006) 227–248.
[17] H. Mantsch, M. Jackson, Molecular spectroscopy in biodiagnostics (from Hippocrates to Herschel and beyond), J. Mol. Struct. 347 (1995) 187–206.
[18] D. Simonova, I. Karamancheva, Application of Fourier transform infrared spectroscopy for tumor diagnosis, Biotechnol. Biotechnol. Equip. 27 (6) (2013) 4200–4207.
[19] M.J. Walsh, S.E. Holton, A. Kajdacsy-Balla, R. Bhargava, Attenuated total reflectance Fourier-transform infrared spectroscopic imaging for breast histopathology, Vib. Spectrosc. 60 (2012) 23–28.
[20] P. Lasch, L. Chiriboga, H. Yee, M. Diem, Infrared spectroscopy of human cells and tissue: detection of disease, Technol. Cancer Res. Treat. 1 (1) (2002).

[21] M. Diem, M. Romeo, S. Boydston-White, M. Miljković, C. Matthäus, A decade of vibrational micro-spectroscopy of human cells and tissue (1994–2004), Analyst 129 (10) (2004) 880–885.

[22] L.P. Choo-Smith, H.G.M. Edwards, H.P. Endtz, J.M. Kros, F. Heule, H. Barr, J.S. Robinson Jr., H.A. Bruining, G.J. Puppels, Medical applications of Raman spectroscopy: from proof of principle to clinical implementation, Biopolymers 67 (1) (2002) 1–9.

[23] A.A. Bunaciu, V.D. Hoang, H.Y. Aboul-Enein, Vibrational micro-spectroscopy of human tissues analysis: review, Crit. Rev. Anal. Chem. 47 (3) (2016) 194–203.

[24] Z. Movasaghi, S. Rehman, I.U. Rehman, Raman spectroscopy of biological tissues, Appl. Spectrosc. Rev. 42 (5) (2007) 493–541.

[25] J. Anderson, J. Dellomo, A. Sommer, A. Evan, S. Bledsoe, A concerted protocol for the analysis of mineral deposits in biopsied tissue using infrared microanalysis, Urol. Res. 33 (2005) 213–219.

[26] J.A. Buckwalter, M.J. Glimcher, R.R. Cooper, R. Recker, Bone biology. I: structure, blood supply, cells, matrix, and mineralization, Instr. Course Lect. 45 (1996) 371–386.

[27] A.S. Posner, G. Duyckaerts, Infrared study of the carbonate in bone, teeth and francolite, Experientia 10 (1954) 424–425.

[28] A.S. Posner, Bone mineral on the molecular level, Fed. Proc. 32 (1973) 1933–1937.

[29] J.D. Termine, A.S. Posner, Infrared analysis of rat bone: age dependency of amorphous and crystalline mineral fractions, Science 153 (1966) 1523–1525.

[30] A. Boskey, R. Mendelshon, Infrared analysis of bone in health and disease, J. Biomed. Opt. 10 (3) (2005) 031102.

[31] R. Mendelsohn, A. Hassankhani, E. DiCarlo, A. Boskey, FT-IR microscopy of endochondral ossification at 20 μm spatial resolution, Calcif. Tissue Int. 44 (1989) 20–24.

[32] E.P. Paschalis, E. DiCarlo, F. Betts, P. Sherman, R. Mendelsohn, A.L. Boskey, FTIR microspectroscopic analysis of human osteonal bone, Calcif. Tissue Int. 59 (1996) 480–487.

[33] P. Demir, F. Severcan, Monitoring radiation induced alterations in biological systems, from molecules to tissues, through infrared spectroscopy, Appl. Spectrosc. Rev. 51 (10) (2016) 839–863.

[34] D.A. Dias, M.N. Veloso, P.A. Augusto de Castro, C.A. Lima, D.M. Zezell, Biochemical evaluation of bone submitted to ionizing radiation using ATR-FTIR spectroscopy associated to cluster analysis, in: 2015 International Nuclear Atlantic Conference, INAC 2015, São Paulo, Brazil, 2015.

[35] J.A. Buckwalter, H.J. Mankin, Articular cartilage: tissue design and chondrocyte-matrix interactions, Instr. Course Lect. 47 (1998) 477–486.

[36] N. Ramakrishnan, Y. Xia, A. Bidthanapally, Fourier-transform infrared anisotropy in cross and parallel sections of tendon and articular cartilage, J. Orthop. Surg. Res. 3 (2008) 48.

[37] Y. Kobrina, L. Rieppo, S. Saarakkala, J.S. Jurvelin, H. Isaksson, Clustering of infrared spectra reveals histological zones in intact articular cartilage, Osteoarthr. Cartil. 20 (5) (2012) 460–468.

[38] A. Hanifi, C. McGoverin, Y.T. Ou, F. Safadi, R. Spencer, N. Pleshko, Differences in infrared spectroscopic data of connective tissues in transflectance and transmittance modes, Anal. Chim. Acta 779 (2013) 41–49.

[39] N. Verzijl, J. DeGroot, C.B. Zaken, O. Braun-Benjamin, A. Maroudas, R.A. Bank, J. Mizrahi, C.G. Schalkwijk, S.R. Thorpe, J.W. Baynes, J.W.J. Bijlsma, F.P.J.G. Lafeber, J.M. TeKoppele, Crosslinking by advanced glycation end products increases the stiffness of the collagen network in human articular cartilage: a possible mechanism through which age is a risk factor for osteoarthritis, Arthritis Rheumatol. 46 (1) (2002) 114–123.

[40] L. Rieppo, H.T. Kokkonen, K.A.M. Kulmala, V. Kovanen, M.J. Lammi, J. Töyräs, S. Saarakkala, Infrared microspectroscopic determination of collagen cross-links in articular cartilage, J. Biomed. Opt. 22 (3) (2017) 035007.

[41] R. Roy, A. Boskey, L.J. Bonassar, Processing of type I collagen gels using nonenzymatic glycation, J. Biomed. Mater. Res. A 93A (3) (2010) 843–851.

[42] M. Guilbert, G. Said, T. Happillon, V. Untereiner, R. Garnotel, P. Jeannesson, G.D. Sockalingum, Probing non-enzymatic glycation of type I collagen: a novel approach using Raman and infrared biophotonic methods, Biochim. Biophys. Acta 1830 (6) (2013) 3525–3531.

[43] C.d.C.A. Lopes, P.H.J.O. Limirio, V.R. Novais, P. Dechichi, Fourier transform infrared spectroscopy (FTIR) application chemical characterization of enamel, dentin and bone, Appl. Spectrosc. Rev. 53 (9) (2018) 747–769.

[44] L. Bachmann, R. Diebolder, R. Hibst, D.M. Zezell, Infrared absorption bands of enamel and dentin tissues from human and bovine teeth, Appl. Spectrosc. Rev. 38 (2003) 1–14.

[45] L.C. Palmer, C.J. Newcomb, S.R. Kaltz, E.D. Spoerke, S.I. Stupp, Biomimetic systems for hydroxyapatite mineralization inspired by bone and enamel, Chem. Rev. 108 (11) (2008) 4754–4783.

[46] L.E. Soares, A.C. De Carvalho Filho, Protective effect of fluoride varnish and fluoride gel on enamel erosion: roughness, SEM-EDS, and micro-EDXRF studies, Microsc. Res. Tech. 78 (3) (2015) 240–248.

[47] R. Reed, C. Xu, Y. Liu, J.P. Gorski, Y. Wang, M.P. Walker, Radiotherapy effect on nanomechanical properties and chemical composition of enamel and dentine, Arch. Oral Biol. 60 (5) (2015) 690–697.

[48] L. Sun, S. Liang, Y. Sa, Z. Wang, X. Ma, T. Jiang, Y. Wang, Surface alteration of human tooth enamel subjected to acidic and neutral 30% hydrogen peroxide, J. Dent. 39 (10) (2011) 686–692.

[49] F.C. Lopes, R. Roperto, A. Akkus, O. Akkus, R.G. Palma-Dibb, M.D. de Sousa-Neto, Effect of laser activated bleaching on the chemical stability and morphology of intracoronal dentin, Arch. Oral Biol. 86 (2017) 40–45.

[50] F.S. Tabatabaei, S. Tatari, R. Samadi, M. Torshabi, Surface characterization and biological properties of regular dentin, demineralized dentin, and deproteinized dentin, J. Mater. Sci.: Mater. Med. 27 (11) (2016) 164.

[51] https://www.medicinenet.com/script/main/art.asp?articlekey=2580, (Accessible on 25 January 2020).

[52] C. Kendall, M. Isabelle, F. Bazant-Hegemark, J. Hutchings, L. Orr, J. Babrah, R. Baker, N. Stone, Vibrational spectroscopy: a clinical tool for cancer diagnostics, Analyst 134 (6) (2009) 1029–1245.

[53] H. Grönberg, Prostate cancer epidemiology, Lancet 361 (2003) 859–864.

[54] R. Yatani, I. Chigusa, K. Akazaki, G.N. Stemmermann, R.A. Welsh, P. Correa, Geographic pathology of latent prostatic carcinoma, Int. J. Cancer 29 (1982) 611–616.

[55] I.I. Patel, J. Trevisan, P.B. Singh, C.M. Nicholson, R.K. Gopala Krishnan, S.S. Matanhelia, F.L. Martin, Segregation of human prostate tissues classified high-risk (UK) versus low-risk (India) for adenocarcinoma using Fourier-transform infrared or Raman microspectroscopy coupled with discriminant analysis, Anal. Bioanal. Chem. 401 (2011) 969–982.

[56] J.R.N. Torres, J.S. Arcieri, F.R. Teixeira, Aspectos epidemiológicos dos pólipos e lesões plano-elevadas colorretais, Rev. Bras. Coloproctol. 30 (4) (2011) 419–429.

[57] P. Lasch, W. Haensch, E.N. Lewis, L.H. Kidder, C. Naumann, Characterization of colorectal adenocarcinoma sections by spatially resolved FT-IR microspectroscopy, Appl. Spectrosc. 56 (1) (2002) 1–9.

[58] A. Kallenbach-Thieltges, F. Großerüschkamp, A. Mosig, M. Diem, A. Tannapfel, K. Gerwert, Immunohistochemistry, histopathology and infrared spectral histopathology of colon cancer tissue sections, J. Biophotonics 6 (1) (2013) 88–100.

[59] L.F. de Azevedo Villela, L.C.C. de Freitas, M. Brandão, L. Bachmann, R. Iwakura, Characterization of thyroid tissue using infrared spectroscopy, Arch. Head Neck Surg. 47 (1) (2018) e0874.

[60] C. Murali Krishna, G.D. Sockalingum, A.R. Bhat, L. Venteo, P. Kushtagi, M. Pluot, M. Manfait, FTIR and Raman microspectroscopy of normal, benign, and malignant formalin-fixed ovarian tissues, Anal. Bioanal. Chem. 387 (2007) 1649–1656.

[61] R.S. Shaikh, T.K. Dora, S. Chopra, A. Maheshwari, D.K. Kedar, R. Bharat, C.M. Krishna, In vivo Raman spectroscopy of human uterine cervix: exploring the utility of vagina as an internal control, J. Biomed. Opt. 19 (8) (2014) 087001.

[62] https://www.who.int/cancer/detection/breastcancer/en/index1.html, (Accessible on 30 January 2020).

[63] J. Depciuch, E. Kaznowska, I. Zawlik, R. Wojnarowska, M. Cholewa, P. Heraud, J. Cebulski, Application of Raman spectroscopy and infrared spectroscopy in the identification of breast cancer, Appl. Spectrosc. 70 (2) (2016) 251–263.

[64] World Health Organization, Latest Global Cancer Data: Cancer Burden Rises to 18.1 Million New Cases and 9.6 Million Cancer Deaths in 2018, International Agency for Research on Cancer—World Health Organization, 2018. Press Release No. 263 (12 September 2018).

[65] J. Lv, L. Zhang, J. Feng, Y. Liu, Z. Wang, M. Zhao, R. Shi, Optical observation of lung cancer with attenuated total reflectance—Fourier transform infrared microscope (ATR-FTIR) and confocal Raman microscope, Spectrosc. Lett. 44 (2011) 312–317.

[66] X. Sun, Y. Xu, J. Wu, Y. Zhang, K. Sun, Detection of lung cancer tissue by attenuated total reflection-Fourier transform infrared spectroscopy—a pilot study of 60 samples, J. Surg. Res. 179 (2013) 33–38.

Part III

Pharmaceutical analysis applications

Chapter 5

Chemical drug analysis

The term "quality control" pertains to the sum of all procedures, ranging from simple chemical experiments to more complicated pharmacopoeial monographs, implemented for ensuring the identity and purity of a particular drug [1]. Since the 1960s, the quality control of pharmaceuticals has been considered as an essential operation of the pharmaceutical industry when new and better medicinal agents being produced at an accelerated rate [2].

Marketed drugs are expected to be safe and therapeutically active formulations. In order to have a consistent and predictable effectiveness, a drug must be formulated (usually with excipients) in suitable dosage forms for administration to the patients by various routes for disease diagnosis and treatment [3]. It means that the matrix of pharmaceutical formulations may have effect on not only their bioavailability but also the performance of their analytical methods.

At the present time, the principal application of vibrational spectroscopy in the major pharmacopeias (e.g., United States Pharmacopeia [4], European Pharmacopeia [5] and British Pharmacopeia [6]) is limited to the IR identification of excipients and active pharmaceutical ingredients (APIs) by comparing spectrally a sample against a reference standard. On the other hand, Raman spectroscopy has a much lower profile with regulatory authorities as compared to IR methods, e.g., a collection of Raman spectra for widely used pharmaceutical excipients was published only in a research article for reference [7].

In the literature, however, there has been a great number of papers dedicated to chemical drug analysis using vibrational spectroscopic techniques [8–19] that can be categorized into three main groups: (i) drug quantification and formulation characterization, (ii) polymorphic analysis and (iii) counterfeiting drug analysis (as detailed discussion in this chapter).

Drug quantification and formulation characterization

As a rule, IR quantification of a drug is feasible when the Lambert-Beer law is applicable in the case of transmission measurements of liquids. As an example, an FTIR method was developed for the determination of diazepam (7-chloro-1,3-dihydro-1-methyl-5-phenyl-2H-1,4-benzodiazepin-2-one) [20], a benzodiazepine compound [21] often prescribed for the treatment of severe anxiety disorders, short-term management of insomnia, and suppression of alcohol withdrawal syndrome. This method involved (i) off-line extracting from

https://doi.org/10.1016/B978-0-12-818827-9.00006-8

© 2020 Elsevier Inc. All rights reserved.

pharmaceuticals with chloroform by sonication, (ii) measuring peak area for the extracts in the spectral interval between 1672 and 1682 cm^{-1} using a baseline correction defined between 1850 and 1524 cm^{-1} (Fig. 5.1). These FTIR data were statistically comparable to those obtained from the UV spectrometry reference method specified by USP [22].

Using IR spectroscopy, APIs in semisolid and/or solid dosage forms could be also directly examined in reflectance mode. For instance, FTIR spectroscopy was able to quantify piroxicam, a nonsteroidal antiinflammatory drug [23], in tablets and ointments when coupled with chemometrics (PLS and PCR) [24]. In other studies, diclofenac sodium (the sodium salt of 2-[(2,6-dichlorophenyl) aminophenyl]-acetic acid), a potent analgesic and antiinflammatory agent [25], was successfully determined in tablets and capsules (relative standard errors of prediction ≤3%) by using different techniques of vibrational spectroscopy combined with PLS, i.e., DRIFT (diffuse reflectance infrared Fourier transform)

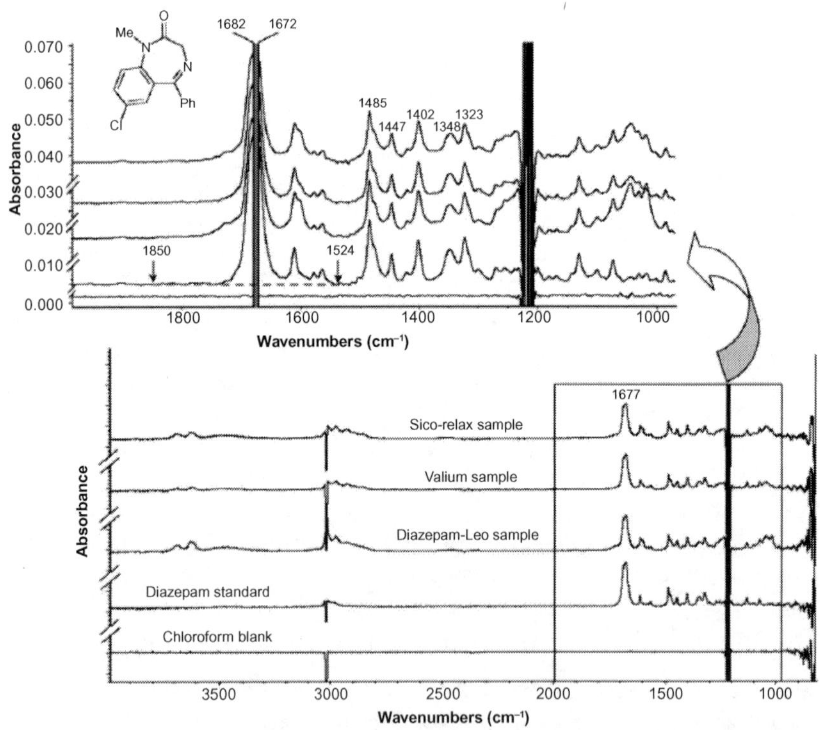

FIG. 5.1 FTIR spectra of a diazepam standard in chloroform at a concentration level of 1 $mg\,g^{-1}$ and extracts of each one of the three pharmaceuticals assayed. Instrumental conditions: 25 scans averaged per spectra, 4 cm^{-1} nominal resolution using a background of the cell filled with chloroform. Note: the spectra were shifted on the absorbance axis to clearly show their bands. *(Reproduced with permission from J. Moros, S. Garrigues, M. de la Guardia, Quality control Fourier transform infrared determination of diazepam in pharmaceuticals, J. Pharm. Biomed. Anal. 43 (2007) 1277–1282.)*

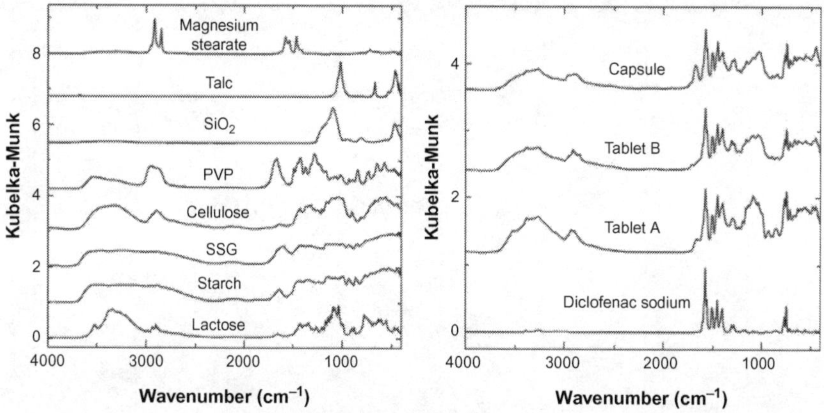

FIG. 5.2 DRIFTS spectra of excipients *(left)*, diclofenac sodium and its analyzed preparations *(right)*; the spectra are offset for clarity; *PVP*, polyvinylpyrrolidone; *SSG*, sodium starch glycolate. *(Reproduced with permission from R. Szostak, S. Mazurek, The influence of sample area on diclofenac sodium quantification by diffuse reflectance IR spectroscopy, Talanta 84 (2011) 583–586.)*

spectroscopy (Fig. 5.2) [26], FTIR ATR (Fourier transform infrared attenuated total reflection), and FT-Raman spectroscopy (Fig. 5.3) [27]. It was pointed out for DRIFTS accessories (the spectrometer beam area on the surface of the sample was approximately sevenfold smaller for Collector II accessory as compared to Seagull) that it was possible to reduce quantification errors associated with the use of a smaller beam spot by collecting spectra for seven times while randomly changing the sample position. To obtain acceptable quantification errors, in the case of ATR accessory, spectral measurement had to be repeated several times.

PLS-based FTIR spectroscopy was also investigated for quantitative assessment of cefixime in oral suspension samples using the fingerprinting region 1485–887 cm^{-1} [28], roxithromycin, and erythromycin ethylsuccinate in tablets from different manufacturers in China [29], ephedrine as the minor component in solid-state mixtures with pseudoephedrine [30], ibuprofen in urine and tablets using the spectral region 1807–1461 cm^{-1} [31], ibuprofen in sustained-release capsules choosing the absorption range 7500–6100 cm^{-1} [32].

In clinical use, all bronchodilating beta(2)-adrenoceptor agonists (e.g., terbutaline) are available as racemates of the derivatives of adrenaline. It occurs that pharmacologic effects of such agonists reside in the *R*-enantiomer, with the *S*-enantiomer being inactive and likely responsible for induction of airway hyperreactivity [33]. In a research, the combination of DRIFTS and artificial neural networks (ANNs) was proposed as a means for establishing the level of *S*-enantiomer present as impurity in *R*-terbutaline [34]. The best ANN model to determine the percentage of *S*-terbutaline enantiomer in the mixture was a radial basis function network with 23 selected inputs, one hidden layer with

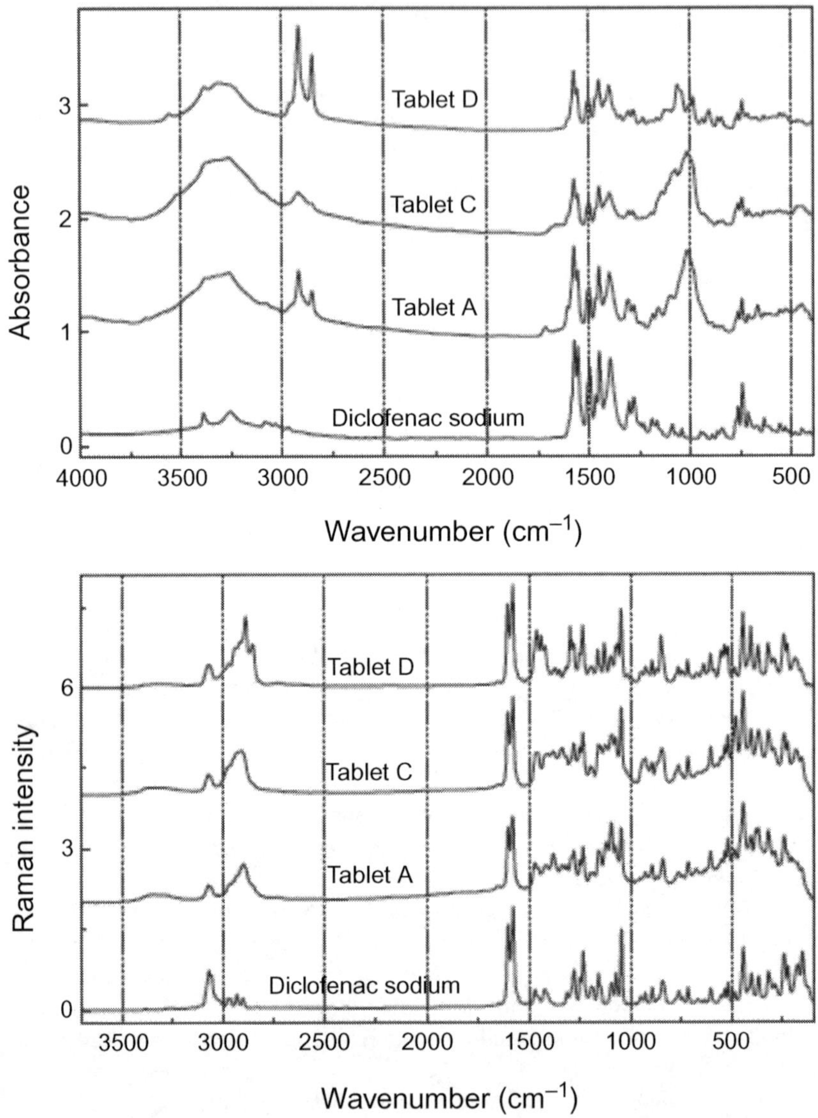

FIG. 5.3 FTIR ATR *(top)* and FT-Raman *(bottom)* spectra of diclofenac sodium and its analyzed preparations; the spectra are offset for clarity. *(Reproduced with permission from R. Szostak, S. Mazurek, Comparison of infrared attenuated total reflection and Raman spectroscopy in the quantitative analysis of diclofenac sodium in tablets, Vib. Spectrosc. 57(1) (2011) 157–162.)*

10 neurons and one output neuron. According to the sensitivity reports of selected inputs for the ANN model, the most important peaks were found to be at 1227 cm^{-1} (C–N stretch of amines) and 2346 cm^{-1} (N–H stretch of secondary amino group) (Fig. 5.4). A quantifiable range of 17.43–100% was reported for *R*-stereoisomer dispersed as 5% mixture in KBr.

The oral route remains the most convenient for drug administration, with tablets emerging as the most common solid dosage form, given its respective consumption by patients [35]. To detect physical changes in an API and formulated product, dissolution test is a requirement for any solid dosage form throughout all the stages of development for product release and stability testing [36]. The usage of conventional dissolution test, nevertheless, is somewhat awkward because it only monitors the amount of released drug and does not yield information on drug release mechanism.

In the 2000s, the combination of FTIR spectroscopic imaging and macro-ATR-IR approach was suggested for studying polymer/drug formulations in contact with aqueous solution as a function of time [37]. Based on this enhanced chemical visualization, the crystallization of ibuprofen molecularly dispersed in poly (ethylene glycol) (PEG) was spectroscopically imaged during the dissolution process in water, showing that inclusion complex of ibuprofen with cyclodextrins could prevent drug crystallization (Figs. 5.5 and 5.6). This imaging method was also implemented to investigate the dissolution of a solid dispersion of nifedipine (a poorly water-soluble drug) in PEG. In other words, it consisted of several steps such as (i) transferring a small amount of powdered sample to the surface of the ATR crystal (inverted zinc selenide (ZnSe) prism) to be heated to 60°C; (ii) placing a cover glass slide over the molten sample with a spacer

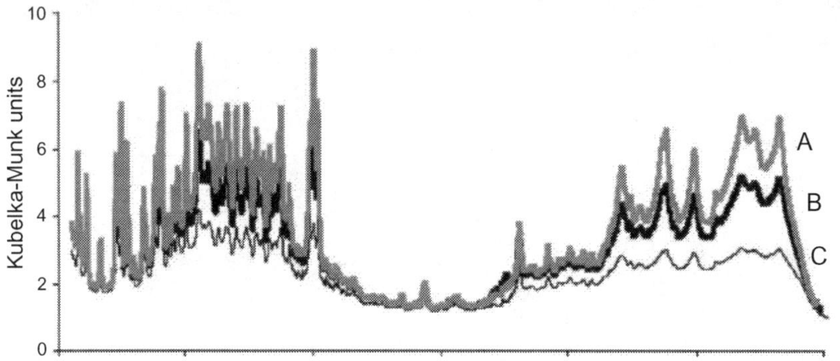

FIG. 5.4 DRIFT spectra of pure *S*-terbutaline (A), 50:50 (w/w) racemic mixture (B) and pure *R*-terbutaline (C). *(Reproduced with permission from S. Agatonovic-Kustrin, R. Alany, Application of diffuse reflectance infrared Fourier transform spectroscopy combined with artificial neural networks in analyzing enantiomeric purity of terbutaline sulfate bulk drug, Anal. Chim. Acta 449 (2001) 157–165.)*

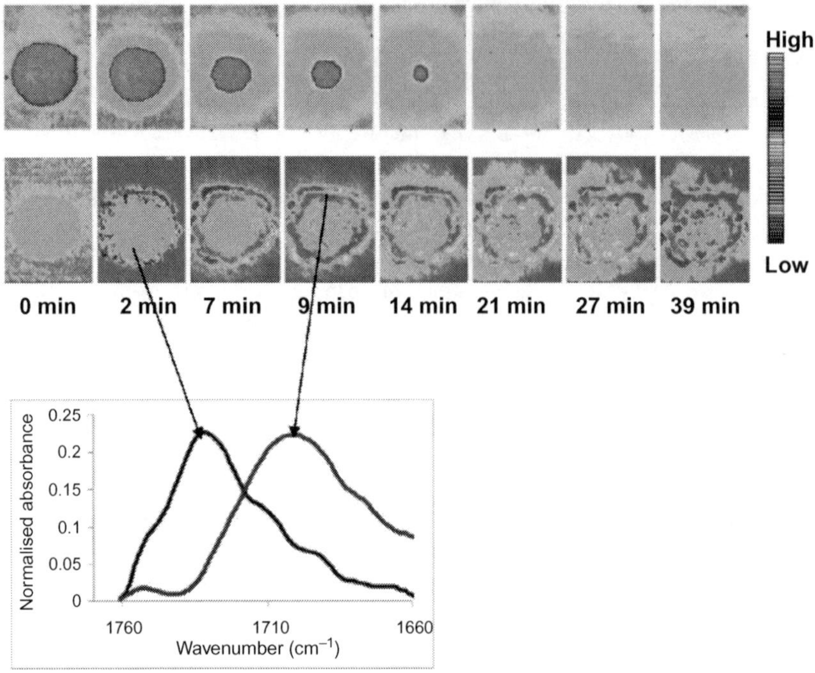

FIG. 5.5 Macro-ATR-IR images of PEG/ibuprofen formulation show distribution of PEG and ibuprofen as a function of time during contact with water. The images (size $3.8 \times 5.3\,mm^2$) are based on the distribution of the integral absorbance of the IR band of PEG-8000 (*top row*, spectral band integration from 1170 to $1020\,cm^{-1}$) and based on the distribution of the integral absorbance of the ν(C=O) band of ibuprofen (*bottom row*, integration from 1760 to $1665\,cm^{-1}$) acquired sequentially as a function of time. *The right-hand column* is the relative concentration scale. The inset shows representative IR spectra in the ν(C=O) region of ibuprofen to demonstrate spectral difference between molecularly dispersed and crystalline drug. *(Reproduced with permission from S.G. Kazarian, K.L.A. Chan, "Chemical photography" of drug release, Macromolecules 36 (2003) 9866–9872.)*

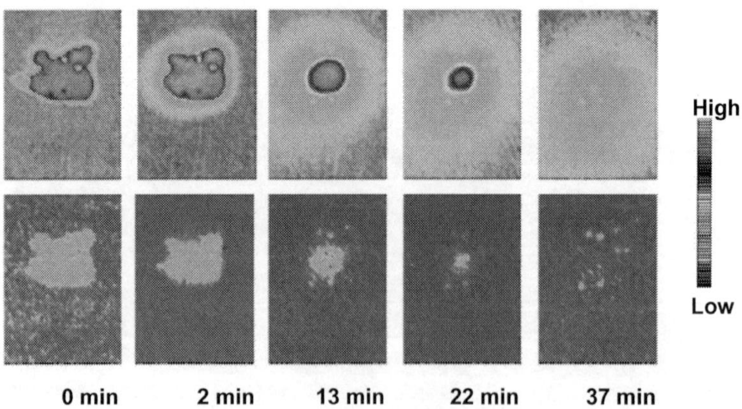

FIG. 5.6 Macro-ATR-IR images of methyl-β-cyclodextrin/ibuprofen showing distribution of methyl-β-cyclodextrin *(top row)* and ibuprofen *(bottom row)* as a function of time during contact with water. The images $3.8 \times 5.3\,mm^2$ demonstrate almost total dissolution of ibuprofen after 40 min contact with water. *(Reproduced with permission from S.G. Kazarian, K.L.A. Chan, "Chemical photography" of drug release, Macromolecules 36 (2003) 9866–9872.)*

to create a gap of a few hundred micrometers between the cover glass and ATR crystal; and (iii) allowing the sample to solidify at 40°C before adding water from the side of the film to study dissolution of the polymer/drug formulation. It was revealed that the formation of a significant amount of crystalline drug (with a drug loading of at least 10 wt%) within the polymer matrix could diminish the dissolution rate of the drug [38]. The field of view (FOV) of approximately $3.8 \times 5.3\,mm^2$ achieved by macro-ATR-IR images herein was in the same order of magnitude as ordinary tablets, implying that the investigated tablets need be neither thin nor very small.

In another study, an innovative design of compaction cell was proposed for in situ FTIR imaging of tablet dissolution, i.e., the tablet formulation could be directly compacted on a diamond ATR accessory for a subsequent FTIR imaging when water flowing around the compacted tablet (Fig. 5.7) [39]. This cell was applied to study the dissolution of a tablet containing hydroxypropyl methylcellulose (HPMC) and caffeine, showing the displacement of low solubility particles inside a swollen polymer matrix. It is noted that this diamond ATR system integrates a FOV of ca. $0.8 \times 1.1\,mm^2$ with a spatial resolution of about 15 μm, which could be obtained without the implementation of an IR microscope [40]. In case of requiring a larger FOV, it is unaffordable though to use a bigger size diamond crystal. As a logical alternative, the ATR system could be

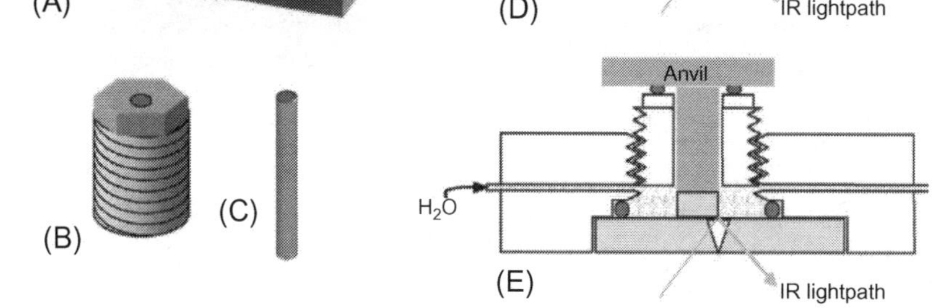

FIG. 5.7 Design and operating principle of the compaction cell. (A) Base plate that can be bolted down to the supercritical fluid analyzer (SFA) ATR system. (B) The threaded ring can be fitted in the base plate and forms a die, in which powder can be compacted using tablet punch (C). (D) Cross-section through this cell (1: tablet, 2: punch, 3: ring, 4: base plate, 5: diamond ATR plate, 6: water tube). (E) When the tablet is formed, the ring can be lifted using a spanner, allowing water flow around the compacted tablet. *(Reproduced with permission from J. van der Weerd, K.L.A. Chan, S.G. Kazarian, An innovative design of compaction cell for in situ FT-IR imaging of tablet dissolution, Vib. Spectrosc. 3 (2004) 9–13.)*

constructed from a ZnSe crystal, which has a refractive index similar to that of diamond but is not nearly as hard as diamond.

Combining this FTIR imaging approach coupled with PLS calibration and the conventional dissolution test (i.e., a UV detector was incorporated into the compaction cell to monitor the amount of drug dissolved in the effluent of the cell) could provide quantitative information of all components in both the tablet and liquid phase [41]. The validity of this procedure was proved by similar results obtained from UV spectra of the effluent and FTIR imaging data of the tablet composed of HPMC and niacinamide (Fig. 5.8). It is noteworthy to mention that the dissolution test could be simply monitored in situ with a NIR transflectance probe (e.g., PLS regression was used to determine the dissolution profile of folic acid tablets from NIR spectra acquired in the wavenumber range of 6310–5540 cm^{-1}) [42].

ATR-based measurements are only meaningful if a good contact exists between the sample and ATR crystal. This is due to the fact that the penetration depth of IR light into the sample is short and the migration of water through the void between the sample and ATR crystal can happen in tablet dissolution studies. In a dissolution study of realistically sized HPMC tablets, microscopic ATR-IR imaging was employed to scrutinize the water intake into in situ compacted and precompacted tablets as a function of compaction pressure [43]. A rigorous analysis of the imaging datasets showed that the intake of water into HPMC tablet occurred at a constant speed of roughly 4 μm/min, implying no leakage ("creeping") of water into the space between the tablet and ATR crystal. It was also hardly affected by either the range of compaction pressures under study or the type of ATR crystal used (diamond or ZnSe) (Fig. 5.9). Swelling was found to efficiently double the radius of the HPMC tablet, making its characterization incomplete by using the microscopic system as well as macroscopic diamond crystal owing to a restriction in FOV.

In another development, NIR spectroscopy and perturbation-correlation moving window two-dimensional correlation spectroscopy (PCMW-2DCS) [44] were utilized to investigate water interaction with disaccharides (i.e., α-lactose monohydrate (La) and trehalose) in pharmaceutical tablets [45]. The water penetration in particle layer was theoretically described by the Lucas-Washburn theory, indicating that the penetration rate and amount of water were principally dependent upon the porosity of the tablet rather than the type of excipient. PCMW-2DCS data revealed a two-staged process, i.e., (i) interaction between water and disaccharide (due to water adsorption and disaccharide hydration number) and (ii) diffusion-controlled disaccharide dissolution (due to saturated interaction) (e.g., Fig. 5.10).

In parallel with dissolution test, ATR-FTIR spectroscopic imaging could be also employed to analyze in situ the spatial distribution of different components in tablets [46]. During tablet formation, the distribution of caffeine was found to be strongly affected by the composition of polymer matrix used, i.e., caffeine acts as magnesium stearate (a lubricant) for lactose making it more

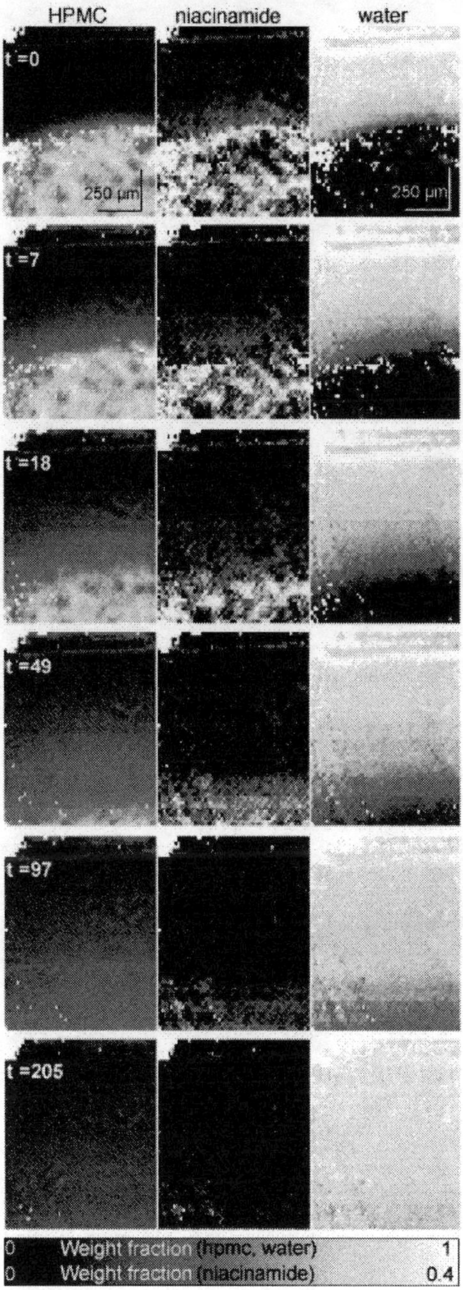

FIG. 5.8 Results of PLS on a number of imaging sets. The three columns represent the amounts (fraction w/w) of HPMC, niacinamide, and water. Every row of images is based on a single data set. The time between the start of the experiment and the acquisition of this data set is indicated in every row. The tablet is only partly visible as its size (Ø 3 mm) exceeds the field of view (FOV) of 820×1140 µm. The *white pixels* locate the outliers that were rejected from further data processing. *(Reproduced with permission from J. Van der Weerd, S.G. Kazarian, Combined approach of FTIR imaging and conventional dissolution tests applied to drug release, J. Control. Release 98 (2004) 295–305.)*

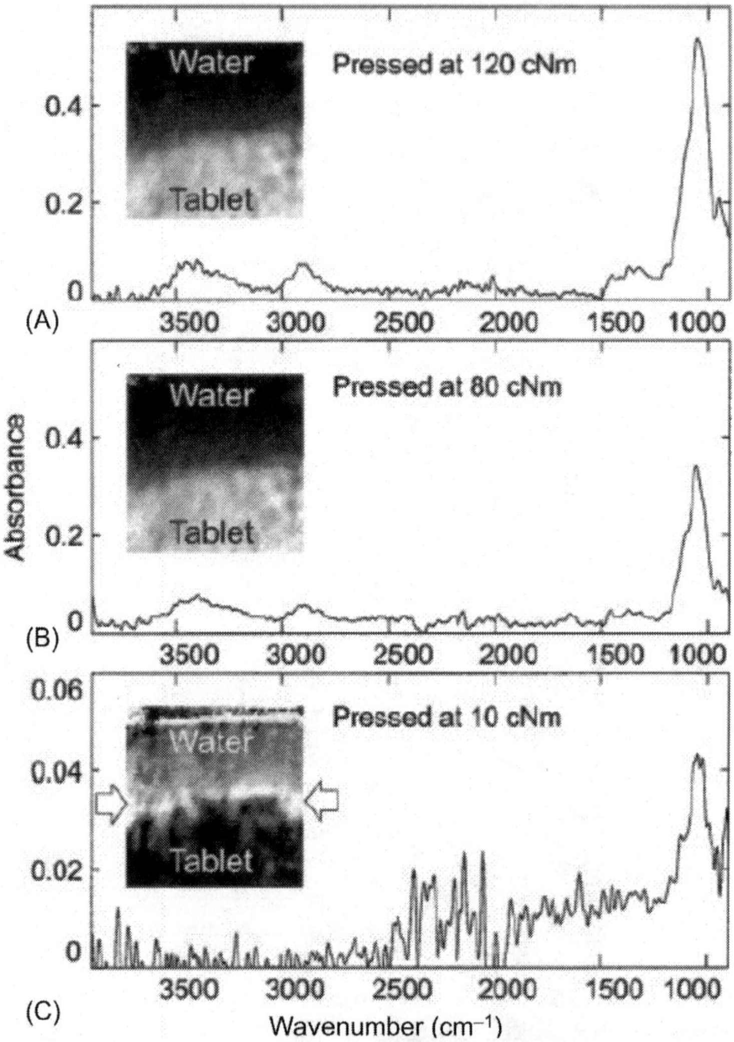

FIG. 5.9 Representative spectra of tablets compacted/pressed at the indicated pressures (A–C). The insets show the absorbance of the HPMC band at 1049 cm^{-1} in the corresponding dataset (*dark color* indicates a low absorbance; *lighter color* indicates higher absorbance). The inset in the lowest spectrum shows a negligible absorption by the tablet, except for the hydrated part adjacent to the water fraction, as indicated by *arrows*. *(Reproduced with permission from J. Van der Weerd, S.G. Kazarian, Validation of macroscopic ATR-FTIR imaging to study dissolution of swelling pharmaceutical tablets, Appl. Spectrosc. 58 (2004) 1413–1419.)*

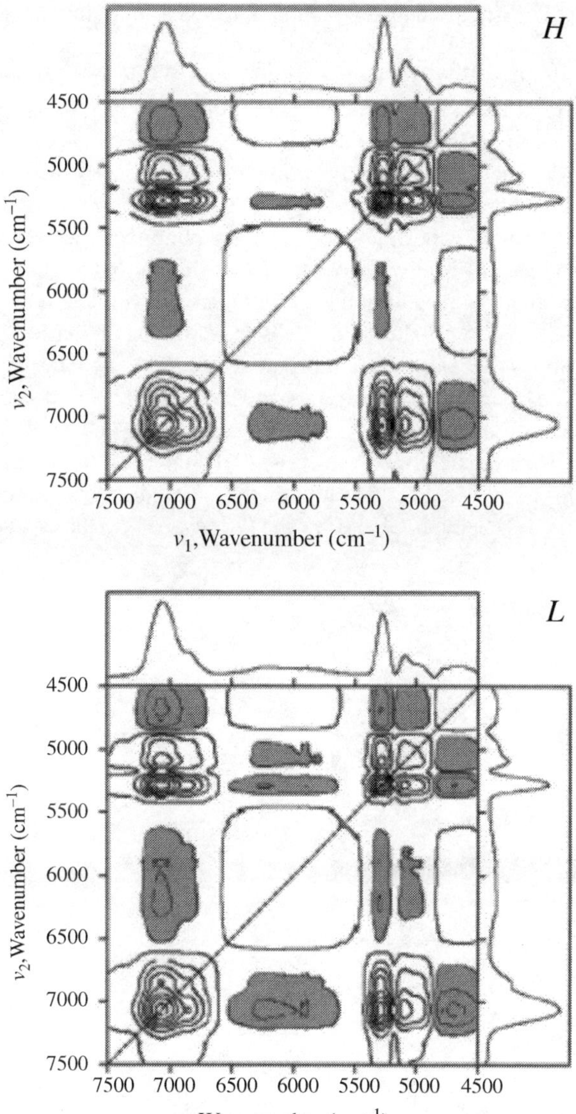

FIG. 5.10 Synchronous generalized 2D correlation spectral maps calculated from the NIR spectra of La-H and La-L with penetrating water. *White and gray areas* indicate positive (include zero) and negative correlation, respectively. The spectra shown at the top and right-hand side of the 2D spectra are the autocorrelation spectra. H and L stand for high and low porosity of the tablets, respectively. *(Reproduced with permission from Y. Hattori, M. Otsuka, NIR spectroscopic study of the dissolution process in pharmaceutical tablets, Vib. Spectrosc. 57 (2011) 275–281.)*

compactable. Statistical analysis of in situ spectroscopic imaging data, in this research, proved that HPMC had superior compaction properties than microcrystalline cellulose. Because the validity of ATR-FTIR imaging technique was limited by a relatively short IR penetration depth, X-ray tomography could be used as a complementary technique to produce a density map of tablets (in particular, for a system containing only two species of distinctly different densities) (Fig. 5.11).

In another context, the homogeneity of a pharmaceutical cream containing imiquimod as API [47] was investigated by using IR imaging spectroscopy and chemometrics (such as principal component analysis (PCA) and multivariate curve resolution with alternating least squares (MCR-ALS)) [48]. After a 3-month period of accelerated stability at 45°C, PCA exploratory analysis indicated that the composition of crystals appeared was different from that of the emulsion. The conjunction of IR imaging spectroscopy and MCR-ALS showed that these crystals (basically presented by C–H and N–H stretches) were a salt formed by carboxylic acid and imiquimod. Fig. 5.12 displays four different pure spectra obtained by using MCR-ALS calculations, i.e., (a) the crystals contained

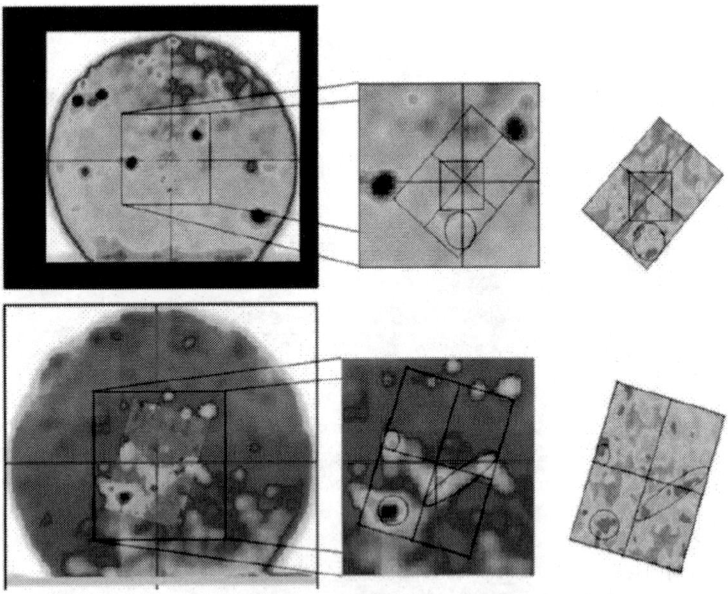

FIG. 5.11 Images showing the results from FTIR spectroscopy compared with images from X-ray microtomography. The *top* set of images shows the results for the 100–125 μm particle size. The *bottom* set shows the data for the 125–150 μm particle size. The FTIR data are shown on the right-hand side, the key area of the X-ray tomography is shown in the middle, and the FTIR data are layered over the X-ray data for comparison on the left-hand side. *(Reproduced with permission from P. Wray, K.L. Andrew Chan, J. Kimber, S.G. Kazarian, Compaction of pharmaceutical tablets with different polymer matrices studied by FTIR imaging and X-ray microtomography, J. Pharm. Sci. 97(10) (2008) 4269–4277.)*

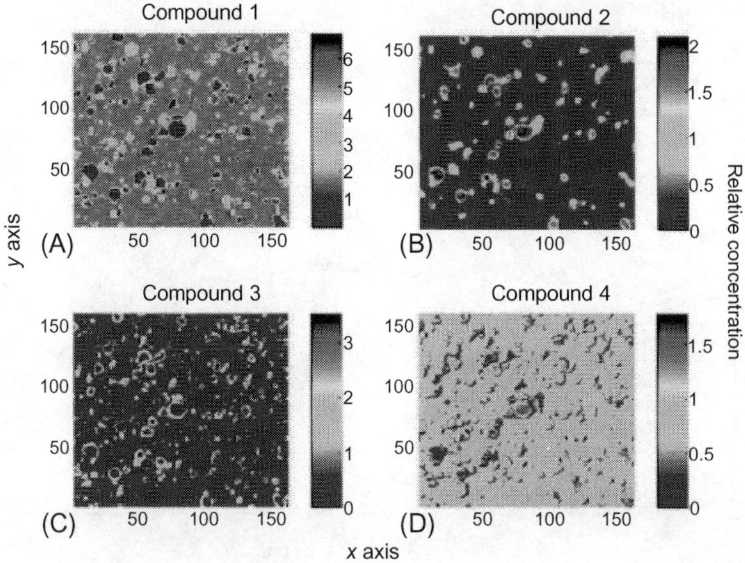

FIG. 5.12 Relative concentration maps of first (A), second (B), third (C), and fourth (D) pure compounds recovered by MRC-ALS. *(Reproduced with permission from R.L. Carneiro, R.J. Poppi, Infrared imaging spectroscopy and chemometric tools for in situ analysis of an imiquimod pharmaceutical preparation presented as cream, Spectrochim. Acta A Mol. Biomol. Spectrosc. 118 (2014) 215–220.)*

no water; (b) the crystals contained a saturated carbon chain and amine group; (c) an interphase compound or another crystalline structure formed by API, API solvent, and water; (d) a water-free compound.

More interestingly, the stability of cosmetic or pharmaceutical "oil in water" emulsions could be even predicted by using FTIR spectroscopy along with classical methods (conductivity, viscosity, pH, texture analysis) [49]. During the aging process, modifications of chemical functions were uncovered by FITR spectrometric indices, including a decrease in the unsaturation index (associated with to the $\nu_{C=CH}$ band at 3006 cm^{-1}) and an increase in the carbonyl index (associated with the $\nu_{C=O}$ band at 1746 cm^{-1}) (e.g., Fig. 5.13). Through deconvolution, the broadening phenomenon of the carbonyl band was probably caused by the appearance of free fatty acids as a consequence of emulsion hydrolysis. Unlike other techniques, FTIR spectroscopy did not truly predict the aging process because few alterations were measured at 40°C. So, this should be complemented by using conductimetry (the most sensitive technique to assess the physical alterations during emulsion's aging).

In another case, surface-enhanced Raman scattering (SERS) chemical imaging (CI) was reported for the comparison of pharmaceutical tablets with trace amounts of API [50]. This method was exceptionally suitable in evaluating tablets prepared by different manufacturing technologies, of which any significant difference could not be detected by ordinary Raman mapping (without

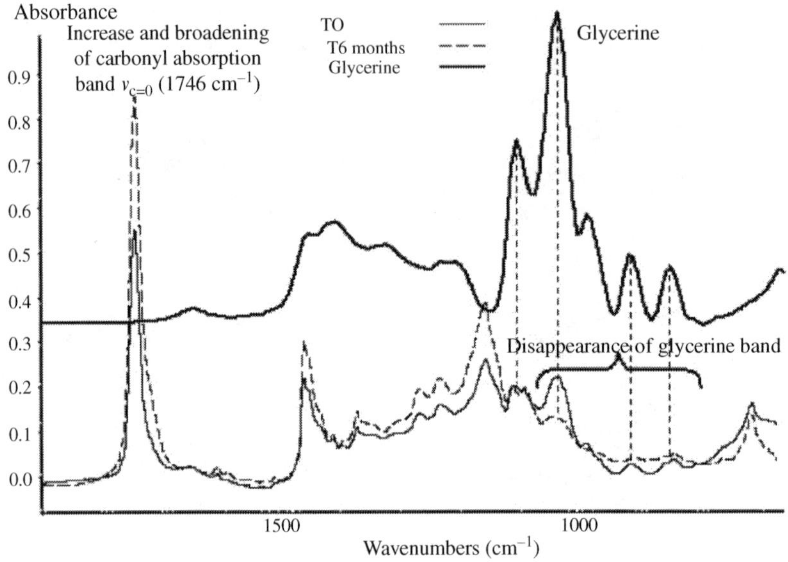

FIG. 5.13 FTIR spectra of an emulsion (containing apricot cores (20%, v/v), Montanov 68 (5%, w/v), glycerin (5%, v/v), methylparaben/propylparaben 50:50 (0.3%, w/v), and demineralized water q.s. (100%, v/v)) before and after a thermal stress during 6 months at 50°C. *(Reproduced with permission from H. Masmoudi, Y. Le Dréau, P. Piccerelle, J. Kister, The evaluation of cosmetic and pharmaceutical emulsions aging process using classical techniques and a new method: FTIR, Int. J. Pharm. 289(1–2) (2005) 117–131.)*

SERS). In the presence of silver colloids, the overall acquisition time for each SERS-CI map—with a higher number of pixels as compared to Raman chemical imaging—was about 20 min. SERS-CI enabled the detection of API's true spatial distribution in the tablets produced by dry (D, direct compression) and wet (W, cosolution and solvent evaporation, then compression) technologies as displayed in Figs. 5.14 and 5.15. Despite that, due to its high spectral variability (governed by colloid size and shape) a chemometric approach was thereby proposed for data treatment, i.e., all spectra were baseline corrected by using Eiler's asymmetric least squares, followed by multivariate curve resolution-alternating least squares data decomposition to find all the various SERS positive (in this case, API-related) loadings.

Polymorphic analysis

Polymorphism is the characteristic, for solid materials, of being able to exit in more than one crystalline form with different conformations or arrangements of the constituents in the crystal lattice [51]. As early as the 1960s, the pharmaceutical applications of polymorphism were noticed since polymorphic forms of a given drug could differ in physicochemical properties (such as solubility,

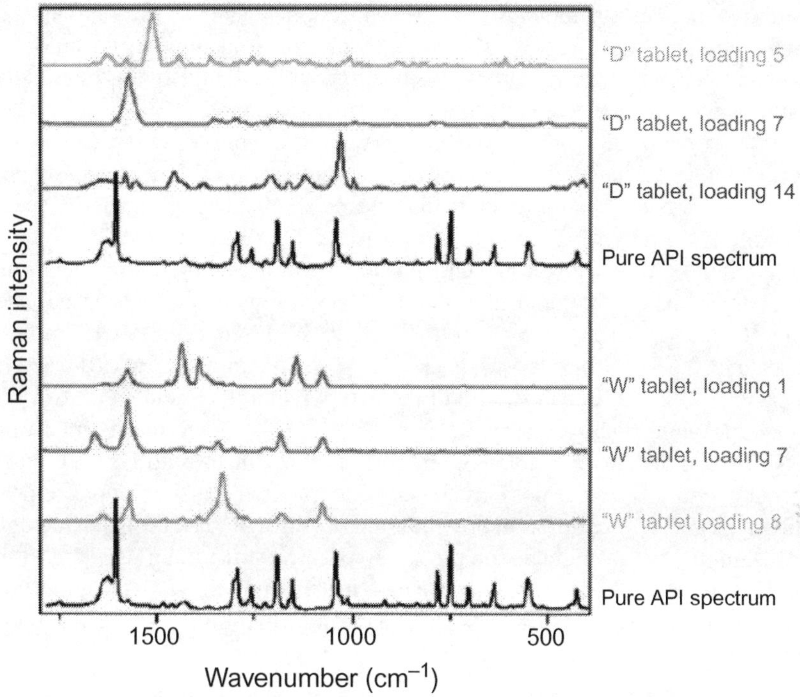

FIG. 5.14 Selected API-related MCR-ALS loadings obtained from SER-CI datasets of "D" tablet and "W" tablet, compared to the pure API spectrum. *(Reproduced with permission from T. Firkala, A. Farkas, B. Vajna, I. Farkas, G. Marosi, Investigation of drug distribution in tablets using surface enhanced Raman chemical imaging, J. Pharm. Biomed. Anal. 76 (2013) 145–151.)*

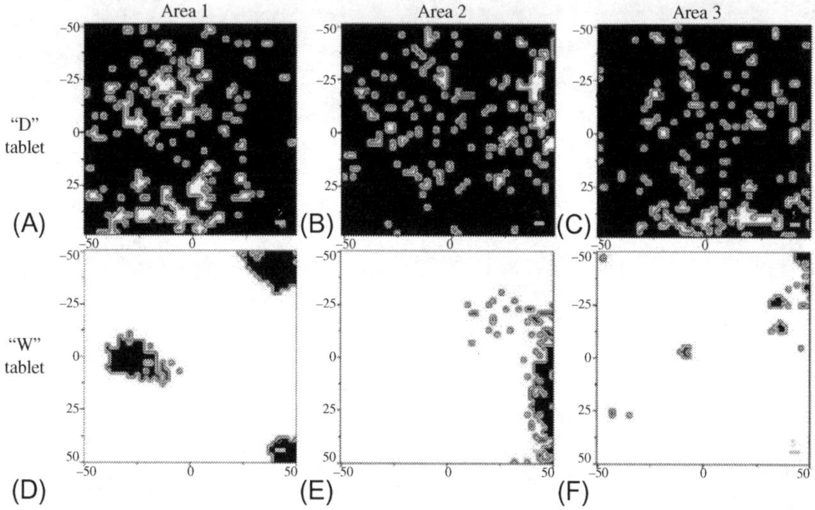

FIG. 5.15 API distribution by SER-CI in randomly selected locations of "D" and "W" tablet (A–F). *(Reproduced with permission from T. Firkala, A. Farkas, B. Vajna, I. Farkas, G. Marosi, Investigation of drug distribution in tablets using surface enhanced Raman chemical imaging, J. Pharm. Biomed. Anal. 76 (2013) 145–151.)*

melting point, density, hardness, crystal shape, optical and electrical properties, vapor pressure, etc.) [52]. Thus, the choice of a proper polymorph would determine whether a pharmaceutical preparation is chemically and/or physically stable as well as therapeutically and/or toxically acceptable.

In the 1990s, bulk pharmaceutical preparations of spironolactone (a diuretic steroidal aldosterone having variable and incomplete oral behavior due to poor water solubility and dissolution rate [53]) were examined by using DRIFT spectroscopy for different polymorphic forms [54]. In a subsequent study, the differentiation of solvated spironolactone samples (prepared by crystallization in absolute methanol, acetonitrile, absolute ethanol, ethyl acetate, benzene, and chloroform) was realized by employing both FT-Raman and FT-IR diffuse reflectance spectroscopy [55]. Under high pressures up to about 50 kbar, the behavior of IR and Raman spectra of the two polymorphs (I and II) of the synthetic spironolactone was very different [56]. It was shown that Form I (the less thermodynamically stable polymorph) did not transform to Form II under pressure; and both forms possibly underwent structural transformations to new polymorphs, but over different pressure ranges (Figs. 5.16 and 5.17). Although it was impossible to comment on the nature of these structural transitions without high-pressure X-ray crystallographic data, this finding was still important for the tableting process of spironolactone during which some mechanical pressure was applied.

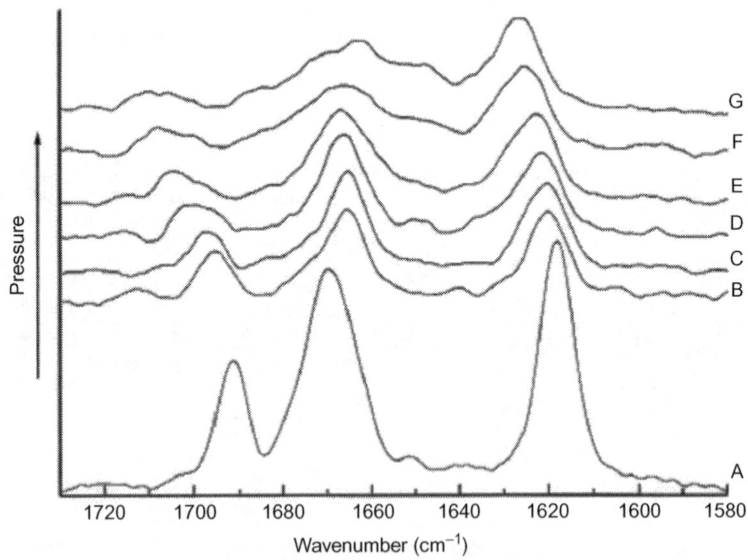

FIG. 5.16 Wavenumber versus pressure plots for selected IR bands of Form I. A (ambient), B (9.3), C (12.7), D (18.0), E (23.1), F (27.5), and G (37.7 kbar). *(Reproduced with permission from G.L. Pisegna, D.F.R. Gilson, I.S. Butler, High-pressure infrared and Raman studies of polymorphism in pharmaceutical compounds: spironolactone, Forms I and II, J. Mol. Struct. 1078 (2014) 146–150.)*

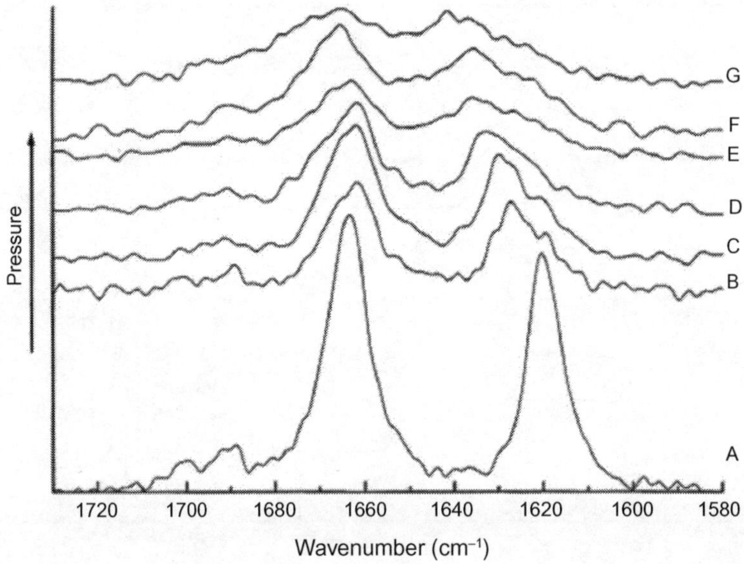

FIG. 5.17 Wavenumber versus pressure plots for selected IR bands of Form II. A (ambient), B (9.3), C (12.7), D (18.0), E (23.1), F (27.5), and G (37.7 kbar). *(Reproduced with permission from G.L. Pisegna, D.F.R. Gilson, I.S. Butler, High-pressure infrared and Raman studies of polymorphism in pharmaceutical compounds: spironolactone, Forms I and II, J. Mol. Struct. 1078 (2014) 146–150.)*

Using Raman and FT-IR spectroscopy, vibrational spectra were recorded for anhydrous and hydrated polymorphs of the antiviral drug lamivudine to track the dynamic process of its polymorphic transformation (Figs. 5.18 and 5.19) [57]. The interaction between crystalline water and lamivudine was demonstrated to have an important impact on the molecular vibration modes of lamivudine polymorphs. Hydrated and anhydrous lamivudine were characteristically presented by two Raman peaks at 783 and 798 cm^{-1}. The time-dependent dehydration process of hydrated lamivudine could be deduced by fitting these two characteristic peaks' normalized areas and heating time with single exponential functions.

In the course of the milling process, the solid-state polymorphic conversion of famotidine, an H2 histamine receptor antagonist [58], could be nondestructively determined by confocal Raman microspectroscopy combined with a thermal analyzer [59]. From the mapping spectra, the blend uniformity of milled samples was estimated by evaluating the consistency of the Raman peak intensity ratio of the 2920 cm^{-1} band (for form A) and 2897 cm^{-1} band (for form B) (Fig. 5.20). The polymorphic conversion of famotidine from form B to form A was discovered to be strongly induced with milling time and synergistically accelerated by heating (i.e., the thermal-dependent critical temperatures

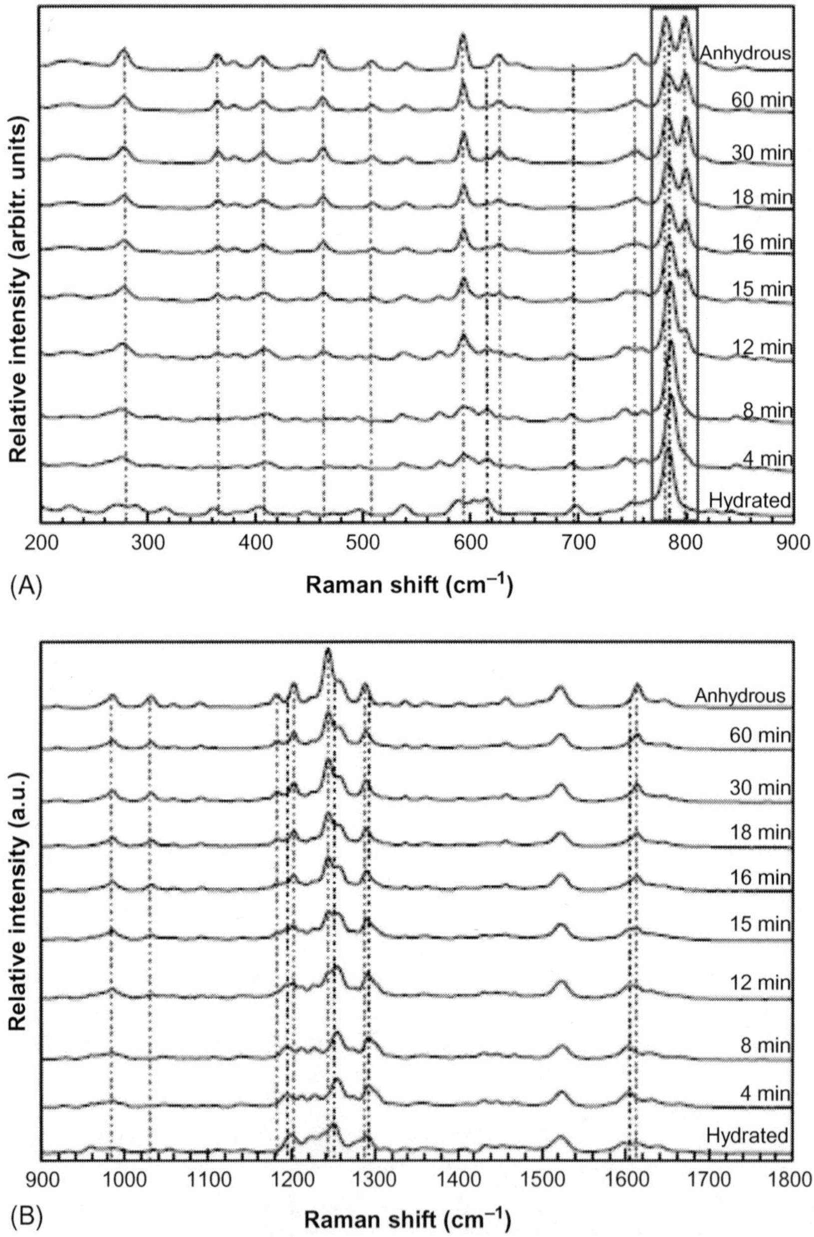

FIG. 5.18 Heating time-dependent changes in the representative Raman spectra of hydrated lamivudine samples (different heating times are shown next to the spectra) in different spectral regions (A) 150–900 cm^{-1}; (B) 900–1800 cm^{-1}. Spectrum of anhydrous lamivudine is also shown at the top of figure for comparison. The characterized vibrational peaks for hydrated and anhydrous lamivudine polymorphs are marked with *black and red dotted lines*, respectively. *(Reproduced with permission from Y. Du, H. Zhang, J. Xue, W. Tang, H. Fang, Q. Zhang, Y. Li, Z. Hong, Vibrational spectroscopic study of polymorphism and polymorphic transformation of anti-viral drug Lamivudine, Spectrochim. Acta A Mol. Biomol. Spectrosc. 137 (2015) 1158–1163.)*

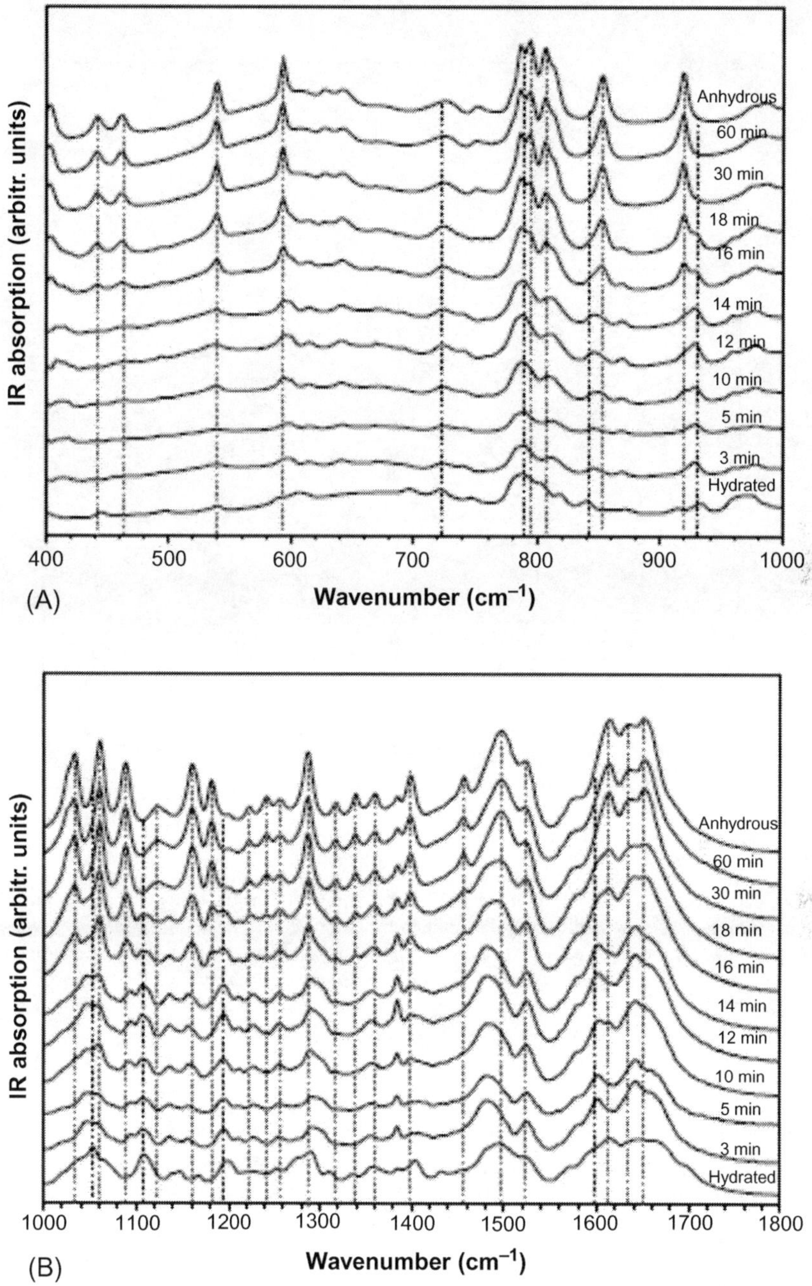

FIG. 5.19 Heating time-dependent changes in the representative FT-IR spectra of hydrated lamivudine samples (different heating times are shown next to the spectra) in different spectral regions (A) 400–1000 cm^{-1}; (B) 1000–1800 cm^{-1}. Spectrum of anhydrous lamivudine is also shown at the top of figure for comparison. The characterized vibrational peaks for hydrated and anhydrous lamivudine polymorphs are marked with *black and red dotted lines*, respectively. *(Reproduced with permission from Y. Du, H. Zhang, J. Xue, W. Tang, H. Fang, Q. Zhang, Y. Li, Z. Hong, Vibrational spectroscopic study of polymorphism and polymorphic transformation of anti-viral drug—Lamivudine, Spectrochim. Acta A Mol. Biomol. Spectrosc. 137 (2015) 1158–1163.)*

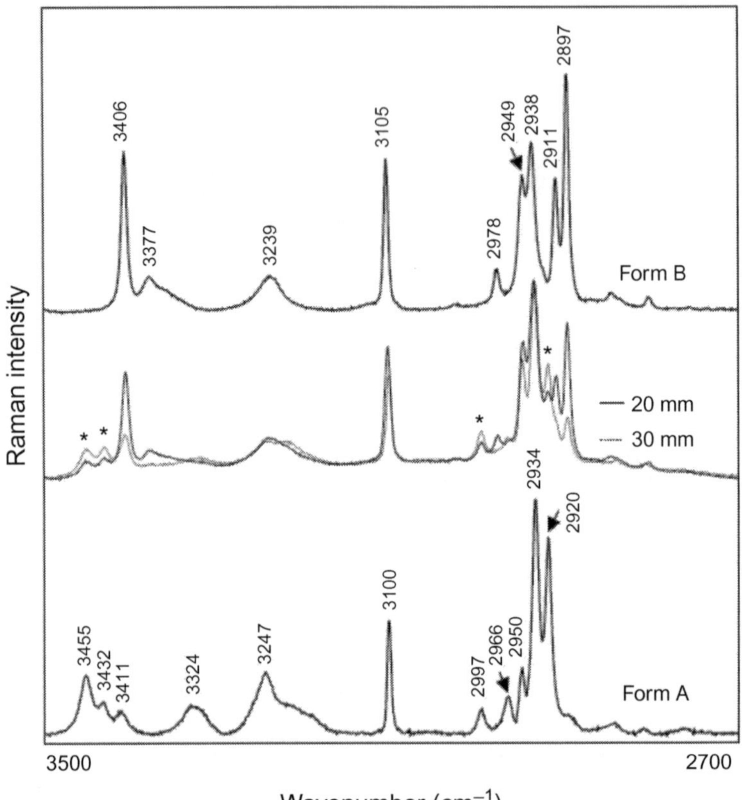

FIG. 5.20 Raman spectra of polymorphic forms A and B of famotidine, as well as the milled samples of intact famotidine form B after milling for 20 or 30 min. *(Reproduced with permission from W.T. Cheng, S.Y. Lin, M.J. Li, Raman microspectroscopic mapping or thermal system used to investigate milling-induced solid-state conversion of famotidine polymorphs, J. Raman Spectrosc. 38 (2007) 1595–1601).*

for sharply enhancing the content of famotidine form A were 130°C and 110°C for 10 min- and 20–30 min-milled samples).

To discriminate the two anhydrate phases of commercial relevance of olanzapine, an antipsychotic agent [60], FT-Raman and IR (mid and near) spectra of these polymorphs were analyzed by comparing with the calculated vibrational spectra using ab initio methods (Fig. 5.21) [61]. According to the geometrical optimizations, the crystalline structure of form (1) exhibited the most stable conformer (Fig. 5.22). While form (1) had vibrational features characteristic of a NH…N hydrogen bond, being the primary intermolecular interaction of this structure, the changes in the vibrational spectra of form (2) were connected to atoms directly or indirectly involved in the NH hydrogen bond. It was suggested that the weakening of the intermolecular interaction was responsible for a lower

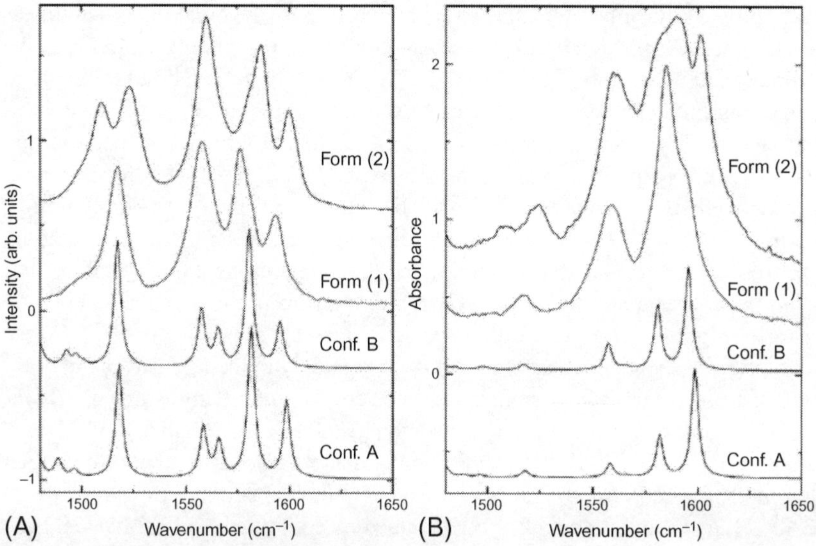

FIG. 5.21 Comparison of (A) Raman and (B) infrared spectra of the polymorphs of olanzapine in the ν(C=C) and ν(C=N) double bond region with the calculated spectra of conformers A and B. *(Reproduced with permission from A.P. Ayala, H.W. Siesler, R. Boese, G.G. Hoffmann, G.I. Polla, D.R. Vega, Solid state characterization of olanzapine polymorphs using vibrational spectroscopy, Int. J. Pharm. 326 (2006) 69–79.)*

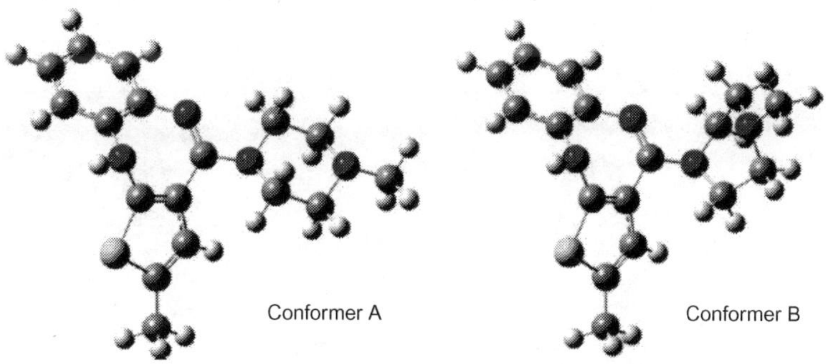

FIG. 5.22 Olanzapine conformers obtained from density functional theory (DFT) geometrical optimizations. *(Reproduced with permission from A.P. Ayala, H.W. Siesler, R. Boese, G.G. Hoffmann, G.I. Polla, D.R. Vega, Solid state characterization of olanzapine polymorphs using vibrational spectroscopy, Int. J. Pharm. 326 (2006) 69–79).*

density of form (2) and could provide additional support to the monotropic relationship proposed for the olanzapine forms.

In a different research, the polymorphic transformation of $\alpha \rightarrow \gamma$ form of pyrazinamide was detected in situ between 145°C and 146°C by temperature-dependent Raman spectroscopy [62]. Using quantum chemical calculations based on DFT, this phase change was characterized by the breaking of two linear N–H ⋯ O type hydrogen bonds (associated with C=O stretching vibration in the α dimer) and formation of one linear N–H ⋯ N type hydrogen bond (along with a weak intramolecular C–H ⋯ O type hydrogen bond in the γ dimer). The combination of DFT theoretical simulation and experimental vibrational spectroscopy was utilized for investigation into tautomeric polymorphism of 2-thiobarbituric acid as well [63]. Differences of 2-thiobarbituric acid tautomeric polymorphs (forms I, II, and IV) were more easily distinguished by terahertz time-domain spectroscopy (more sensitive to intermolecular interaction within crystalline unit cells) than Raman spectroscopy (active mostly from intramolecular interaction of various functional groups within a specific molecule).

Being investigated by Raman spectroscopy mainly in the low-wavenumber region, the polymorphism of sulindac (a nonsteroidal antiinflammatory drug) was understandable in terms of the amorphization method on recrystallization and crystalline form stability [64]. Low-wavenumber data revealed the enantiotropic relationship between Forms I and IV of sulindac in the temperature range from −20 to 40°C (i.e., Form IV could be only obtained by cooling Form I, which had been crystallized from the devitrification of quenched liquid). It was shown that (i) Form IV was an intermediate crystalline state between Forms II and I in the ordering process from the amorphous state toward the stable Form II (commercial form); and (ii) both the temperature of crystallization and physical stability of Form I were dependent on the technique used for preparing amorphous sulindac (Fig. 5.23).

Using three known polymorphs of ganciclovir as model compounds, ATR-FTIR spectroscopy in coupling with PLS was justified to be feasible for

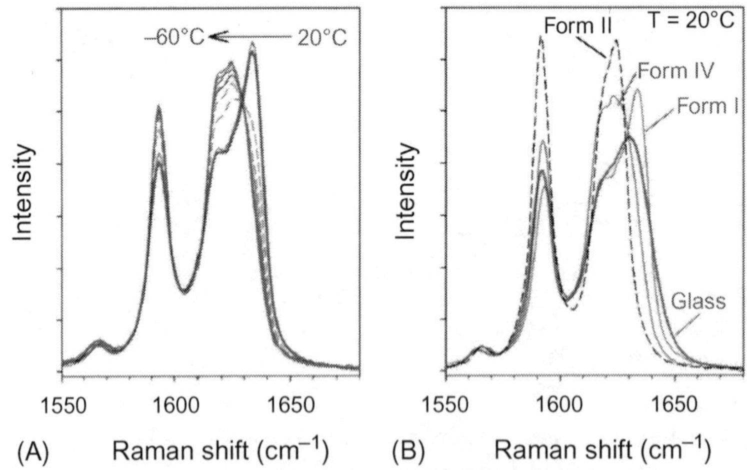

FIG. 5.23 Raman spectra in the fingerprint region collected (A) during Form I to Form IV transformation; (B) in the different states of sulindac at 20°C. *(Reproduced with permission from M. Latreche, J.F. Willart, L. Paccou, Y. Guinet, A. Hédoux, Polymorphism versus devitrification mechanism: low-wavenumber Raman investigations in sulindac, Int. J. Pharm. 567 (2019) 118476.)*

qualitative and quantitative analysis of both binary and ternary mixtures of polymorphs [65]. The ability of vibrational spectroscopy combined with multivariate analysis for quantification of different solid-state forms (including amorphous form) in mixtures was further proved in other studies, e.g., PLS-based Raman and IR spectroscopy for mixtures of α, β, and amorphous indomethacin [66]; ATR-IR, NIR, and Raman spectroscopy combined with PLS algorithm with standard normal variate preprocessing for sulfathiazole polymorphs (forms I, III, and V) in ternary mixtures [67]; second-derivative DRIFT spectroscopy and artificial neural network modeling for mebendazole polymorphs A–C in powder mixtures [68]; DRIFT spectroscopy together with multivariate statistical process control analysis, soft independent modeling of class analogy (SIMCA), orthogonal signal correction preprocessing and PLS regression methods for polymorphic composition of sulfathiazole [69]; near-infrared chemical imaging (NIR-CI), NIR, Raman, and ATR-IR spectroscopy in combination with PLS regression models and different data preprocessing algorithms (such as normalization, standard normal variate, multiplicative scatter correction, first to third derivatives) for three polymorphic forms (I, II, III) of furosemide in ternary mixtures [70].

Counterfeiting drug analysis

Poor-quality, and especially counterfeit, drugs can seriously pose potential health risks to patients and damage the pharmaceutical industry. With regard to a global health commentary, the prevalence and risks of drug counterfeiting have become more and more sophisticated based on data published accessible via Pubmed from 2007 to 2016 [71].

The internationally recognized definition of a counterfeit medicine was approved by the World Health Organization (WHO) as follows [72]: "*A counterfeit medicine is one which is deliberately and fraudulently mislabelled with respect to identity and/or source. Counterfeiting can apply to both branded and generic products and counterfeit products may include products with the correct ingredients or with the wrong ingredients, without active ingredients, with insufficient active ingredient or with fake packaging*." According to the European Medicines Agency (EMA), counterfeit drugs are being detected all over the world that cover a wide variety of therapeutic agents for the treatment of life-threatening conditions (e.g., malaria, tuberculosis, and HIV/AIDS) in the developing countries as well as expensive lifestyle medicines in the wealthy countries (e.g., hormones, steroids, antihistamines, anticancer drugs, antiviral drugs) [73].

To search for illegal drugs, the analytical platform, consisting of accurate mass determination with liquid chromatography time-of-flight mass spectrometry (LC-QTOF-MS) in combination with nuclear magnetic resonance (NMR) spectroscopy, has been verified to be an excellent tool [74]. However, this platform cannot be favorably applied, in particular on site, for screening analysis of counterfeit drugs on account of a costly, time-consuming, and labor-intensive process.

Since the 2000s, NIR spectroscopy combined with chemometrics was suggested to be simple and rapid means for facing the challenge of the counterfeit pharmaceutical market [75]. This suggestion was made with a widespread utilization of NIR equipment among analytical laboratories, allowing on-site usage of inexpensive portable NIR spectrometers [76] together with setting up mobile laboratories equipped with NIR spectrometers [77].

The feasibility of NIR measurements in diffuse reflectance mode as an express analytical means for detection of counterfeit drugs was demonstrated by Rodionova's group [78]. This NIR approach together with PCA was able to detect differences between genuine and counterfeit drugs for 250-mg film-coated antimicrobial drug tablets and 40-mg uncoated antispasmodic drug tablets. In addition, multivariate hyperspectral image analysis could provide a useful diagnostic tool for identifying nonhomogeneous spatial regions of drug formulation as illustrated in Fig. 5.24. This research group also used FT-NIR

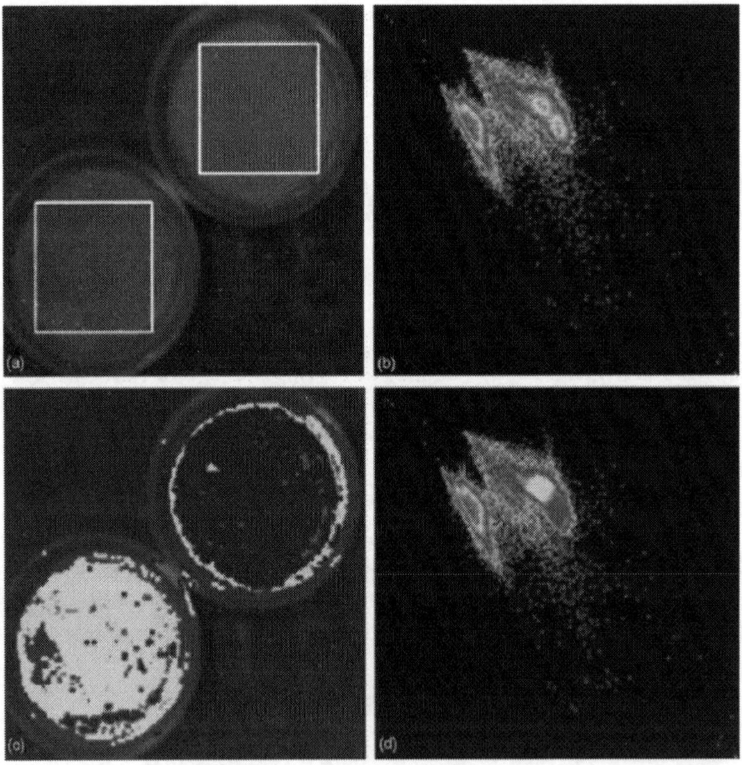

FIG. 5.24 Data Set 3 (antimicrobial drug, crushed tablets without coating). Multivariate image analysis: (A) T1 score image of crushed tablet ingredients; (B) T1–T3 score plot showing ingredient clustering; (C) class masks for two ingredients; (D) image mapping of pixels selected by masks. *(Reproduced with permission from O.Ye. Rodionova, L.P. Houmøller, A.L. Pomerantsev, P. Geladi, J. Burger, V.L. Dorofeyev, A.P. Arzamastsev, NIR spectrometry for counterfeit drug detection. A feasibility study, Anal. Chim. Acta 549 (2005) 151–158).*

spectrometer fitted with a handheld diffuse reflectance fiber-optic probe to measure the spectra of taurine (a nonessential sulfur-containing amino acid) in closed polyethylene (PE) bags in the 4000–10,000 cm^{-1} region [79]. It was revealed that the first overtone around 5770 cm^{-1} of PE significantly shifted the API peak to the mid-IR region and PE's combination bands around 4300 cm^{-1} amplified the corresponding API peak (Fig. 5.25.). To avoid misclassification, the influence of varying thickness of the PE package caused by folds had to be taken into account for constructing calibration sets.

In a different study, Puchert et al. proposed a four-stage concept for counterfeit drug identification combining single-point near infrared spectroscopy (NIRS) and near-infrared chemical imaging (NIR-CI) with statistical variance analysis (Fig. 5.26) [80]. While single-point NIRS was highly efficient for differentiation of innovator and generic tablets (e.g., with compositional differences in major excipients), NIR-CI turned out to be better than single-point NIRS for counterfeit identification thanks to its capability of making the whole tablet surface ascertainable in relation to spatial uniformity of API and major excipients (described by the skewness of pixel distribution given by image statistics) (Fig. 5.27). To overcome a disadvantage of this approach (i.e., the partial loss of spatial information), a method was also developed through (i) summation and unfolding of multidimensional predicted classification scores, which results in a Linear Image Signature (LIS) containing spatial information despite linearization and (ii) multivariate LIS data analysis (LIS-MVA) that reveals

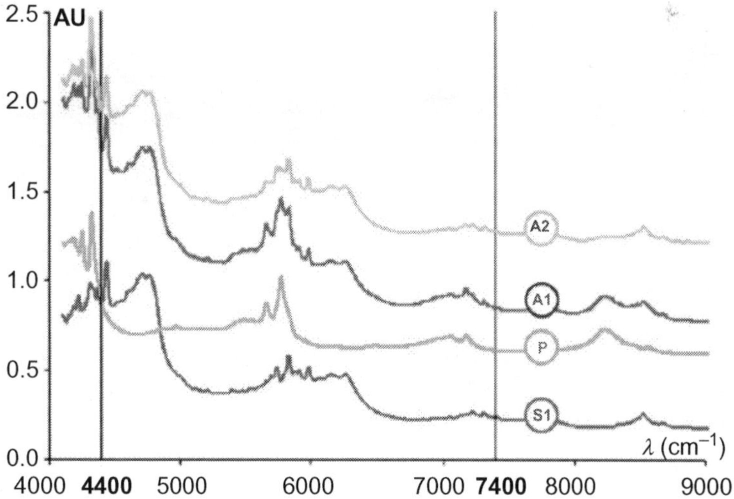

FIG. 5.25 Spectrum S1 obtained from sample 1 without PE bag (substance), P is a spectrum of PE bag, A1 is a spectrum of sample 1 in PE bag, A2 is a spectrum of sample 2 in PE bag. Spectral region between vertical lines is used for further data processing.*(Reproduced with permission from O.Ye. Rodionova, Ya.V. Sokovikov, A.L. Pomerantsev, Quality control of packed raw materials in pharmaceutical industry, Anal. Chim. Acta 642 (2009) 222–227.)*

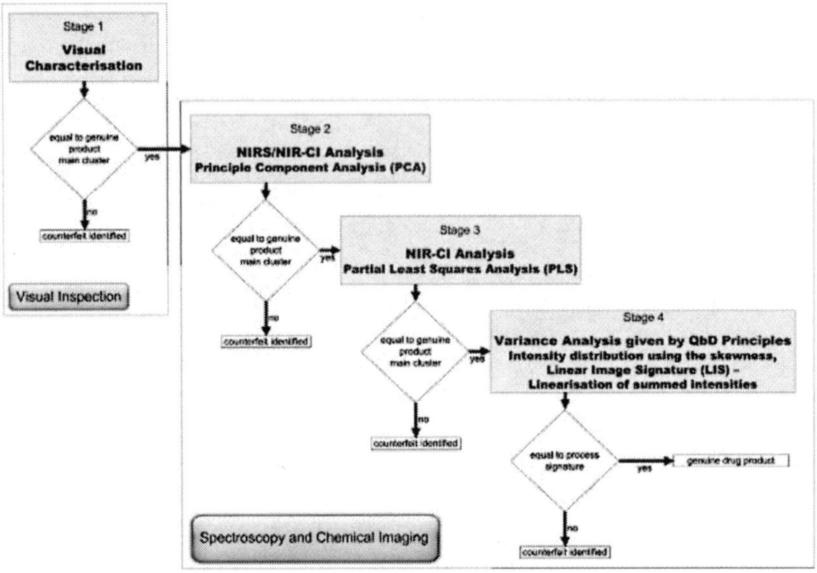

FIG. 5.26 General approach for the identification of counterfeit tablets. *(Reproduced with permission from T. Puchert, D. Lochmann, J.C. Menezes, G. Reich, Near-infrared chemical imaging (NIR-CI) for counterfeit drug identification—a four-stage concept with a novel approach of data processing (Linear Image Signature), J. Pharm. Biomed. Anal. 51 (2010) 138–145.)*

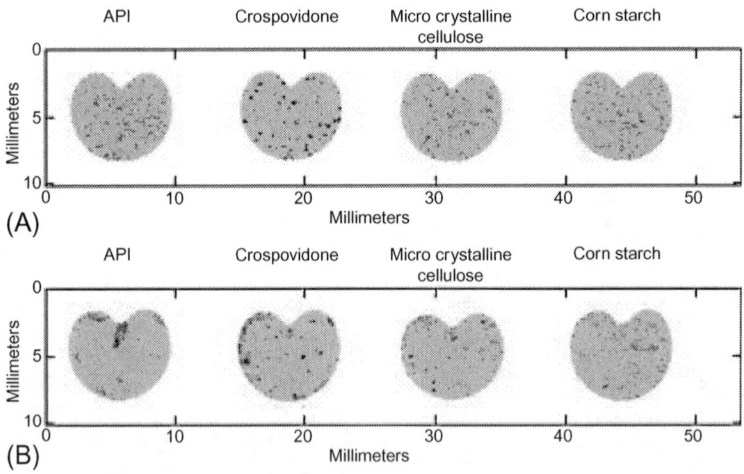

FIG. 5.27 Exemplary PLS predicted images: (A) genuine tablet; (B) counterfeit tablet—*red pixels* indicate higher concentration and *blue pixels* indicate lower concentration; images result from histogram plots centered to the mean and normally distributed; predicted concentration threshold: 3SD. *(Reproduced with permission from T. Puchert, D. Lochmann, J.C. Menezes, G. Reich, Near-infrared chemical imaging (NIR-CI) for counterfeit drug identification—a four-stage concept with a novel approach of data processing (Linear Image Signature), J. Pharm. Biomed. Anal. 51 (2010) 138–145.)*

higher variability in pixel distribution and predicted concentrations as comparing counterfeit to genuine tablets.

The complexity of the classification may generate the heterogeneity of classes (i.e., one product family having different spectra signatures due to different formulations) and/or the similarity of classes (different product families having similar spectra due to close formulations). Using a large database representing the whole tablet portfolio of a firm, a NIRS method was introduced for identifying 29 different product families of pharmaceutical tablets, one family containing one or more formulation(s), e.g., different dosages [81]. The dataset of 7120 spectra (obtained from 53 formulations, at least 5 batches (from all possible production sites), 5 tablets per batch, 2 repetitions per tablet, and 2 benchtop instruments) was first subjected to PCA and then different supervised techniques, namely, SVM (SVC Kernel linear and Radial Basis Function), K-Nearest Neighbors (KNN), and Discriminant Analysis (DA) to produce an outstanding classification rate of 100% of correct answer. DA was selected herein for the routine analysis of suspected tablets with the Mahalanobis distance as acceptance criterion for identification. In an analogous way, the identification of 29 pharmaceutical tablet families was successfully done by using an innovative strategy with handheld NIR spectrometers [82]. This consisted of comparing the spectrum of a suspected counterfeit with a reference database using the "One vs Rest" classification approach in combination with SVM for the short-wavelength NIR dataset or LDA for the classical NIR dataset. Acceptance criteria for counterfeit detection were: a class name check and a correlation distance limit. With this strategy, 100% of the genuine samples tested were correctly identified, and all the challenging samples (counterfeits and generics) were rejected.

To detect counterfeit drugs with identical API composition, the use of hyperspectral imaging was investigated in the range of 400–2500 nm [83]. PCA was performed, in this study, for effective visualization of ingredient distribution, while Gray-Level Co-Occurrence Matrix analysis enabled quantifying the homogeneity of distribution of tablet ingredients (Fig. 5.28). It was indicated that the range of 1000–2500 nm is definitely advantageous over that of 400–1000 nm because higher wavelength is less sensitive to nonuniform illumination and radiation scattering at the edges of the tablet and its embossment (e.g., a rounded shape of the Viagra tablet).

Since being approved for use in erectile dysfunction in 1998, Viagra (sildenafil citrate, a phosphodiesterase-5 (PDE5) inhibitor) was imitated and counterfeited posing significant safety risks to human health [84]. To describe NIR spectroscopy as a fast-screening method for Viagra, a total of 103 samples were analyzed to (i) check the homogeneity of a batch, (ii) distinguish imitations and counterfeits from authentic samples when combined with Wavelength Correction algorithm with the threshold value for similarity set at 0.998, and (iii) identify sildenafil citrate in the presence of excipients by using second-derivative spectra and specific absorption bands (unidentified in only two

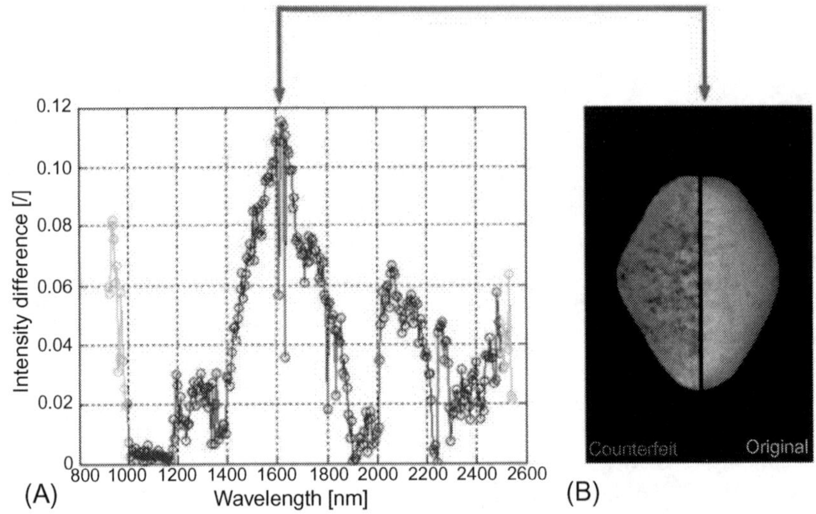

FIG. 5.28 Spectrum showing the differences in reflectance of original Viagra tablets and counterfeit tablets (A) and the images of the original Viagra and counterfeit drug at the wavelength for which the maximum mean difference in reflectance was recorded for the original and falsified drug (B). *(Reproduced with permission from S. Wilczyński, R. Koprowski, M. Marmion, P. Duda, Barbara Błońska-Fajfrowska, The use of hyperspectral imaging in the VNIR (400–1000nm) and SWIR range (1000–2500nm) for detecting counterfeit drugs with identical API composition, Talanta 160 (2016) 1–8.)*

samples due to the coexistence of another pharmacological active substance) [85]. It was also proved that directional hemispherical reflectance analysis in the NIR range could differentiate between genuine and counterfeit Viagra tablets with reference to the integrated reflectance of a surface at two different angles of incidence (20 and 60°C) and six discrete spectral bands from 0.9 to 12 μm [86].

In another research, the detection of counterfeit Viagra tablets could be also reliably realized by designing an automated spectral interpretation that includes Raman spectra recorded between 1150 and 700 cm^{-1} and a combined approach of principal components analysis (PCA) and hierarchical cluster analysis (HCA) [87]. Of 18 tablets examined for this study, Raman spectroscopy was able to detect 9 counterfeit ones (containing less amount of active ingredient or other inactive compounds) (Fig. 5.29).

Along with chemometric tools (Multivariate Curve Analysis—Alternate Least Squares for qualitative analysis and Direct Classical Least Squares for direct quantitative analysis without prior calibration), Raman chemical imaging was applied for determination of real (authentic and falsified) samples of Viagra (yielding relative errors ($-15 \div +24\%$) as compared to HPLC quantitative data) and also discrimination of three salts of clopidogrel in generic Plavix samples [88].

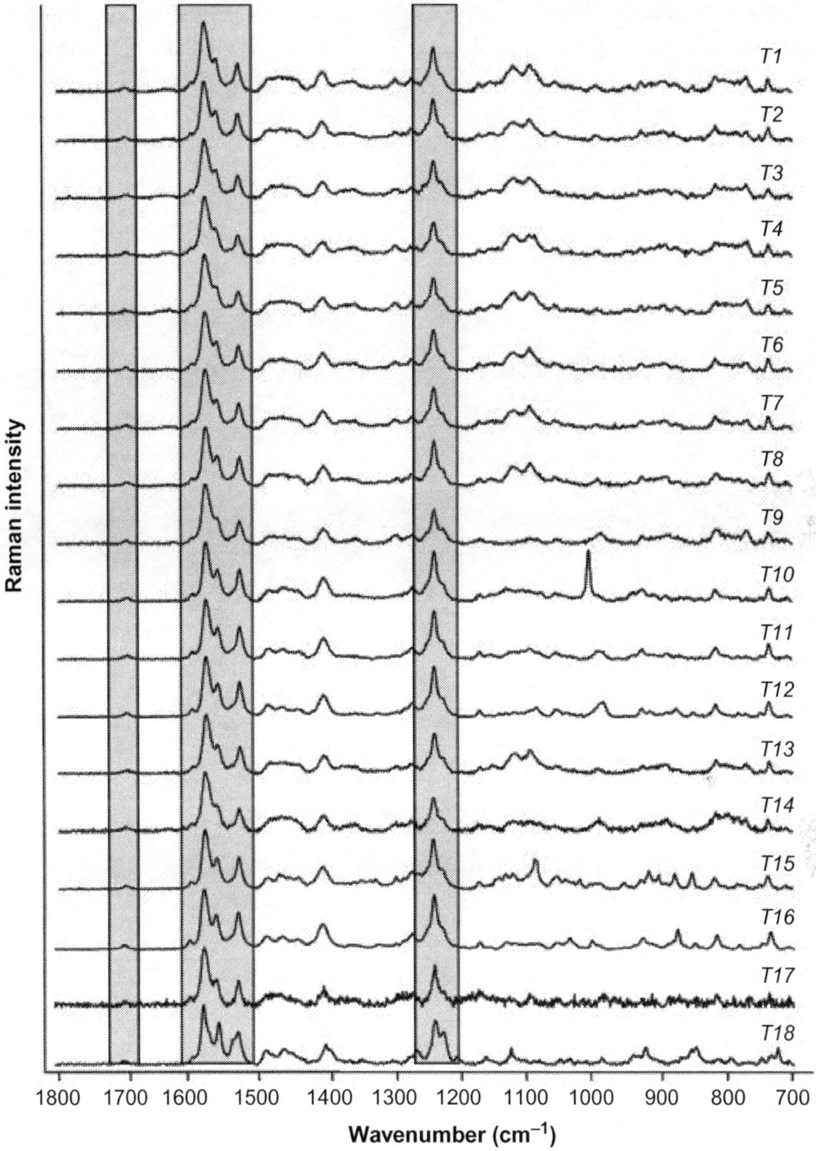

FIG. 5.29 Raman spectra of the 18 tablets (5× objective, 10 accumulations of 30 s, 700–1800 cm^{-1}, 785-nm laser wavelength, after baseline correction). *(Reproduced with permission from M. de Veij, A. Deneckere, P. Vandenabeele, D. de Kaste, L. Moens, Detection of counterfeit Viagra® with Raman spectroscopy, J. Pharm. Biomed. Anal. 46 (2008) 303–309.)*

As stated in another investigation, both genuine and counterfeit Cialis tablets (i.e., tadalafil, another PDE5 inhibitor) were able, by using Raman microscopy and multivariate curve resolution, to be analyzed for identification of excipients, quantification of API, and spatial distribution of both excipients and API in each tablet (Figs. 5.30 and 5.31) [89].

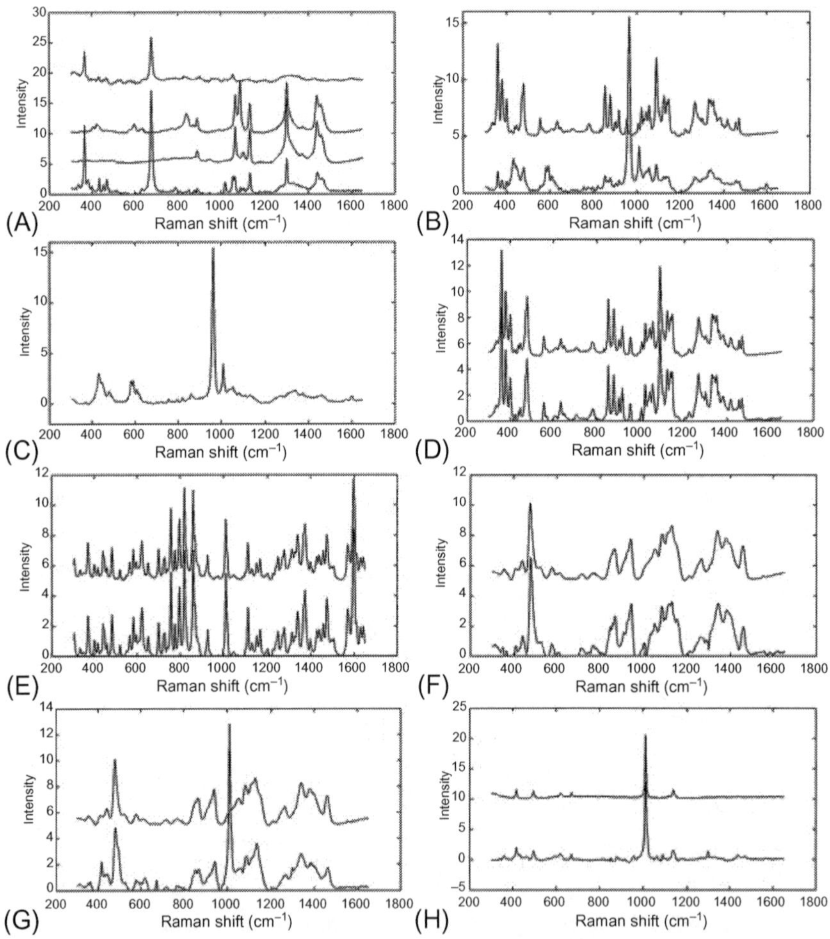

FIG. 5.30 Resolved spectra of counterfeit sample A. (A) First component compared to talc *(top)*, sodium lauryl sulfate *(second from top)*, and magnesium stearate *(second from bottom)*. (B) Second component compared to lactose monohydrate *(top)*. (C) Corrected second component. (D) Third component compared to lactose monohydrate *(top)*. (E) Fifth component compared to tadalafil *(top)*. (F) Sixth component compared to corn starch *(top)*. (G) Fourth component compared to corn starch *(top)*. (H) Corrected fourth component compared to calcium sulfate *(top)*. *(Reproduced with permission from K. Kwok, L.S. Taylor, Analysis of counterfeit Cialis® tablets using Raman microscopy and multivariate curve resolution, J. Pharm. Biomed. Anal. 66 (2012) 126–135.)*

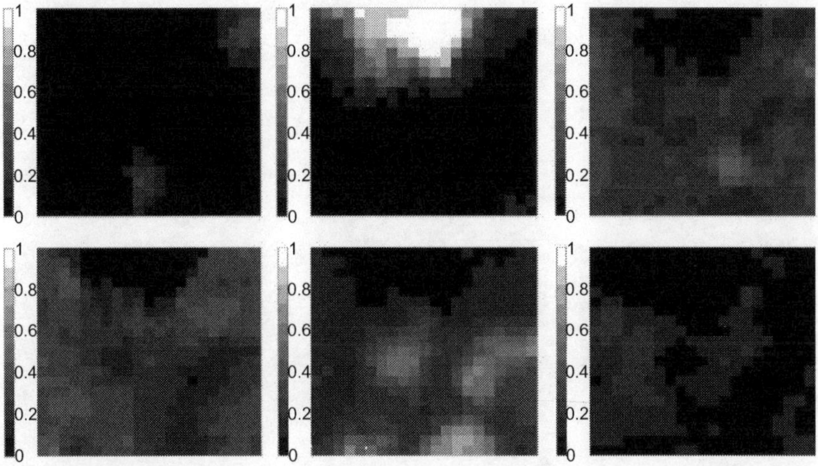

FIG. 5.31 Raman images of resolved components of counterfeit sample A. *Top*: first to third components *(left to right)*. *Bottom*: fourth to sixth components *(left to right)*. First component: talc, sodium lauryl sulfate, magnesium stearate; second component: lactose monohydrate, unknown; third component: lactose monohydrate; fourth component: starch, calcium sulfate; fifth component: tadalafil; sixth component: starch. *(Reproduced with permission from K. Kwok, L.S. Taylor, Analysis of counterfeit Cialis® tablets using Raman microscopy and multivariate curve resolution, J. Pharm. Biomed. Anal. 66 (2012) 126–135.)*

The employment of Raman spectroscopy for screening imitation or fake Cialis tablets was additionally justified by other analytical techniques (such as ^{1}H NMR, 2D DOSY ^{1}H NMR, LC-DAD and LC-MS) [90]. Intriguingly, counterfeit Cialis tablets could be recognized by employing Raman microscopy and two-dimensional correlation spectroscopy to examine their packaging [91]. It was uncovered that two (white and yellow) color regions could be selectively chosen for the analysis, i.e., for the white color region, the genuine package was made of cellulose (box), calcium carbonate (white pigment), and a styrene-based compound (coating binder), while the counterfeit package consisted of cellulose, calcium carbonate (box), a styrene based compound, and an unknown white colorant; for the yellow color region, both the genuine and counterfeit packages contained an unknown yellow colorant, which could be chemically dissimilar for each (Figs. 5.32 and 5.33).

In a systemic study, both the benefits and limitations of dispersive Raman spectroscopy backscattering mode combined with PCA were evaluated for characterization of genuine and counterfeit drugs based on a large set of model tablets [92]. It was confirmed that the discrimination was possible for tablet samples with different coating materials, API contents, and excipients. In contrast, it was impossible to discriminate tablet samples with a variation in compression force, mixing quality and granulation. As exemplified, the change in Raman signals was monitored for commercial effervescent tablets being stored in five different conditions (Fig. 5.34).

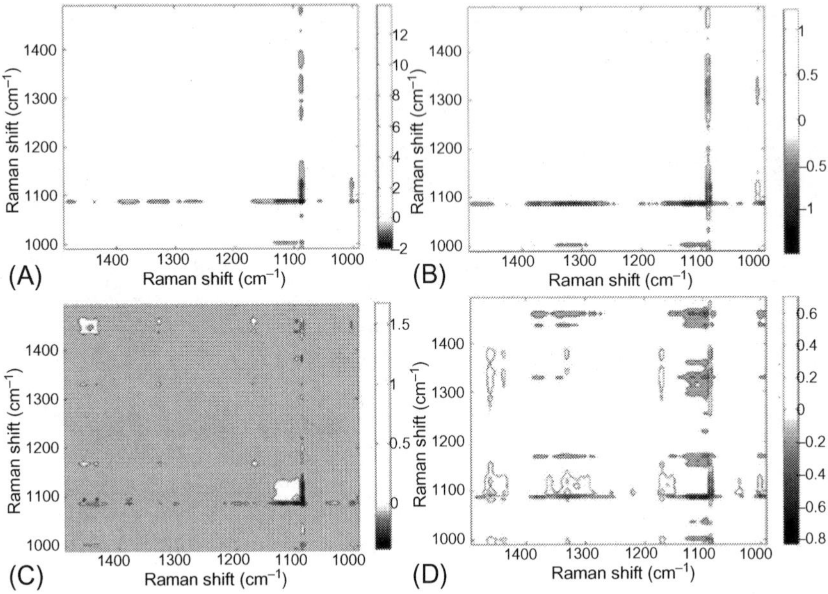

FIG. 5.32 Synchronous (A and C) and asynchronous (B and D) maps of white color region of the package (A and B: genuine; C and D: counterfeit). *(Reproduced with permission from K. Kwok, L.S. Taylor, Analysis of the packaging enclosing a counterfeit pharmaceutical tablet using Raman microscopy and two-dimensional correlation spectroscopy, Vib. Spectrosc. 61 (2012) 176–182.)*

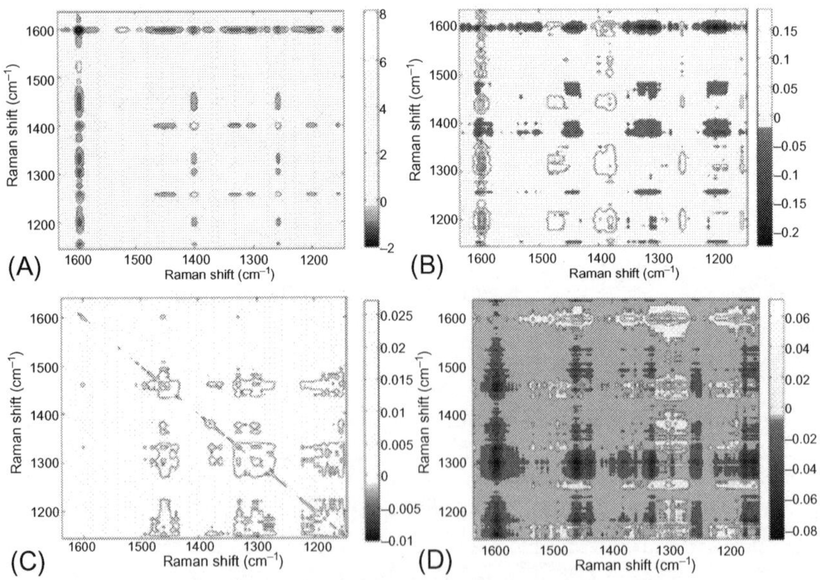

FIG. 5.33 Synchronous (A and C) and asynchronous (B and D) maps of yellow color region of the package (A and B: genuine; C and D: counterfeit). *(Reproduced with permission from K. Kwok, L.S. Taylor, Analysis of the packaging enclosing a counterfeit pharmaceutical tablet using Raman microscopy and two-dimensional correlation spectroscopy, Vib. Spectrosc. 61 (2012) 176–182.)*

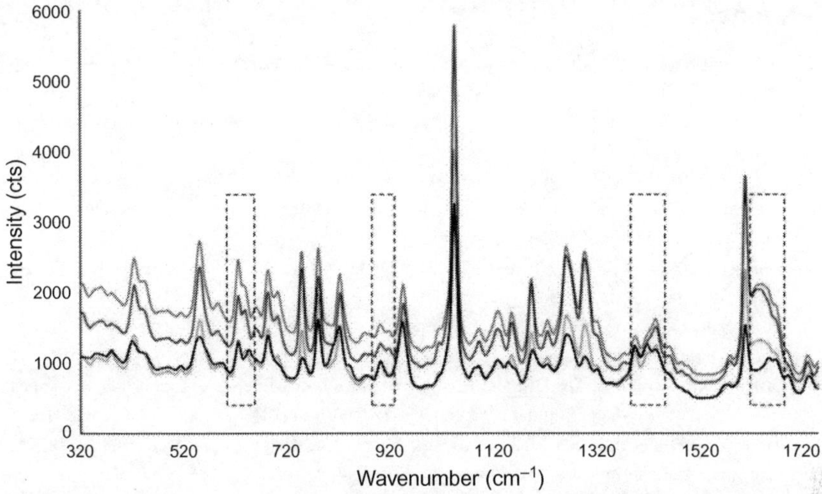

FIG. 5.34 Raman spectra of ASS + C-ratiopharm effervescent tablets (containing acetylsalicylic acid) stored at normal conditions in the spectral range 320–1720 cm^{-1}: Comparison of raw data after one *(blue)*, two *(red)*, three *(green)*, and four *(black)* days of storage. The occurring spectral changes are highlighted with *dotted lines*. *(Reproduced with permission from S. Neuberger, C. Neusüß, Determination of counterfeit medicines by Raman spectroscopy: systematic study based on a large set of model tablets, J. Pharm. Biomed. Anal. 112 (2015) 70–78.)*

In 2008, Acomplia (rimonabant, a selective CB1 endocannabinoid receptor antagonist indicated for obesity treatment) was withdrawn from the market as a result of unacceptable side effects such as depression and suicidal behavior [93]. Unfortunately, it could be still ordered via the Internet, resulting in the delivery of Acomplia counterfeit and imitation products. It was underlined that the use of only chromatographic techniques (LC-DAD-MSn) was not sufficient when investigating rimonabant in illegal medicines [94]. For this purpose, spectroscopic (both Raman and IR) and X-ray diffraction experiments provided valuable information on rimonabant polymorphism, i.e., two suspected samples contained an unapproved rimonabant polymorph as affirmed by Raman, IR, and X-ray diffraction data. The corroboration of Raman and NIR spectroscopy was also evidenced by their capacity to rapidly screen genuine and counterfeit products of the cholesterol-lowering medicine Lipitor [95]. Based on partial least squares discriminant analysis (PLS-DA) models, NIR or Raman spectra could be furthermore applied for distinguishing between atorvastatine and lovastatine as the API used in the counterfeit drugs tested, despite of spectral differences observed by Raman microscopy (for the coating and tablet core) and NIR spectroscopy (due to water adsorption from the atmosphere after unpacking a tablet blister).

References

[1] https://www.who.int/medicines/areas/quality_safety/quality_assurance/control/en/, (Accessed January 7, 2020).

[2] L. Levi, G.C. Walker, L.I. Pugsley, Quality control of pharmaceuticals, Can. Med. Assoc. J. 91 (15) (1964) 781–785.

[3] L.V. Allen Jr., H.C. Ansel, H.C. Anseil, Section I: Introduction to drugs, drug dosage forms, and drug delivery systems, in: Ansel's Pharmaceutical Dosage Forms and Drug Delivery Systems, 11th ed., Wolters Kluwer, Philadelphia, 2018.

[4] USP 41—NF 36, The United States Pharmacopeia and National Formulary, United States, Pharmacopeial Convention Inc., 2018.

[5] European Pharmacopoeia (Ph. Eur.), European Directorate for the Quality of Medicines & HealthCare (EDQM), 10th ed., 2016.

[6] British pharmacopoeia, The British Pharmacopoeia Commission Secretariat of the Medicines and Healthcare Products Regulatory Agency (MHRA), TSO (The Stationery Office), 2019.

[7] M. de Veij, P. Vandenabeele, T. De Beer, J.P. Remonc, L. Moens, Reference database of Raman spectra of pharmaceutical excipients, J. Raman Spectrosc. 40 (2009) 297–307.

[8] A.A. Bunaciu, H.Y. Aboul-Enein, Vibrational spectroscopy applications of drug analysis, in: J.C. Lindon, G.E. Tranter, D.W. Koppenaal (Eds.), The Encyclopedia of Spectroscopy and Spectrometry, third ed., vol. 4, Academic Press, Oxford, 2017, pp. 575–581.

[9] A.A. Bunaciu, H.Y. Aboul-Enein, S. Fleschin, Applications of Fourier transform infrared spectrophotometry in pharmaceutical drugs analysis, Appl. Spectrosc. Rev. 45 (2010) 206–219.

[10] R. Deidda, P.Y. Sacre, M. Clavaud, L. Coïc, H. Avohou, P. Hubert, E. Ziemons, Vibrational spectroscopy in analysis of pharmaceuticals: critical review of innovative portable and handheld NIR and Raman spectrophotometers, TrAC Trends Anal. Chem. 114 (2019) 251–259.

[11] C.L. Anderton, Vibrational Spectroscopy in Pharmaceutical Analysis, Blackwell, Publishing, Oxford, 2003.

[12] S. Wartewig, R.H.H. Neubert, Pharmaceutical applications of mid-IR and Raman spectroscopy, Adv. Drug Deliv. Rev. 57 (2005) 1144–1170.

[13] Y. Roggo, P. Chalus, L. Maurer, C. Lema-Martinez, A. Edmond, N. Jent, A review of near infrared spectroscopy and chemometrics in pharmaceutical technologies, J. Pharm. Biomed. Anal. 44 (2007) 683–700.

[14] C. Gendrin, Y. Roggo, C. Collet, Pharmaceutical applications of vibrational chemical imaging and chemometrics: a review, J. Pharm. Biomed. Anal. 48 (2008) 533–553.

[15] C. De Bleye, P.F. Chavez, J. Mantanus, R. Marini, P. Hubert, E. Rozet, E. Ziemons, Critical review of near-infrared spectroscopic methods validations in pharmaceutical applications, J. Pharm. Biomed. Anal. 69 (2012) 125–132.

[16] B. Van Eerdenbrugh, L.S. Taylor, Application of mid-IR spectroscopy for the characterization of pharmaceutical systems, Int. J. Pharm. 417 (2011) 3–16.

[17] D.E. Bugay, W.P. Findlay, Pharmaceutical Excipients Characterization by IR Raman and NMR Spectroscopy, Marcel Dekker, New York, 1999.

[18] S.C. Pinzaru, I. Pavel, N. Leopold, W. Kiefer, Identification and characterization of pharmaceuticals using Raman and surface-enhanced Raman scattering, J. Raman Spectrosc. 35 (2004) 338–346.

[19] A.A. Bunaciu, H.Y. Aboul-Enein, V.D. Hoang, Vibrational spectroscopy in polymorphic analysis, TrAC Trends Anal. Chem. 69 (2015) 14–22.

[20] J. Moros, S. Garrigues, M. de la Guardia, Quality control Fourier transform infrared determination of diazepam in pharmaceuticals, J. Pharm. Biomed. Anal. 43 (2007) 1277–1282.

[21] U.S. Drug Enforcement Administration, https://www.dea.gov/factsheets/benzodiazepines, (Accessed January 7, 2020).

[22] The United States Pharmacopeia, Twentieth Revision, United States Pharmacopeial Convention, Inc., Rockville, MD, 1980.

[23] R.N. Brogden, R.C. Heel, T.M. Speight, G.S. Avery, Piroxicam: a review of its pharmacological properties and therapeutic efficacy, Drugs 22 (3) (1981) 165–187.

[24] A.A. Bunaciu, S. Fleschin, H.Y. Aboul-Enein, A new method for a quantitative determination of piroxicam in pharmaceutical formulations using FT-IR spectrometry, Antiinflamm. Antiallergy Agents Med. Chem. 11 (2012) 262–266.

[25] A.F. Sallmann, The history of diclofenac, Am. J. Med. 80 (Suppl. 4B) (1986) 29–33.

[26] R. Szostak, S. Mazurek, The influence of sample area on diclofenac sodium quantification by diffuse reflectance IR spectroscopy, Talanta 84 (2011) 583–586.

[27] R. Szostak, S. Mazurek, Comparison of infrared attenuated total reflection and Raman spectroscopy in the quantitative analysis of diclofenac sodium in tablets, Vib. Spectrosc. 57 (1) (2011) 157–162.

[28] A.A. Kandhro, A.H. Laghari, S.A. Mahesar, R. Saleem, A. Nelofar, S.T. Khan, S.T.H. Sherazi, Application of attenuated total reflectance Fourier transform infrared spectroscopy for determination of cefixime in oral pharmaceutical formulations, Spectrochim. Acta A Mol. Biomol. Spectrosc. 115 (2013) 51–56.

[29] Y.C. Feng, C.Q. Hu, Construction of universal quantitative models for determination of roxithromycin and erythromycin ethylsuccinate in tablets from different manufacturers using near infrared reflectance spectroscopy, J. Pharm. Biomed. Anal. 41 (2) (2006) 373–384.

[30] Y.K. Dijiba, A. Zhang, T.M. Niemczyk, Determinations of ephedrine in mixtures of ephedrine and pseudoephedrine using diffuse reflectance infrared spectroscopy, Int. J. Pharm. 289 (1–2) (2005) 39–49.

[31] A.R. Khaskheli, S.T.H. Sirajuddin, S.A. Sherazi, S.A. Mahesar, N.H.K. Kandhro, A.M. Mallah, Estimation of ibuprofen in urine and tablet formulations by transmission Fourier transform infrared spectroscopy by partial least square, Spectrochim. Acta A Mol. Biomol. Spectrosc. 102 (2013) 403–407.

[32] X. Wang, Q. Fu, J. Sheng, X. Yang, J. Jia, W. Du, Construction of a universal quantitative model for ibuprofen sustained-release capsules from different manufacturers using near-infrared diffuse reflection spectroscopy, Vib. Spectrosc. 53 (2010) 214–217.

[33] B. Waldeck, Enantiomers of bronchodilating beta2-adrenoceptor agonists: is there a cause for concern? J. Allergy Clin. Immunol. 103 (1999) 742–748.

[34] S. Agatonovic-Kustrin, R. Alany, Application of diffuse reflectance infrared Fourier transform spectroscopy combined with artificial neural networks in analysing enantiomeric purity of terbutaline sulphate bulk drug, Anal. Chim. Acta 449 (2001) 157–165.

[35] M. Helliwell, D. Taylor, Solid oral dosage forms, Prof. Nurse 8 (5) (1993) 313–317.

[36] Y. Chen, Z. Gao, J.Z. Duan, Dissolution testing of solid products, in: Y. Qiu, Y. Chen, G. Zhang, L. Yu, R.V. Mantri (Eds.), Developing Solid Oral Dosage Forms—Pharmaceutical Theory And Practice, Academic Press, 2016, pp. 355–380. (Chapter 13).

[37] S.G. Kazarian, K.L.A. Chan, "Chemical photography" of drug release, Macromolecules 36 (2003) 9866–9872.

[38] K.L.A. Chan, S.G. Kazarian, FTIR spectroscopic imaging of dissolution of a solid dispersion of nifedipine in poly (ethylene glycol), Mol. Pharm. 1 (2004) 331–335.

[39] J. van der Weerd, K.L.A. Chan, S.G. Kazarian, An innovative design of compaction cell for in situ FT-IR imaging of tablet dissolution, Vib. Spectrosc. 3 (2004) 9–13.
[40] K.L.A. Chan, S.G. Kazarian, New opportunities in micro- and macro-attenuated total reflection infrared spectroscopic imaging: spatial resolution and sampling versatility, Appl. Spectrosc. 57 (2003) 381–389.
[41] J. Van der Weerd, S.G. Kazarian, Combined approach of FTIR imaging and conventional dissolution tests applied to drug release, J. Control. Release 98 (2004) 295–305.
[42] M.C. Sarraguça, R. Matias, R. Figueiredo, P.R.S. Ribeiro, A. Teixeira Martins, J.A. Lopes, Near infrared spectroscopy to monitor drug release in situ during dissolution tests, Int. J. Pharm. 513 (1–2) (2016) 1–7.
[43] J. Van der Weerd, S.G. Kazarian, Validation of macroscopic ATR-FTIR imaging to study dissolution of swelling pharmaceutical tablets, Appl. Spectrosc. 58 (2004) 1413–1419.
[44] S. Morita, H. Shinzawa, I. Noda, Y. Ozaki, Perturbation-correlation moving-window two-dimensional correlation spectroscopy, Appl. Spectrosc. 60 (4) (2006) 398–406.
[45] Y. Hattori, M. Otsuka, NIR spectroscopic study of the dissolution process in pharmaceutical tablets, Vib. Spectrosc. 57 (2011) 275–281.
[46] P. Wray, K.L. Andrew Chan, J. Kimber, S.G. Kazarian, Compaction of pharmaceutical tablets with different polymer matrices studied by FTIR imaging and X-ray microtomography, J. Pharm. Sci. 97 (2008) 4269–4277.
[47] A.J. Wagstaff, C.M. Perry, Topical imiquimod: a review of its use in the management of anogenital warts, actinic keratoses, basal cell carcinoma and other skin lesions, Drugs 67 (15) (2007) 2187–2210.
[48] R.L. Carneiro, R.J. Poppi, Infrared imaging spectroscopy and chemometric tools for in situ analysis of an imiquimod pharmaceutical preparation presented as cream, Spectrochim. Acta A Mol. Biomol. Spectrosc. 118 (2014) 215–220.
[49] H. Masmoudi, Y. Le Dréau, P. Piccerelle, J. Kister, The evaluation of cosmetic and pharmaceutical emulsions aging process using classical techniques and a new method: FTIR, Int. J. Pharm. 289 (1–2) (2005) 117–131.
[50] T. Firkala, A. Farkas, B. Vajna, I. Farkas, G. Marosi, Investigation of drug distribution in tablets using surface enhanced Raman chemical imaging, J. Pharm. Biomed. Anal. 76 (2013) 145–151.
[51] https://dictionary.cambridge.org/dictionary/english/polymorphism, (Accessed December 25, 2019).
[52] J. Haleblian, W. McCrone, Pharmaceutical applications of polymorphism, J. Pharm. Sci. 58 (8) (1969) 911–929.
[53] H. Seo, M. Tsuruoka, T. Hashimoto, T. Fujinaga, M. Otagiri, K. Uekama, Enhancement of oral bioavailability of spironolactone by β- and γ-cyclodextrin complexations, Chem. Pharm. Bull. 31 (1) (1983) 286–291.
[54] G.A. Neville, H.D. Beckstead, H.F. Shurvell, Utility of Fourier-transform Raman and Fourier-transform infrared diffuse reflectance spectroscopy for determination of polymorphic spironolactone samples, J. Pharm. Sci. 81 (1992) 1141–1146.
[55] H.D. Beckstead, G.A. Neville, H.F. Shurvell, Differentiation of solvated spironolactone samples by FT-Raman and FT-IR diffuse reflectance spectroscopy, Fresenius J. Anal. Chem. 345 (1993) 727–732.
[56] G.L. Pisegna, D.F.R. Gilson, I.S. Butler, High-pressure infrared and Raman studies of polymorphism in pharmaceutical compounds: spironolactone, forms I and II, J. Mol. Struct. 1078 (2014) 146–150.

[57] Y. Du, H. Zhang, J. Xue, W. Tang, H. Fang, Q. Zhang, Y. Li, Z. Hong, Vibrational spectroscopic study of polymorphism and polymorphic transformation of anti-viral drug Lamivudine, Spectrochim. Acta A Mol. Biomol. Spectrosc. 137 (2015) 1158–1163.

[58] H.D. Langtry, S.M. Grant, K.L. Goa, Famotidine, Drugs 35 (1989) 551–590.

[59] W.T. Cheng, S.Y. Lin, M.J. Li, Raman microspectroscopic mapping or thermal system used to investigate milling-induced solid-state conversion of famotidine polymorphs, J. Raman Spectrosc. 38 (2007) 1595–1601.

[60] B. Fulton, K.L. Goa, Olanzapine—a review of its pharmacological properties and therapeutic efficacy in the management of schizophrenia and related psychoses, Drugs 53 (1997) 281–298.

[61] A.P. Ayala, H.W. Siesler, R. Boese, G.G. Hoffmann, G.I. Polla, D.R. Vega, Solid state characterization of olanzapine polymorphs using vibrational spectroscopy, Int. J. Pharm. 326 (2006) 69–79.

[62] P. Sharma, R. Nandi, D. Gangopadhyay, A. Singh, R.K. Singh, Temperature dependent polymorphism of pyrazinamide: an in situ Raman and DFT study, Spectrochim. Acta A Mol. Biomol. Spectrosc. 1905 (2018) 177–180.

[63] Q. Wang, J. Xue, Y. Wang, S. Jin, Y. Du, Investigation into tautomeric polymorphism of 2-thiobarbituric acid using experimental vibrational spectroscopy combined with DFT theoretical simulation, Spectrochim. Acta A Mol. Biomol. Spectrosc. 204 (2018) 99–104.

[64] M. Latreche, J.F. Willart, L. Paccou, Y. Guinet, A. Hédoux, Polymorphism versus devitrification mechanism: low-wavenumber Raman investigations in sulindac, Int. J. Pharm. 567 (2019) 118476.

[65] A. Salari, R.E. Young, Application of attenuated total reflectance FTIR spectroscopy to the analysis of mixtures of pharmaceutical polymorphs, Int. J. Pharm. 163 (1998) 157–166.

[66] A. Heinz, M. Savolainen, T. Rades, C.J. Strachan, Quantifying ternary mixtures of different solid-state forms of indomethacin by Raman and near-infrared spectroscopy, Eur. J. Pharm. Sci. 32 (3) (2007) 182–192.

[67] Y. Hu, A. Erxleben, A.G. Ryder, P. McArdle, Quantitative analysis of sulfathiazole polymorphs in ternary mixtures by attenuated total reflectance infrared, near-infrared and Raman spectroscopy, J. Pharm. Biomed. Anal. 53 (3) (2010) 412–420.

[68] K. Kachrimanis, M. Rontogianni, S. Malamataris, Simultaneous quantitative analysis of mebendazole polymorphs A–C in powder mixtures by DRIFTS spectroscopy and ANN modeling, J. Pharm. Biomed. Anal. 51 (3) (2010) 512–520.

[69] K. Pöllänen, A. Häkkinen, M. Huhtanen, S.P. Reinikainen, M. Karjalainen, J. Rantanen, M. Louhi-Kultanen, L. Nyström, DRIFT-IR for quantitative characterization of polymorphic composition of sulfathiazole, Anal. Chim. Acta 544 (2005) 108–117.

[70] S.A. Schönbichler, L.K.H. Bittner, A.K.H. Weiss, U.J. Griesser, J.D. Pallua, C.W. Huck, Comparison of NIR chemical imaging with conventional NIR, Raman and ATR-IR spectroscopy for quantification of furosemide crystal polymorphs in ternary powder mixtures, Eur. J. Pharm. Biopharm. 84 (2013) 616–625.

[71] A. Koczwara, J. Dressman, Global Health Commentary, Poor-quality and counterfeit drugs: a Systematic assessment of Prevalence and risks based on Data Published from 2007 to 2016, J. Pharm. Sci. 106 (10) (2017) 2921–2929.

[72] WHO (World Health Organization), http://www.who.int/medicines/organization.gsm/activities/quality_assurance/cft/counterfeit, (Accessed December 25, 2019).

[73] https://www.ema.europa.eu/en/human-regulatory/overview/public-health-threats/falsified-medicines-overview, (Accessed January 7, 2020).

[74] M. Johansson, D. Fransson, T. Rundlöf, N.H. Huynh, T. Arvidsson, A general analytical platform and strategy in search for illegal drugs, J. Pharm. Biomed. Anal. 100 (2014) 215–229.

[75] O.Y. Rodionova, A.L. Pomerantsev, NIR-based approach to counterfeit-drug detection, Trends Anal. Chem. 29 (8) (2010) 795–803.

[76] F.E. Dowell, E.B. Maghirang, F.M. Fernandez, P.N. Newton, M.D. Green, Detecting counterfeit antimalarial tablets by near-infrared spectroscopy, J. Pharm. Biomed. Anal. 48 (3) (2008) 1011–1014.

[77] J. Shaohong, Roles and functions of Chinese pharmacopoeia and reference standard in China drug administration, in: Proceedings of the International Conference, EDQM New Frontiers in the Quality of Medicines, Strasbourg, France, 13–15 June, 2007, pp. 44–46.

[78] O.Y. Rodionova, L.P. Houmøller, A.L. Pomerantsev, P. Geladi, J. Burger, V.L. Dorofeyev, A.P. Arzamastsev, NIR spectrometry for counterfeit drug detection. A feasibility study, Anal. Chim. Acta 549 (2005) 151–158.

[79] O.Y. Rodionova, Y.V. Sokovikov, A.L. Pomerantsev, Quality control of packed raw materials in pharmaceutical industry, Anal. Chim. Acta 642 (2009) 222–227.

[80] T. Puchert, D. Lochmann, J.C. Menezes, G. Reich, Near-infrared chemical imaging (NIR-CI) for counterfeit drug identification—a four-stage concept with a novel approach of data processing (Linear Image Signature), J. Pharm. Biomed. Anal. 51 (2010) 138–145.

[81] K. Dégardin, A. Guillemain, N. Viegas Guerreiro, Y. Roggo, Near infrared spectroscopy for counterfeit detection using a large database of pharmaceutical tablets, J. Pharm. Biomed. Anal. 128 (2016) 89–97.

[82] A. Guillemain, K. Dégardin, Y. Roggo, Performance of NIR handheld spectrometers for the detection of counterfeit tablets, Talanta 165 (2017) 632–640.

[83] S. Wilczyński, R. Koprowski, M. Marmion, P. Duda, B. Błońska-Fajfrowska, The use of hyperspectral imaging in the VNIR (400–1000 nm) and SWIR range (1000–2500 nm) for detecting counterfeit drugs with identical API composition, Talanta 160 (2016) 1–8.

[84] G. Jackson, S. Arver, I. Banks, V.J. Stecher, Counterfeit phosphodiesterase type 5 inhibitors pose significant safety risks, Int. J. Clin. Pract. 64 (4) (2010) 497–504.

[85] M.J. Vredenbregt, L. Blok-Tip, R. Hoogerbrugge, D.M. Barends, D. de Kaste, Screening suspected counterfeit Viagra® and imitations of Viagra® with near-infrared spectroscopy, J. Pharm. Biomed. Anal. 40 (2006) 840–849.

[86] S. Wilczyński, R. Koprowski, B. Błońska-Fajfrowska, Directional reflectance analysis for identifying counterfeit drugs: preliminary study, J. Pharm. Biomed. Anal. 124 (2016) 341–346.

[87] M. de Veij, A. Deneckere, P. Vandenabeele, D. de Kaste, L. Moens, Detection of counterfeit Viagra® with Raman spectroscopy, J. Pharm. Biomed. Anal. 46 (2008) 303–309.

[88] H. Rebiere, M. Martin, C. Ghyselinck, P.A. Bonnet, C. Brenier, Raman chemical imaging for spectroscopic screening and direct quantification of falsified drugs, J. Pharm. Biomed. Anal. 148 (2018) 316–323.

[89] K. Kwok, L.S. Taylor, Analysis of counterfeit Cialis® tablets using Raman microscopy and multivariate curve resolution, J. Pharm. Biomed. Anal. 66 (2012) 126–135.

[90] S. Trefi, C. Routaboul, S. Hamieh, V. Gilard, M. Malet-Martino, R. Martino, Analysis of illegally manufactured formulations of tadalafil (Cialis®) by 1H NMR, 2D DOSY 1H NMR and Raman spectroscopy, J. Pharm. Biomed. Anal. 47 (2008) 103–113.

[91] K. Kwok, L.S. Taylor, Analysis of the packaging enclosing a counterfeit pharmaceutical tablet using Raman microscopy and two-dimensional correlation spectroscopy, Vib. Spectrosc. 61 (2012) 176–182.

[92] S. Neuberger, C. Neusüß, Determination of counterfeit medicines by Raman spectroscopy: systematic study based on a large set of model tablets, J. Pharm. Biomed. Anal. 112 (2015) 70–78.

[93] D.R. Janero, A. Makriyannis, Cannabinoid receptor antagonists: pharmacological opportunities, clinical experience, and translational prognosis, Expert Opin. Emerg. Drugs 14 (2009) 43–65.

[94] B.J. Venhuis, M.V. Vredenbregt, N. Kaun, J.K. Maurin, Z. Fijałek, D. de Kaste, The identification of rimonabant polymorphs, sibutramine and analogues of both in counterfeit Acomplia bought on the internet, J. Pharm. Biomed. Anal. 54 (2011) 21–26.

[95] P. de Peinder, M.J. Vredenbregt, T. Visser, D. de Kaste, Detection of Lipitor® counterfeits: a comparison of NIR and Raman spectroscopy in combination with chemometrics, J. Pharm. Biomed. Anal. 47 (2008) 688–694.

Chapter 6

Herbal drug analysis

From the ancient history, people have relied on traditional herbal medicine to meet their healthcare needs [1]. The most common reasons for such using a plant or plant part are that products labeled "natural" are thought to be more affordable, accessible, and closely corresponded to the patient's ideology allaying concerns about the side effects of chemical (synthetic) medicines. Although this is not absolutely true, herbal medicines continue to be majorly used for health promotion and chronic conditions, in particular when the treatment is ineffective for advanced cancer and new infectious diseases [2].

In the last decade, about 25% of the drugs prescribed worldwide were derived from plants and 11% of the total 252 drugs in WHO (World Health Organization) essential medicine list were exclusively of plant origin [3]. Based on contributions from 179 member states, the WHO global report 2019 obviously stated that more and more countries are recognizing the role of traditional and complementary medicine in their national health systems [4].

Herbs and plants can be processed and taken in various ways and forms (e.g., whole herb, teas, syrups, essential oils, ointments, salves, rubs, capsules and tablets containing a ground or powdered form of a raw herb or its dried extract) [5]. For herbal medicines, both single and combination preparations contain a myriad of compounds that may be found in a very low concentration range (parts-per-million (ppm) or even parts-per-billion (ppb)), of which many are secondary metabolites [6] and represent their therapeutic effects. The efficacy of these medicines may vary, depending on various factors such as harvest season, geographic origin, plant family, processing, etc. This point makes the quality control of herbal medicines uncontrollable, and is considered as an obstacle for their internationalization and modernization.

Until now, the fingerprint analysis has been internationally recognized as a suitable approach to control the quality of herbal medicines, especially chromatographic, electrophoretic, and their hyphenated techniques are strongly recommended [7–9] In addition to the aforementioned, the chemical analysis of herbal medicines has been also reported in numerous studies with vibrational spectroscopy [10–14] that cover both identification and authentication (as representatively discussed later).

In general, vibrational spectroscopic fingerprinting features could be effectually employed for a quick quality control of herbal medicines. For instance, *Herba Epimedii*, a traditional Chinese medicine well known for kidney

Vibrational Spectroscopy Applications in Biomedical, Pharmaceutical and Food Sciences
https://doi.org/10.1016/B978-0-12-818827-9.00007-X

© 2020 Elsevier Inc. All rights reserved.

nourishing and yang strengthening, was able to be identified by a characteristic FTIR absorption peak at $1259 \pm 1\,\text{cm}^{-1}$ found in both its herbal samples and 70% ethanol extracts [15]. The intensity of this peak was directly correlated with the total content of 4′-methoxyl-prenylflavonols, the major 8-prenyl flavonols in the aerial parts the *Epimedium* species (Fig. 6.1). A correlation value of not less than 0.50, representing the semblance of two IR spectra of herbal sample and icariin, in the range of $1280–1200\,\text{cm}^{-1}$ was established as a screening index.

In another research, multistep macrofingerprint characters were used for an easy identification of lipophilic constituents in Angelica extracted by petroleum ether and wet distillation [16]. In a simultaneous manner, the compositional dissimilarity of two different extracts could be efficiently found by a mutual confirmation of FTIR spectra, their corresponding second derivative spectra (SD-IR), and two-dimensional correlation (2D) IR spectra (Figures 6.2–6.4).

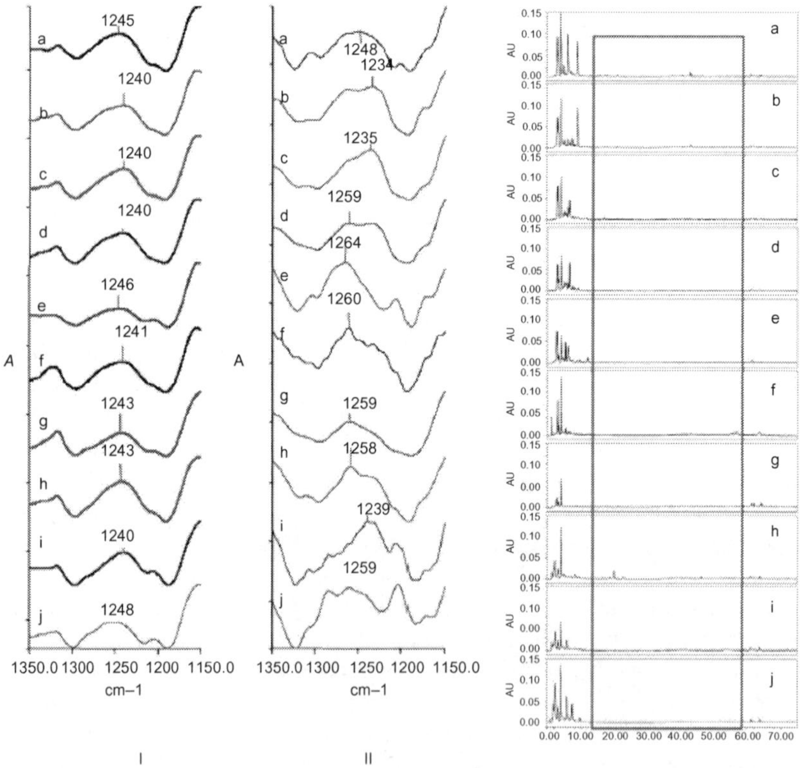

FIG. 6.1 The IR spectra of herbal samples (I) and those of corresponding extract samples (II) and their HPLC chromatograms (III); each letter (a–j) represents 4-digit sample number. *(Reproduced with permission from L.K. Pei, S.Q. Sun, B.L. Guo, W.H. Huang, P.G. Xiao, Fast quality control of Herba Epimedii by using Fourier transform infrared spectroscopy, Spectrochim. Acta A 70 (2008) 258–264.)*

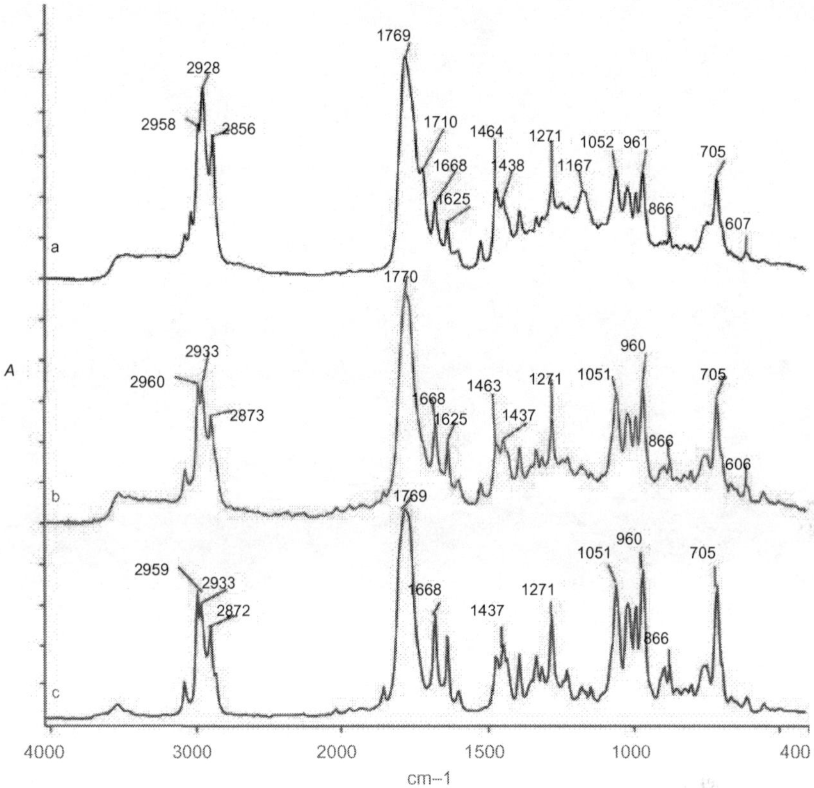

FIG. 6.2 FT-IR spectra of the extracts of: (a) petroleum ether; (b) wet distillation; (c) Z-ligustilide. *(Reproduced with permission from H. Liu, S. Sun, G. Lv, X. Liang, Discrimination of extracted lipophilic constituents of Angelica with multi-steps infrared macro-fingerprinting method, Vib. Spectrosc. 40 (2006) 202–208.)*

Ginseng, a regularly top-selling remedy in herbal medicine, is typically characterized by the presence of ginsenosides and gintonin. It comes from the root part of several plant species in the *Panax* genus, e.g., Korean ginseng (*P. ginseng*), South China ginseng (*P. notoginseng*) and American ginseng (*P. quinquefolius*). Different varieties of ginseng have been used for centuries in Asian and North America to aid a range of medical conditions (such as boosting energy, lowering blood sugar and cholesterol levels, reducing stress, promoting relaxation, treating diabetes, and managing male sexual dysfunction) [17]. It is noted that Korean ginseng is generally more expensive than Chinese one, and it is only able to trade wild and wild-simulated American ginseng roots (a species listed in Appendix II of the Convention on International Trade in Endangered Species of Wild Fauna and Flora [18]) if they were collected from 5-year old plants legally harvested during the designated State harvest season.

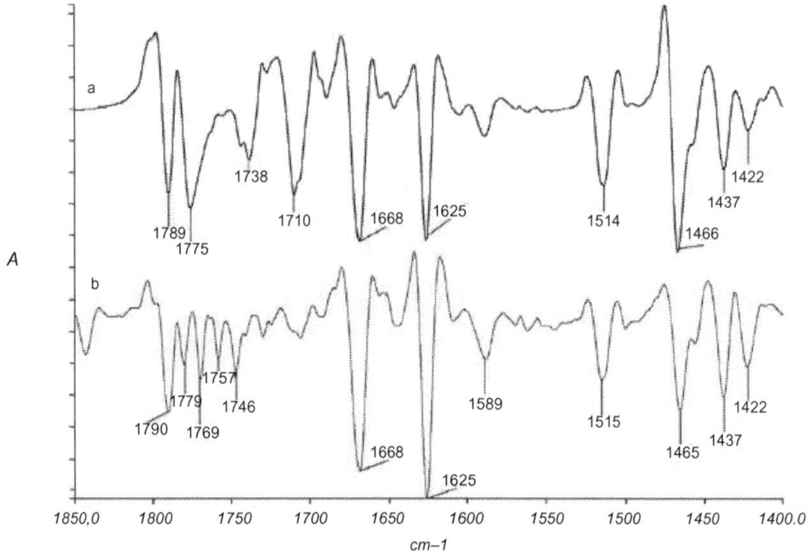

FIG. 6.3 Second derivative spectra of the extracts in the range of 1400–1850 cm^{-1}: (a) petroleum ether and (b) wet distillation. *(Reproduced with permission from K.Y.L. Yap, S.Y. Chan, C.S. Lim, Infrared-based protocol for the identification and categorization of ginseng and its products, Food Res. Int. 40(5) (2007) 643–652.)*

To pinpoint the origin of an unknown ginseng sample, FT-Raman spectroscopy was used [19]. Several spectral features were identified, e.g., a peak at 980 cm^{-1} (ascribed to ring "breathing" (in-plane expansion)) was only found in Chinese ginseng; whereas the C=C stretch at about 1600 cm^{-1} did not appear in Chinese ginseng spectra, but was spectrally seen in American and Korean ginsengs. It was also shown that sliced ginseng samples could produce Raman spectra of much better quality than powdered ones (Fig. 6.5).

In another study, an IR-based protocol was proposed in combination with PCA for ginseng quality surveillance [20]. The spectral region 2000–600 cm^{-1} was identified as characteristic IR fingerprint to authenticate three grades of ginseng root powder (2, 5 and 10) purchased from Hong Kong as well as three commercial products (American ginseng tea, American white ginseng capsules and Korean ginseng capsules) (e.g., Fig. 6.6).

It is speculated that Asian ginseng, American ginseng and Notoginseng contain similar chemical constituents because they own a very close relationship in botanical taxonomy. However, they could be effectively differentiated by using FTIR spectroscopy, i.e., conventional (1D) and two-dimensional (2D) correlation generated by applying thermal perturbation to increase the temperature from 60 to 120°C [21]. For their easy differentiation, comparing the intensity of 1D peaks (located at 1640, 1416, 1372, and 1048 cm^{-1}) and referring to distinctively different features of second derivative FTIR spectra were used. Moreover, their identification could be further based on the positions and intensities of

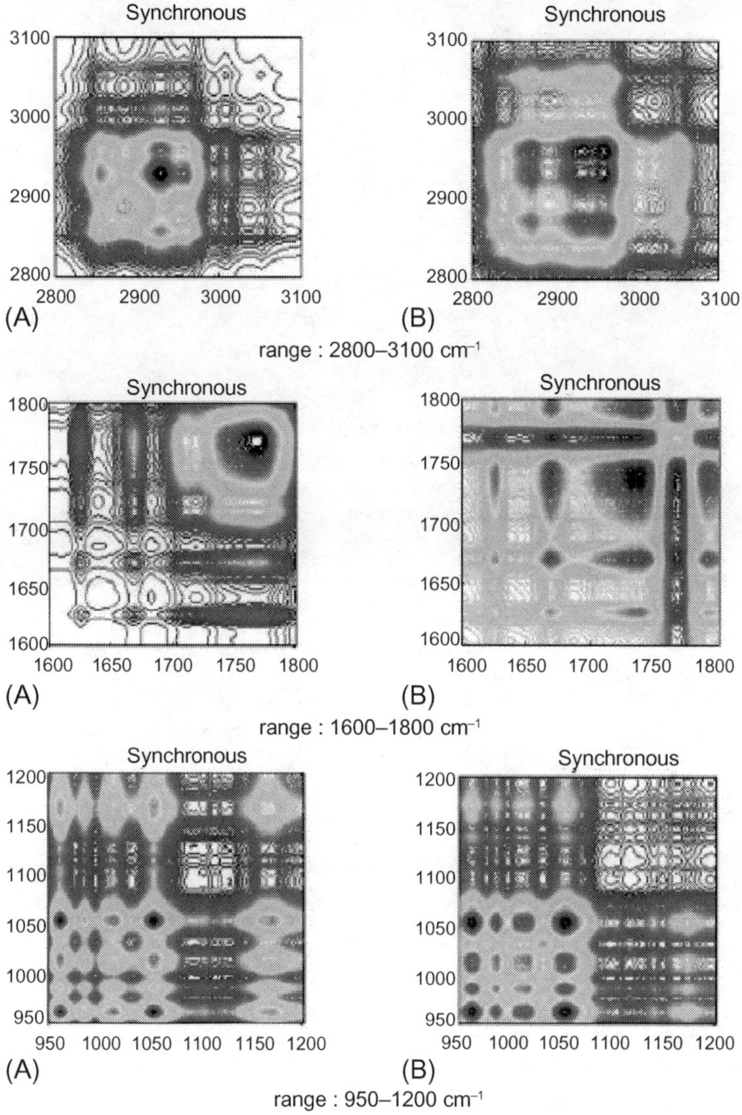

FIG. 6.4 Synchronous 2D-IR spectra of the extracts in the different ranges: (A) petroleum ether and (B) wet distillation. *(Reproduced with permission from H. Liu, S. Sun, G. Lv, X. Liang, Discrimination of extracted lipophilic constituents of Angelica with multi-steps infrared macro-fingerprinting method, Vib. Spectrosc. 40 (2006) 202–208.)*

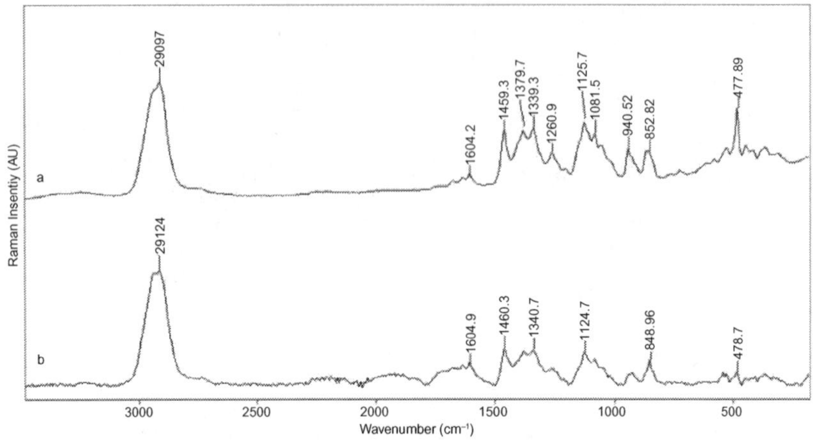

FIG. 6.5 Korean ginseng samples: (a) sliced; (b) sliced and then powdered. *(Reproduced with permission from H.G.M. Edwards, T. Munshi, K. Page, Analytical discrimination between sources of ginseng using Raman spectroscopy, Anal. Bioanal. Chem. 389 (2007) 2203–2215.)*

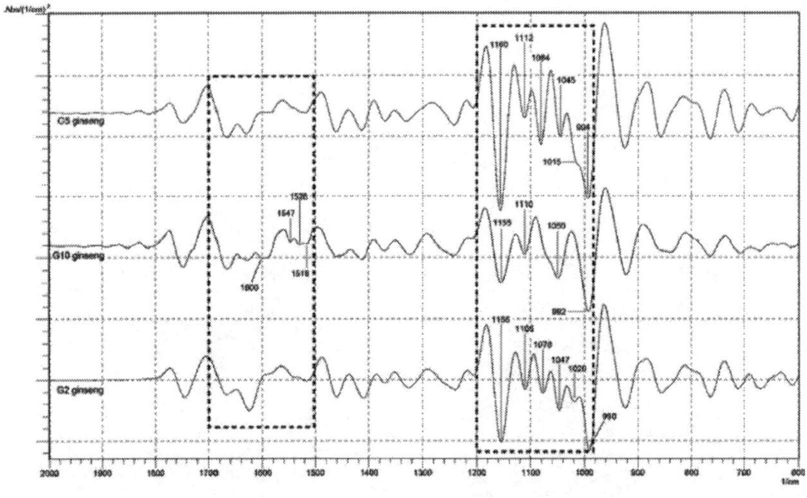

FIG. 6.6 Second-derivative MIR spectra of the G2, G5, and G10 ginsengs. *(Reproduced with permission from K.Y.L. Yap, S.Y. Chan, C.S. Lim, Infrared-based protocol for the identification and categorization of ginseng and its products, Food Res. Int. 40(5) (2007) 643–652.)*

relatively strong autopeaks, positive or negative crosspeaks in their visual and colorful 2D-FTIR spectra (as displayed in Fig. 6.7).

Chrysanthemum is a well-known traditional Chinese medicinal herb commonly used for treatment of wind-heat type cold, headache, dizziness, and dimsightedness [22]. It is suggested that the dried flowers of *Chrysanthemum* containing organic groups (such as alkanes, flavonoids, terpinoids, unsaturated fatty acids, and polysaccharides) can have certain pharmacological effect on

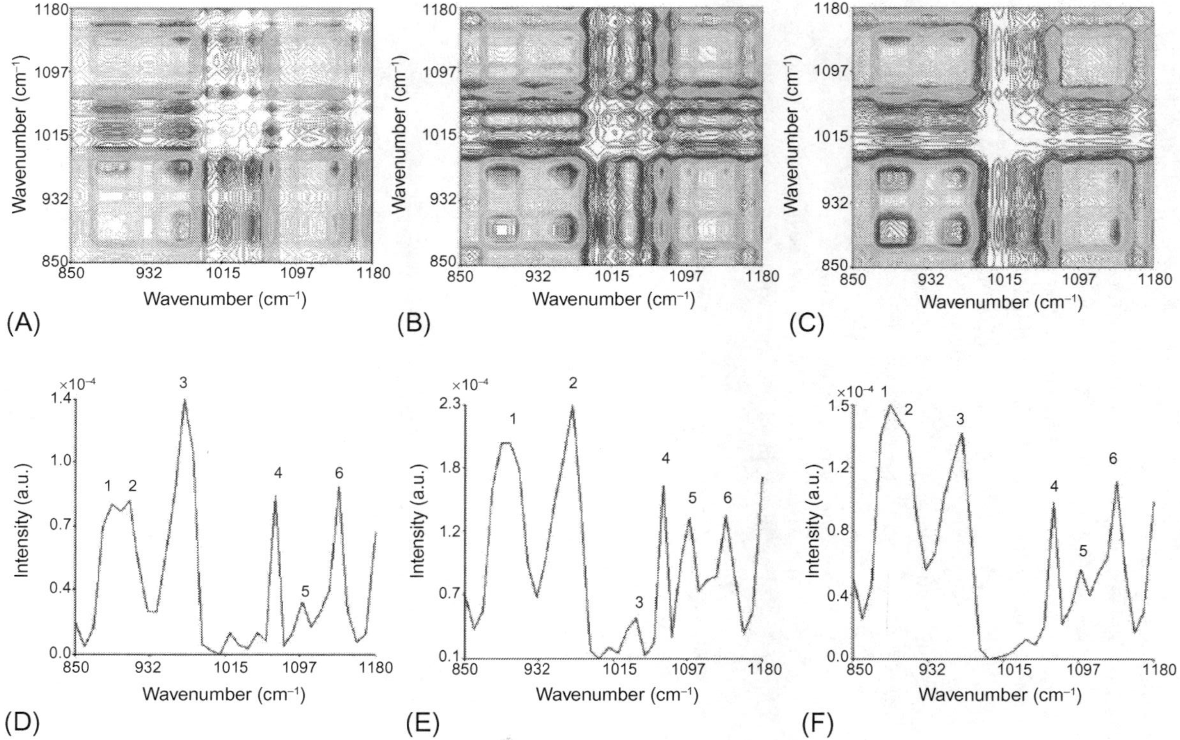

FIG. 6.7 Synchronous 2D-FTIR correlation spectra of (A) Asian ginseng, (B) American ginseng, and (C) Notoginseng in the region of 850–1180 cm^{-1} with their curves of autopeaks in the 2D-FTIR spectra for (D) Asian ginseng, (E) American ginseng, and (F) Notoginseng. *(Reproduced with permission from G.H. Lu, Q. Zhou, S.Q. Sun, K.S.Y. Leung, H. Zhang, Z.Z. Zhao, Differentiation of Asian ginseng, American ginseng and Notoginseng by Fourier transform infrared spectroscopy combined with two-dimensional correlation infrared spectroscopy, J. Mol. Struct. 883–884 (2008) 91–98.)*

cholesterol metabolism as well as antifungal, antibacterial, antivirus, antiinflammatory, antimutagenic, and antineoplastic activities [23]. Using FTIR spectroscopy, it was possible to simultaneously analyze the main chemical constituents in different solvent extracts (water and ethyl acetate) of *Chrysanthemum* samples of seven different regions [24].

Ganoderma lucidum is a Chinese medicinal mushroom, whose fruiting bodies, mycelia, and spores, were traditionally used as a folk medicine in the belief that it could possess many health-promoting properties [25]. To discriminate *G. lucidum* according to cultivation area, NIR diffuse reflectance spectra (raw, first, and second derivative) were compared to develop a robust classification rule [26]. The amount of polysaccharides and triterpenoid saponins in *G. lucidum* samples was found to be considerably different with respect to the geographical origin (Fig. 6.8). In combination with discriminant partial least squares [27], correct classifications were obtained with NIR spectra preprocessed with standard normal transformation second derivative treatment.

Green tea (*Camellia sinensis L.*) is the most popular beverage and widely distributed in China. The atmosphere, in which it is grown, determines much of its flavor and quality. For this reason, FTIR spectroscopy and supervised pattern recognition was attempted for a rapid discrimination of roast green tea sampled in different Chinese provinces (i.e., Anhui, Henan, Jiangsu and Zhejiang) [28]. It was revealed that discrimination rates were all 100% in the training and prediction sets when support vector machine (SVM), a nonlinear algorithm, was used to construct the model based on principal component analysis (PCA) (Fig. 6.9).

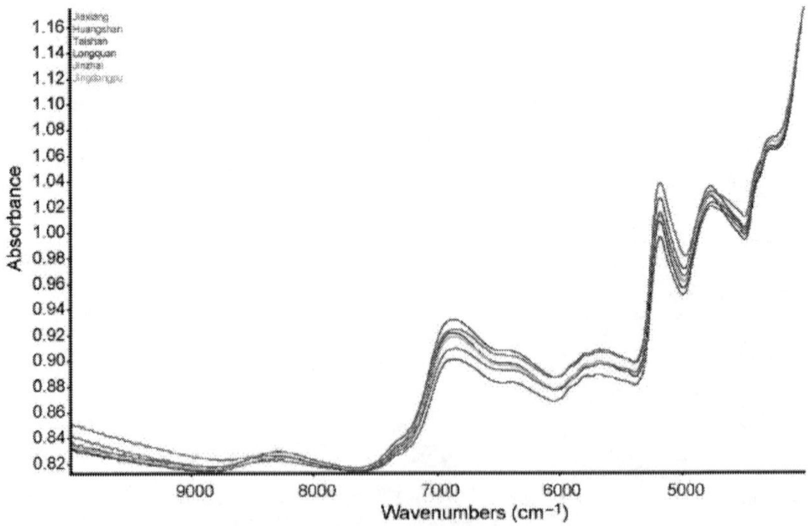

FIG. 6.8 Raw spectra of *Ganoderma lucidum* samples from six different origins. *(Reproduced with permission from Y. Chen, M.Y. Xie, Y. Yan, S.B. Zhu, S.P. Nie, C. Li, Y.X. Wang, X.F. Gong, Discrimination of Ganoderma lucidum according to geographical origin with near infrared diffuse reflectance spectroscopy and pattern recognition techniques, Anal. Chim. Acta 618 (2008) 121–130.)*

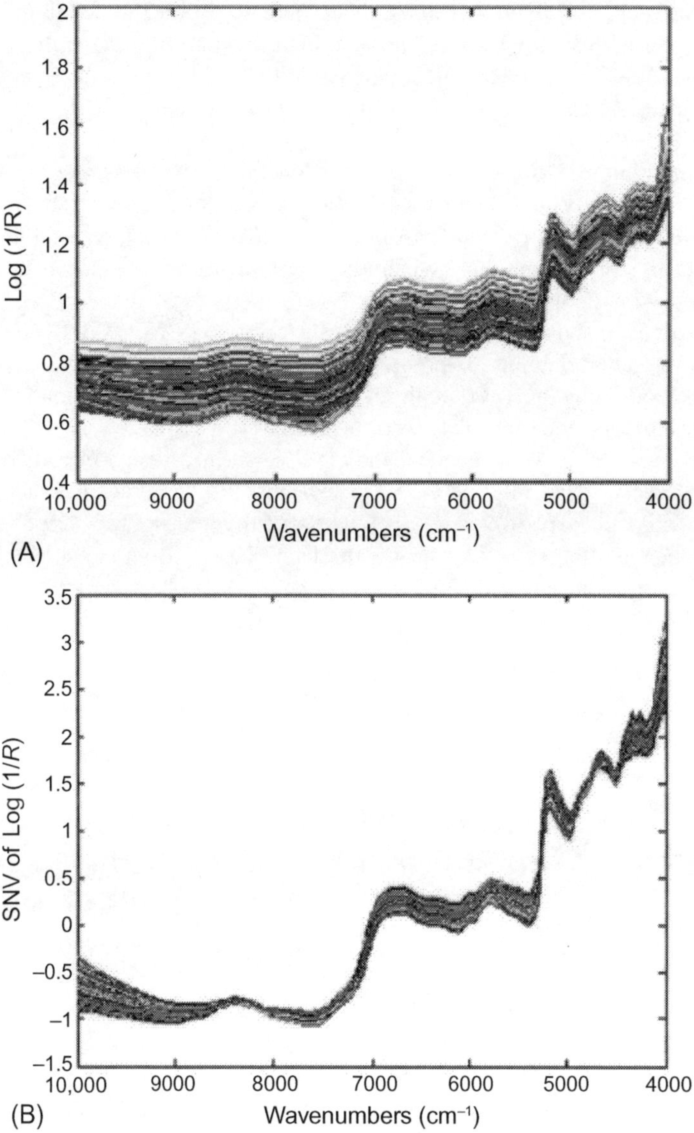

FIG. 6.9 Spectra of tea obtained from (A) raw data and (B) SNV preprocessing data. *(Reproduced with permission from Q.S. Chen, J.W. Zhao, H. Lin, Study on discrimination of roast green tea (Camellia sinensis L.) according to geographical origin by FT-NIR spectroscopy and supervised pattern recognition, Spectrochim. Acta A 72 (2009) 845–850.)*

In another study, FTIR spectra at the frequency region 1800–600 cm^{-1} were well exploited for both classification and identification of *Rhizoma gastrodiae* (Tianma) from different producing areas with the aid of principal component analysis (PCA) and partial least squares-discriminant analysis (PLS-DA) [29] (Fig. 6.10).

Combining FT-NIR spectroscopy and chemometrics was also proved to be able to classify and identify authentic varieties of Hongjingtian in China (*Rhodiola crenulata, Rhodiola fastigiata, Rhodiola kirilowii, Rhodiola brevipetiolata*) [30]. The optimal authenticity classification and identification was obtained with sample set partitioning based on joint *x-y* distances (as dividing method), standard normal variate transformation + Norris-Williams + 2nd derivative transformation (as preprocessing method), competitive adaptive reweighted sampling (as wavelength selection method), and kernel extreme learning machine (as modeling evaluation method).

For many decades, the genus name *Cordyceps* (historically classified in the *Clavicipitaceae*) was referred to any insect-inhabiting ascomycetes fungi [31]. Specifically, *Cordyceps cicadae* (belonging to the genus *Paecilomyces*) is an entomogenous fungi that parasitises on cicada larvae. Both experience-based therapies of traditional Chinese medicine and modern pharmacological studies

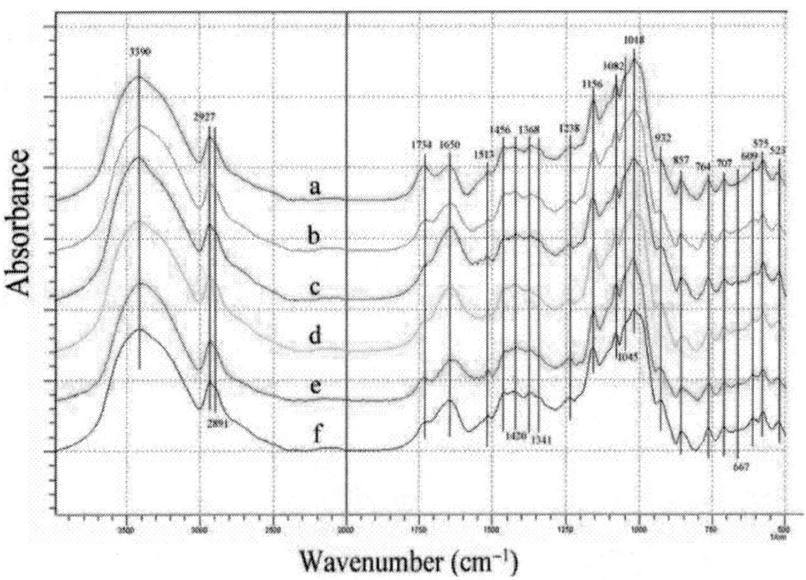

FIG. 6.10 Averaged FT-IR spectra of Tianma from different producing areas in the range of 4000–400 cm^{-1} (a = Anhui Dabieshan Dongma, b = Hubei Dongma, c = Guizhou Tianma, d = Lijiang Hongtianma, e = Zhaotong Wutianma, f = Zhaotong Xiaocaoba Wild Tianma). *(Reproduced with permission from Q. Fan, C. Chen, Y. Lin, C. Zhang, B. Liu, S. Zhao, Fourier transform infrared (FT-IR) spectroscopy for discrimination of Rhizoma gastrodiae (Tianma) from different producing areas, J. Mol. Struct. 1051 (2013) 66–71.)*

showed that *C. cicadae* exhibits a number of biological functions such as vision improvement and protection of renal function [32]. In an effort to discern the false from the genuine, a comprehensive assessment of wild *C. cicadae* from different geographical origins was performed by using TOPSIS (Technique for Order Preference by Similarity to Ideal Solution) based on macroscopic IR fingerprint methods, i.e., FTIR and second-derivative IR spectroscopy (SD-IR) [33]. While FTIR spectra of *C. cicadae* displayed major absorptive peaks of carbohydrates, lipids, and nucleosides at 3291, 2925, 2845, 1651, 1547, 1455, 1080, and 950 cm^{-1}, the difference in proportions and types of chemical composition of different *C. cicadae* samples was further amplified by using SD-IR spectra to resolve overlapped bands and enhance characteristic peaks (Figs. 6.11 and 6.12). The TOPSIS method (initially proposed by Hwang and Yoon to tackle ranking problems in multiobjective decision analysis [34]) was able specifically, in this research, to classify *C. cicadae* from different regions. TOPSIS results were consistent with those from FT-IR spectral correlation coefficients, stating that Anhui samples possessed the strongest intensity of absorption bands.

Aquilariae Lignum Resinatum is the resin-rich wood of *Aquilaria sinensis* (Lour.) Gilg. From old times, this expensive perfume herb has been exploited

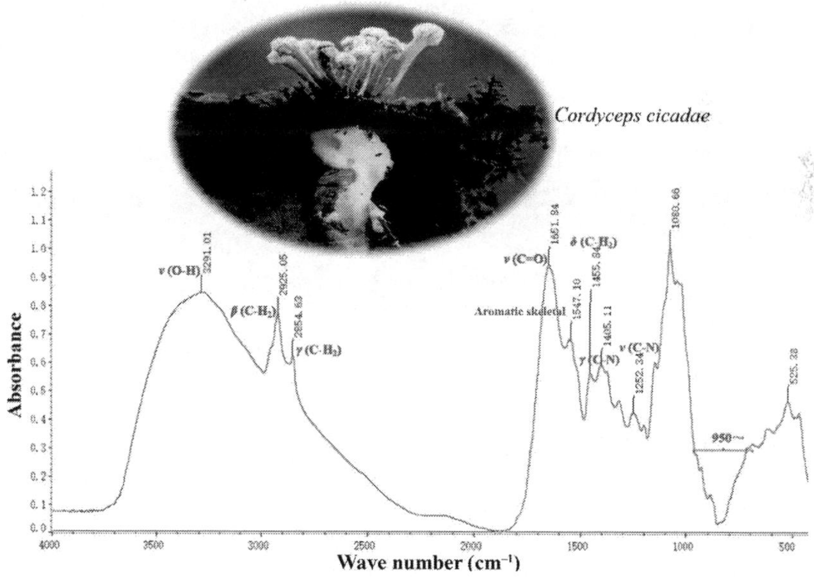

FIG. 6.11 FT-IR spectra characterization of *Cordyceps cicadae*: ν, stretching vibration; δ, bending vibration; γ, symmetrical; β, asymmetrical. *(Reproduced with permission from Y.F. Sun, E. Kmonickova, R.L. Han, W. Zhou, K.B. Yang, H. Lu, Z.Q. Wang, H. Zhao, H. Wang, Comprehensive evaluation of wild Cordyceps cicadae from different geographical origins by TOPSIS method based on the macroscopic infrared spectroscopy (IR) fingerprint, Spectrochim. Acta A: Mol. Biomol. Spectrosc. 214 (2019) 252–260.)*

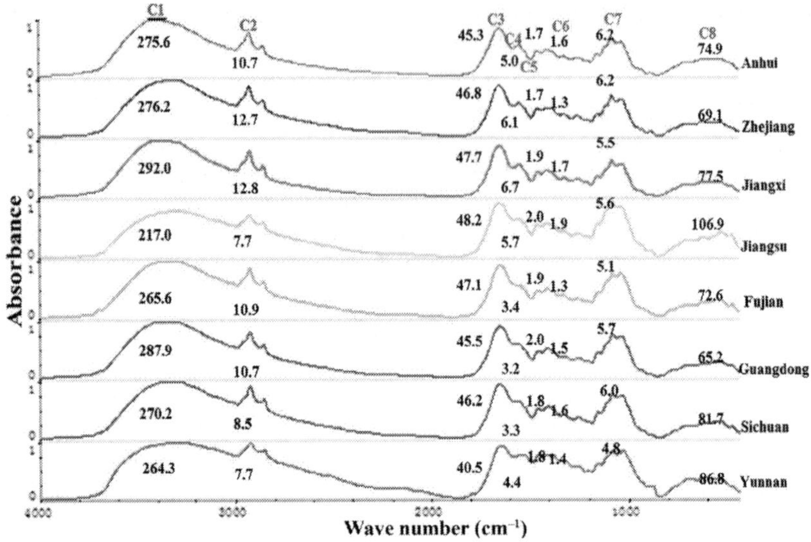

FIG. 6.12 Characteristic peak areas based on the infrared spectra of *Cordyceps cicadae* from different origins. C1–C8: the main characteristic peak areas. *(Reproduced with permission from Y.F. Sun, E. Kmonickova, R.L. Han, W. Zhou, K.B. Yang, H. Lu, Z.Q. Wang, H. Zhao, H. Wang, Comprehensive evaluation of wild Cordyceps cicadae from different geographical origins by TOPSIS method based on the macroscopic infrared spectroscopy (IR) fingerprint, Spectrochim. Acta A: Mol. Biomol. Spectrosc. 214 (2019) 252–260.)*

to treat abdominal distension, chest distress, stomach-cold vomiting, and renal deficiency [35]. With the intention of ensuring the safety and efficacy of *Aquilariae Lignum Resinatum* for medical applications, its major chemical components were characterized by FT-IR and 2D-IR spectroscopy [36]. Besides the common cellulose and lignin compounds, resin was identified as a characteristic constituent of *Aquilariae Lignum Resinatum* (corresponding to an absorption peak near 1658 cm^{-1}). As stated in 2D-IR spectral data, this constituent was more sensitive than cellulose and lignin to the thermal perturbation. IR spectral correlation threshold was set, for identification of *Aquilariae Lignum Resinatum*, to be not less than 0.9886.

Si Wu Tang (SWT), one of the most used traditional Chinese medicine formulae against women's diseases (menstrual discomfort, dysmenorrhea, and other estrogen-caused inconveniences) in Asia [37], is composed of four herbs, i.e., *Radix Paeoniae Alba*, *Rhizoma Chuanxiong*, *Radix Angelicae Sinensis,* and *Radix Rehmanniae Preparata*. Distinction of SWT and non-SWT samples could be rapidly and noninvasively performed by using MIR and NIR benchtop spectrometers and a mobile NIR device coupled with multivariate analysis (PCA with the nonlinear iterative partial least squares algorithm) (e.g., Fig. 6.13) [38].

Tanreqing injection is a patent drug in China, which is made from five kinds of traditional Chinese medicine extracts, namely, *Forsythia Suspense*,

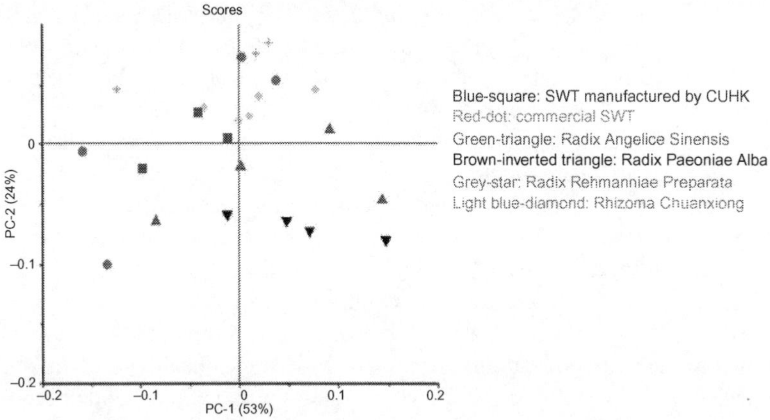

FIG. 6.13 PCA score plot obtained with the mobile NIR spectrometer. *(Reproduced with permission from C.K. Pezzei, M. Watschinger, V.A. Huck-Pezzei, C.B.S. Lau, Z. Zuo, P.C. Leung, C.W. Huck, Infrared spectroscopic techniques for the non-invasive and rapid quality control of Chinese traditional medicine Si-Wu-Tang, Spectrosc. Eur. 28(3) (2016) 16–21.)*

Flos Lonicerae, *Radix Scutellariae*, Bear gall powder, and Cornu gorais. It has been chiefly used in the treatment of acute respiratory tract infections [39]. To ensure a steady run of the manufacturing process, a fast analysis of intermediates of *Tanreqing* injection was developed by using FT-NIR spectroscopy (in transflective mode) combined with PLS [40]. In this study, the concurrent determination of 6 active ingredients (i.e., caffeic acid, chlorogenic acid, baicalin, luteoloside, ursodesoxycholic acid, and chenodeoxycholic acid) in *Tanreqing* injection intermediates was done by using a reference method, HPLC–DAD/ELSD. The established method was applied to analyze three different batches, suggesting that the content of baicalin and ursodesoxycholic acid (the crucial quality indicators of *Tanreqing* injection) could be stable by rigidly controlling three-step operations (pH adjustment, filtration, and hyperfiltration).

For herbal preparations, it is impossible to identify a specific toxic or regulated plant by micro- and macroscopic observation because the plant parts used are often pulverized, mixed with other ingredients, and compressed into tablets or capsules. This real challenge was solved by using ATR-mid IR spectroscopy together with modeling technique (soft independent modeling by class analogy, SIMCA) [41]. Test samples were triturations of the dry extracts and powdered plant materials prepared for each targeted plant in the four blank matrices (lactose and three herbal matrices). For the chosen test set, 21 out of 25 samples were appropriately classified and 6 of 9 targeted plants showed no misclassifications. As a proof of concept, 5 real samples (seized and provided by the Federal Agency for Medicines and Health Products in Belgium) were screened with the proposed strategy; of which 4 samples were ascribed to the correct class as confirmed by a chromatographic fingerprint approach [42].

Gelsemium elegans is a toxic plant indigenous to Southeast Asia. Despite its strong respiratory depressive effect, medicinal plants of the genus *Gelsemium* are still generally used to treat many diseases according to the traditional Chinese medicine "like cures like" [43].

In a study, the combination of multistep IR macrofingerprint analysis (FTIR, SD-IR, 2D-IR) and chemometrics (PCA and SIMCA) was proved to successfully discriminate not only between *G. elegans* and standard herbs but also among different parts (i.e. the stem, leaf, and root that contain different amount of indole alkaloid) of *G. elegans* [44] (c.f. Figures 6.14–6.16).

Since the last decade, the threat of counterfeit drugs has been globally recognized, requiring a close cooperation between drug companies, governments, and international organizations concerned with trade, health, customs, and counterfeiting [45]. With a rapid growth in sales, the intentional adulteration to make herbal medicines more effective has become a problem of more concern, albeit less frequently discovered in the regulated supply chains in comparison with counterfeit chemical drugs [46].

Radix Astragali is one of the most important traditional Chinese medicines used to reinforce "Qi." In commercial herb markets, shredded slices of *Radix Astragali* were found to be mixed with adulterants [47]. To detect potential adulteration of *Radix Astragali*, derived from the root of *Astragalus membranaceus* (Fischer) Bunge var. *mongholicus* (Bunge) Hsiao, FT-IR spectroscopy

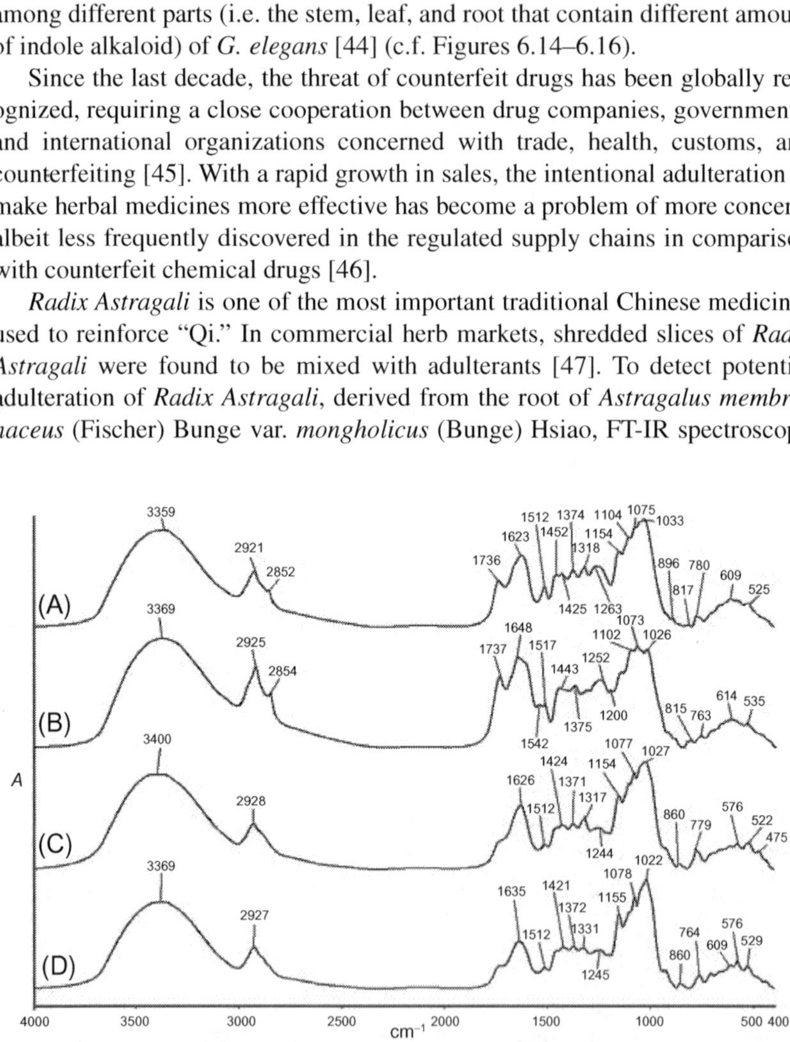

FIG. 6.14 Comparison of FT-IR results for different parts of *Gelsemium elegans*: (a) stem; (b) leaf; (c) root; (d) combination of stem and root. *(Reproduced with permission from C.H. Ng, Y. Chen, Y.S. Ch'ng, C.S. Tan, Z.Q. Yeap, Y.C. Loh, S. Sheng Wu, M.F. Yam, Application of mid-infrared spectroscopy with multivariate analysis for the discrimination of toxic plant, Gelsemium elegans, Vib. Spectrosc. 99 (2018) 13–24, doi:10.1016/j.vibspec.2018.08.013.)*

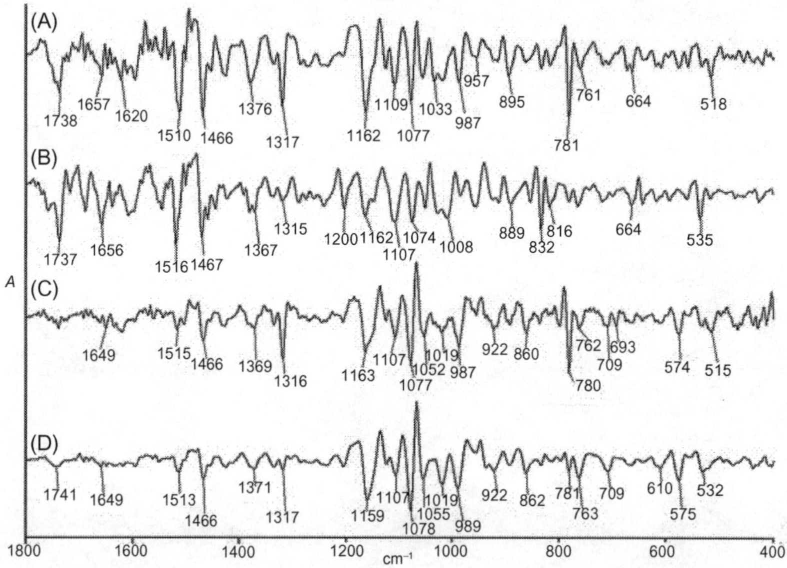

FIG. 6.15 Comparison of SD-IR results for different parts of *Gelsemium elegans*: (a) stem; (b) leaf; (c) root; (d) combination of stem and root. *(Reproduced with permission from C.H. Ng, Y. Chen, Y.S. Ch'ng, C.S. Tan, Z.Q. Yeap, Y.C. Loh, S. Sheng Wu, M.F. Yam, Application of mid-infrared spectroscopy with multivariate analysis for the discrimination of toxic plant, Gelsemium elegans, Vib. Spectrosc. 99 (2018) 13–24, doi:10.1016/j.vibspec.2018.08.013.)*

combined with Mahalanobis distance [48] was employed [49]. Most *Radix Astragali* samples adulterated with different percentages of *Hedysarum polybotrys* Handel-Mazzetti were correctly detected, i.e., in the most informative region 4000–1300 cm^{-1}, the rejection rate could improve up to 88.9% as compared to that based on the whole region (Fig. 6.17). In this study, discriminant partial least squares (DPLS)-based FTIR spectroscopy was also able to discriminate the geographical origin of *Radix Astragali* samples collected from the north of China and extracted with butanone.

Research has shown a significant underrecognition of the adulteration of herbal antidiabetic products with undeclared (both registered and banned) pharmaceuticals such as metformin, gliclazide, rosiglitazone, phenformin, glibenclamide, and glimepiride. Patients orally taking such illicit products may face with the dilemma of being at risk of potentially fatal adverse effects [50]. To directly and rapidly detect herbal antidiabetic medicines dosed with synthetic adulterants, a general detection procedure was developed by using a reverse correlation coefficient method (RCCM) in tandem with comparison of NIR characteristic peaks [51]. It means that any herbal medicine tested as adulterated must meet two criteria: (i) the correlation coefficient between the tested and reference samples is greater than the RCCM threshold and (ii) the NIR spectra of the tested and reference samples contain the same characteristic peaks (as

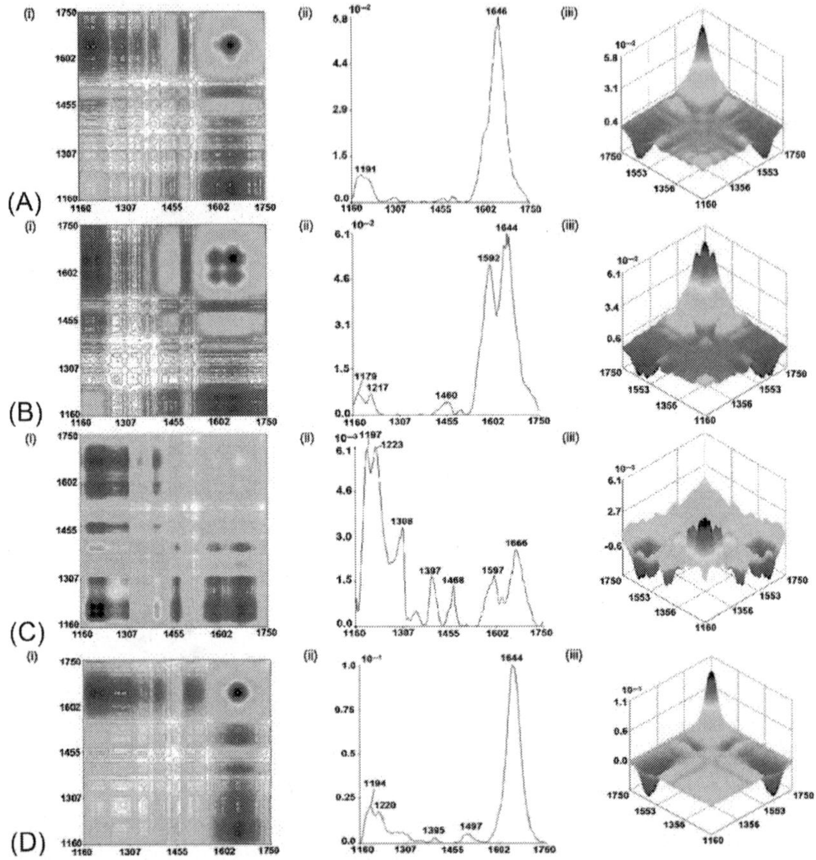

FIG. 6.16 The 2D-correlation IR spectra of each part of the *Gelsemium elegans* plant in the range of 1750–1160 cm^{-1}: (A) stem; (B) leaf; (C) root; (D) combination of stem and root. *(Reproduced with permission from C.H. Ng, Y. Chen, Y.S. Ch'ng, C.S. Tan, Z.Q. Yeap, Y.C. Loh, S. Sheng Wu, M.F. Yam, Application of mid-infrared spectroscopy with multivariate analysis for the discrimination of toxic plant, Gelsemium elegans, Vib. Spectrosc. 99 (2018) 13–24, doi:10.1016/j.vibspec.2018.08.013.)*

illustrated in Fig. 6.18). The minimum detectable concentration of an adulterant proved to be more reasonable than its minimum effective concentration to set up a threshold. The accuracy of this procedure was greater than 80% for 4 pure synthetic antidiabetic drugs (i.e., metformin, gliclazide, glibenclamide and glimepiride) when using 174 batches of laboratory-made samples and 127 batches of herbal antidiabetic medicines for validation.

The prevalence of overweight and obesity is now occurring in both developed and developing countries. Although adverse complication of treating obesity with some herbal medicine was reported [52], some people still seek herbal products for the treatment of weight loss.

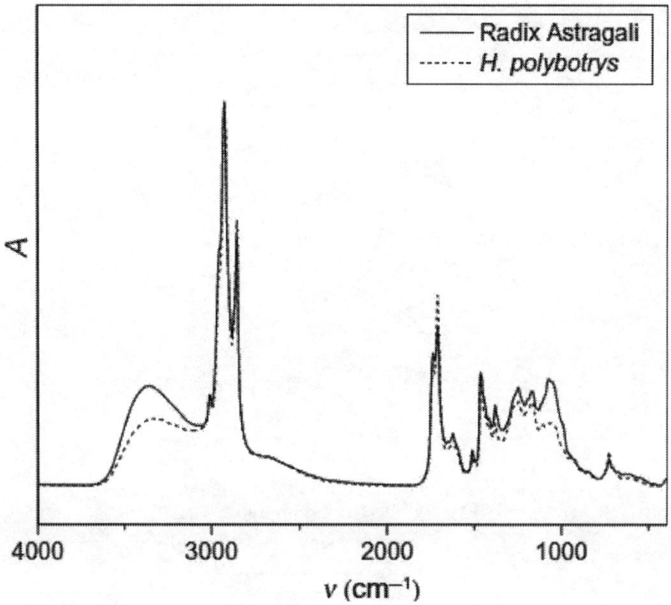

FIG. 6.17 Representative IR spectra of *Radix Astragali* and *H. polybotrys*. *(Reproduced with permission from L. Zhang, L. Nie, Discrimination of geographical origin and adulteration of radix astragali using Fourier transform infrared spectroscopy and chemometric methods, Phytochem. Anal. 21(6) (2010) 609–615.)*

Sibutramine is an appetite suppressant, by inhibiting reuptake of both serotonin and noradrenaline released from hypothalamic neurons [53]. It was demonstrated that NIR spectroscopy could satisfactorily classify herbal medicine samples as adulterated or not adulterated with sibutramine (Fig. 6.19) [54]. Using partial least squares-discriminant analysis (PLS-DA), a correct classification of 100% was obtained for the external validation set; on the other hand, the limit of quantification was 0.8% *w*/w for sibutramine by using PLS for calibration.

In general, the combination of ATR-FTIR spectroscopy and PLS-DA could be used at customs to easily differentiate between nonadulterated and adulterated plant food supplements as well as to get the first idea about the nature of adulterant (if any). This conclusion was drawn from a study on IR spectroscopic detection and identification of multiple adulterants in plant food supplements (Fig. 6.20) [55]. Second-derivative preprocessed Mid-IR data were selected as the best suited for the purpose (i.e., a correct classification rate of 98.3% was achieved for an external test set). Strikingly, noise was added to the data (when fusing Mid-IR and NIR spectra) and it could lead to some false negatives and false positives. PLS-DA was also unable to model the classes of the combination preparations, and systematically classified the samples containing more than one adulterant to the class of the highest-dose active pharmaceutical ingredient (API) per dosage unit.

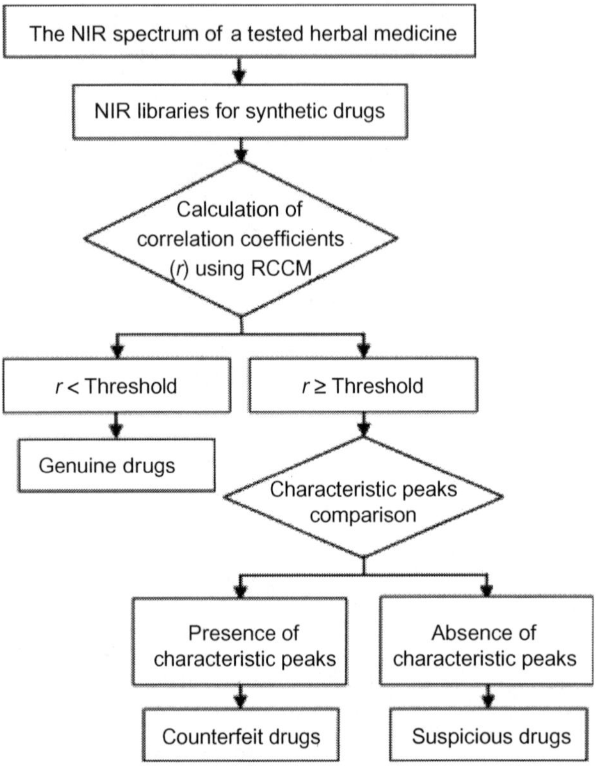

FIG. 6.18 Procedure of RCCM in tandem with characteristic peaks comparison method to detect herbal medicines adulterated with synthetic drugs. *(Reproduced with permission from Y. Feng, D. Lei, C. Hu, Rapid identification of illegal synthetic adulterants in herbal anti-diabetic medicines using near infrared spectroscopy, Spectrochim. Acta A: Mol. Biomol. Spectrosc. 125 (2014) 363–374.)*

Honey is a thick, golden, and sweet liquid produced by industrious bees using the nectar from flowers. It can be used in both traditional and modern ways for treating human diseases [56], e.g., a potential demulcent treatment for cough as recommended by the World Health Organization [57]. Unfortunately, it may be smoothly adulterated with various cheaper sweeteners (such as refined cane sugar beet sugar, high fructose corn syrup, and maltose syrup) for higher commercial profits. To detect adulterants (i.e., high fructose corn syrup and maltose syrup) in honey, Raman spectroscopy was used in coupling with PLS-LDA [58]. By using adaptive iteratively reweighted penalized least squares to remove background of spectral data [59], the classification of honey adulterants using PLS-LDA gave a total accuracy of 84.4% (Fig. 6.21).

In another instance, a surface-enhanced Raman scattering (SERS) analysis method was developed for the simultaneous determination of two or three kinds of illegally added drugs (inclusive of phenformin hydrochloride, rosiglitazone maleate, pioglitazone hydrochloride, metformin hydrochloride, and sibutramine

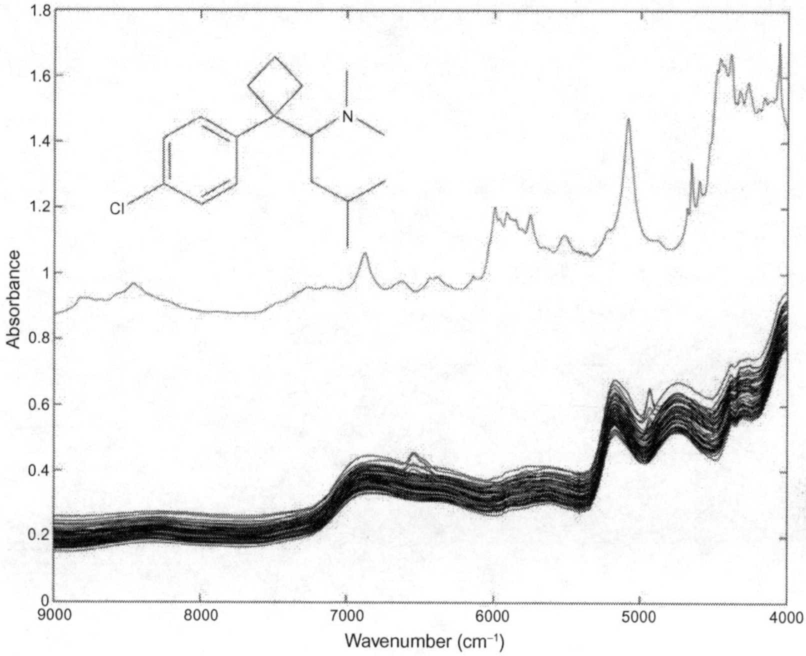

FIG. 6.19 Raw spectra of different herbal samples and a pure standard of sibutramine *(gray)*. *(Reproduced with permission from N.C. da Silva, R.S. Honorato, M.F. Pimentel, S. Garrigues, M.L. Cervera, M. de la Guardia, Near infrared spectroscopy detection and quantification of herbal medicines adulterated with Sibutramine, J. Forensic Sci. 60(5) (2015) 1199–1205.)*

hydrochloride) in Chinese traditional patent medicines [60]. It was indicated that the pH level was extremely important for its cumulative impact on protonation, surface charge, and repulsion of an analyte and nanoparticles (as shown in Figs. 6.22 and 6.23).

It was also reported that a convenient and rapid platform for detecting dye adulteration of medicinal herbs could be developed by combining a silver nanoparticle wiper with SERS technology [61]. To create the wiper, a SERS-active substrate was formed by trapping silver nanoparticles in filter paper. Dye molecules could be then transferred onto the resulting substrate by merely wiping it over the wetted medicinal herb (Fig. 6.24). This method was able to well detect 9 dyes with detection limits ranging from 10^{-6} to 5×10^{-8} g/mL, which were lower than the minimum concentrations required to visibly dye any colorless herb (Fig. 6.25). More importantly, the excellent performance of this SERS-based approach was demonstrated by detecting erythrosine B sodium salt, malachite green, and Rhodamine B on the surface of herbs that had undergone simulated dye adulteration, obviously verifying its potential for field-based applications.

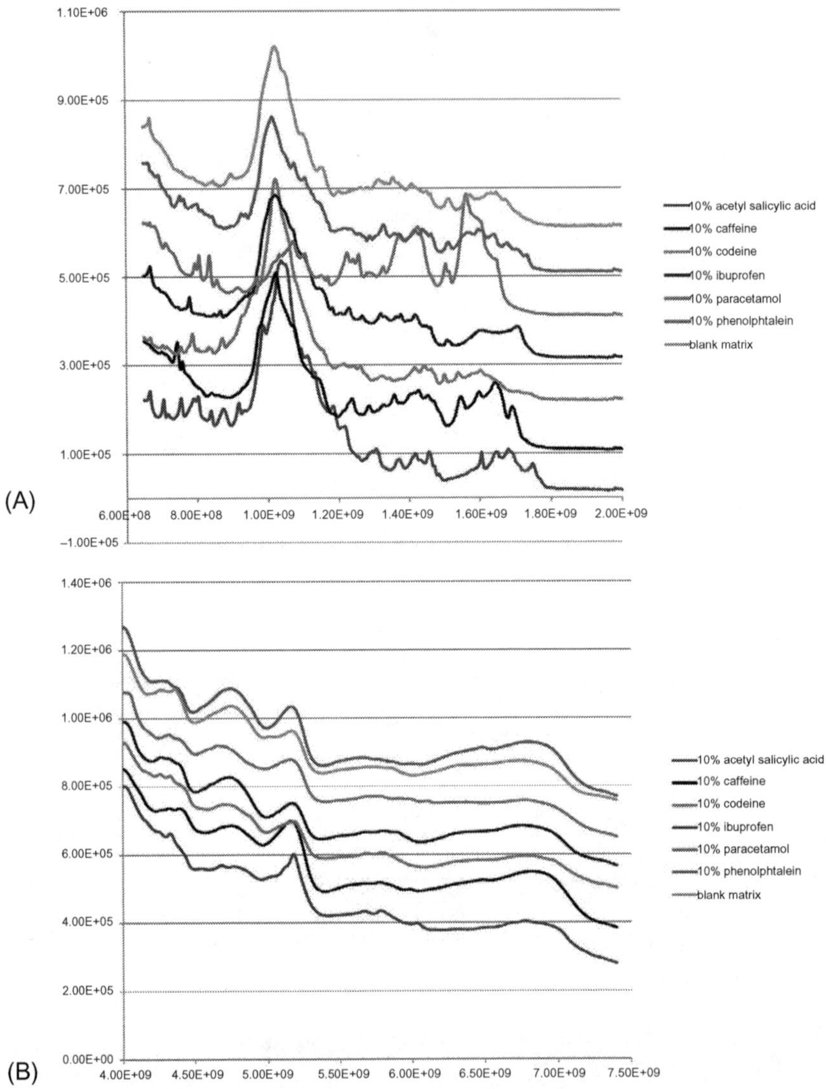

FIG. 6.20 (A) Mid-IR spectra and (B) NIR spectra obtained for each API in randomly selected 10% trituration. *(Reproduced with permission from E. Deconinck, C. Aouadi, J.L. Bothy, P. Courselle, Detection and identification of multiple adulterants in plant food supplements using attenuated total reflectance—infrared spectroscopy, J. Pharm. Biomed. Anal. 152 (2018) 111–119.)*

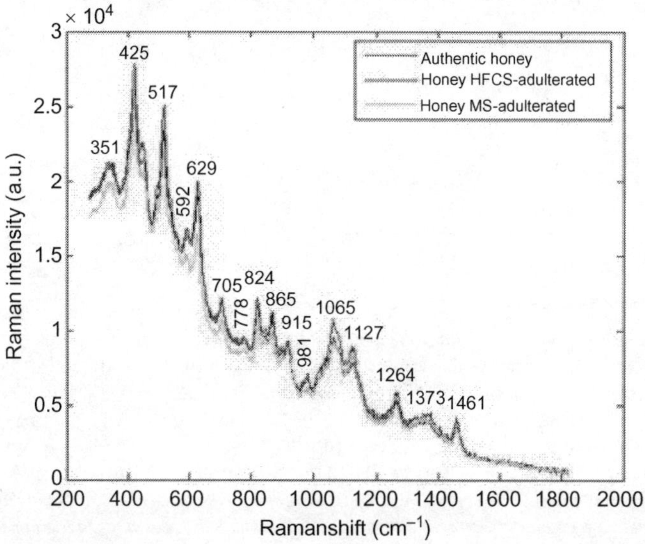

FIG. 6.21 Raw Raman spectra of a randomly selected authentic honey sample and the same honey sample adulterated with high fructose corn syrup (40%, w/w) and maltose syrup (40%, w/w). *(Reproduced with permission from S. Li, Y. Shan, X. Zhu, X. Zhang, G. Ling, Detection of honey adulteration by high fructose corn syrup and maltose syrup using Raman spectroscopy, J. Food Compos. Anal. 28 (2012) 69–74.)*

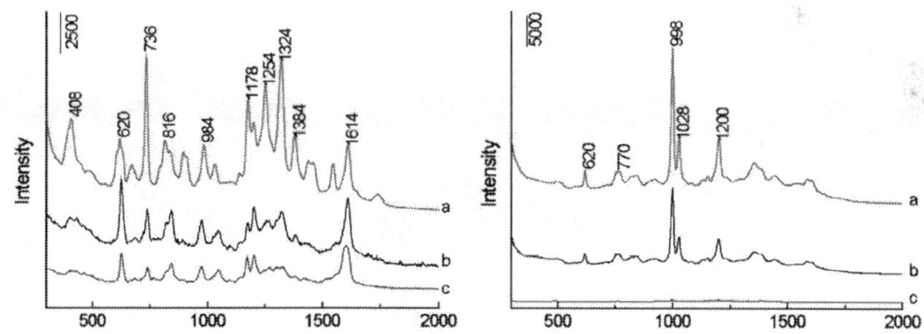

FIG. 6.22 SERS spectra of drugs on Ag colloidal under different pH conditions. (A) and (B) are the SERS spectra of rosiglitazone maleate and phenformin hydrochloride at a concentration of 1×10^{-5} mol L^{-1}, respectively. For (A), lines "a–c" represent rosiglitazone maleate in acidic, alkaline, and neutral conditions, respectively. For (B), lines "a–c" represent phenformin hydrochloride in acidic, alkaline, and neutral conditions, respectively. *(Reproduced with permission from Y. Zhang, X.F. Huang, W.F. Liu, Z.N. Cheng, C.P. Chen, L.H. Yin, Analysis of drugs illegally added into Chinese traditional patent medicine using surface-enhanced Raman scattering, Anal. Sci. 29 (2013) 985–990.)*

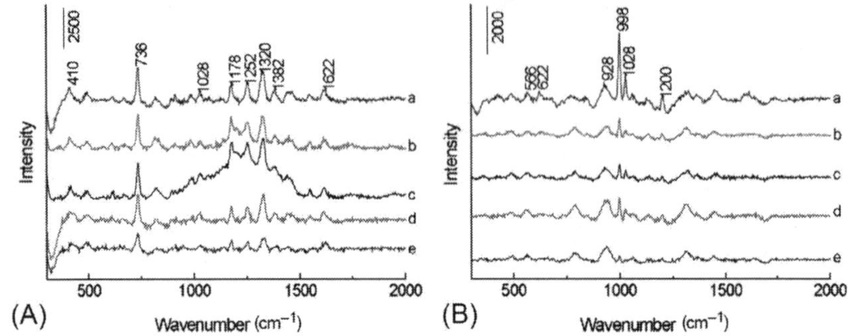

FIG. 6.23 SERS spectra of rosiglitazone maleate (A) and phenformin hydrochloride (B) in different Chinese traditional patent medicines; the detections were operated in acidic and alkaline conditions, respectively. For (A), lines "a–e" represent an adding proportion of 1.0% in Jiangtangshu, 0.8% in Tangniaole, 0.5% in Ganluxiaoke, 0.3% in Jiangtangning, 0.1% in Jiangtang. For (B), lines "a–e" represent an adding proportion of 1.0% in Fengjiaotangtai, 0.8% in Tangxinsukang, 05% in Xiaokejiangtang, 0.3% in Jiangtangshu, 0.1% in Jiangtang, respectively. *(Reproduced with permission from Y. Zhang, X.F. Huang, W.F. Liu, Z.N. Cheng, C.P. Chen, L.H. Yin, Analysis of drugs illegally added into Chinese traditional patent medicine using surface-enhanced Raman scattering, Anal. Sci. 29 (2013) 985–990.)*

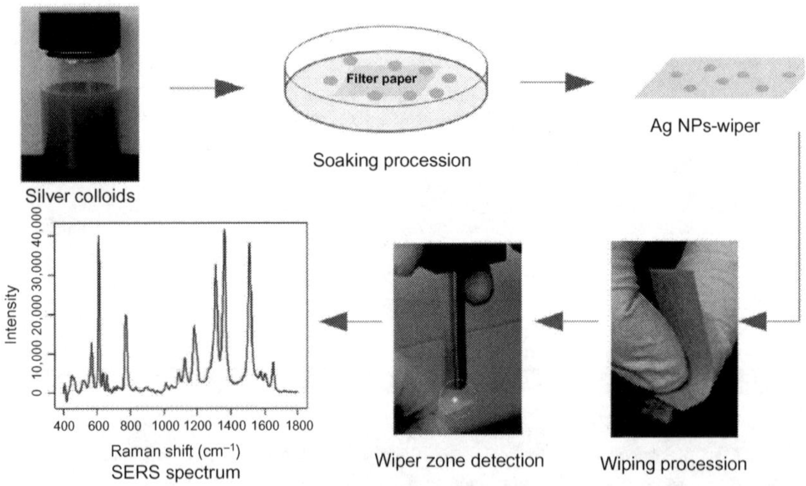

FIG. 6.24 Schematic showing the fabrication of the paper-based wiper and SERS examination of dyes obtained from the surfaces of medicinal herbs. *(Reproduced with permission from D. Li, Q. Zhu, D. Lv, B. Zheng, Y. Liu, Y. Chai, Silver-nanoparticle-based surface-enhanced Raman scattering wiper for the detection of dye adulteration of medicinal herbs, Anal. Bioanal. Chem. 407 (2015) 6031–6039.)*

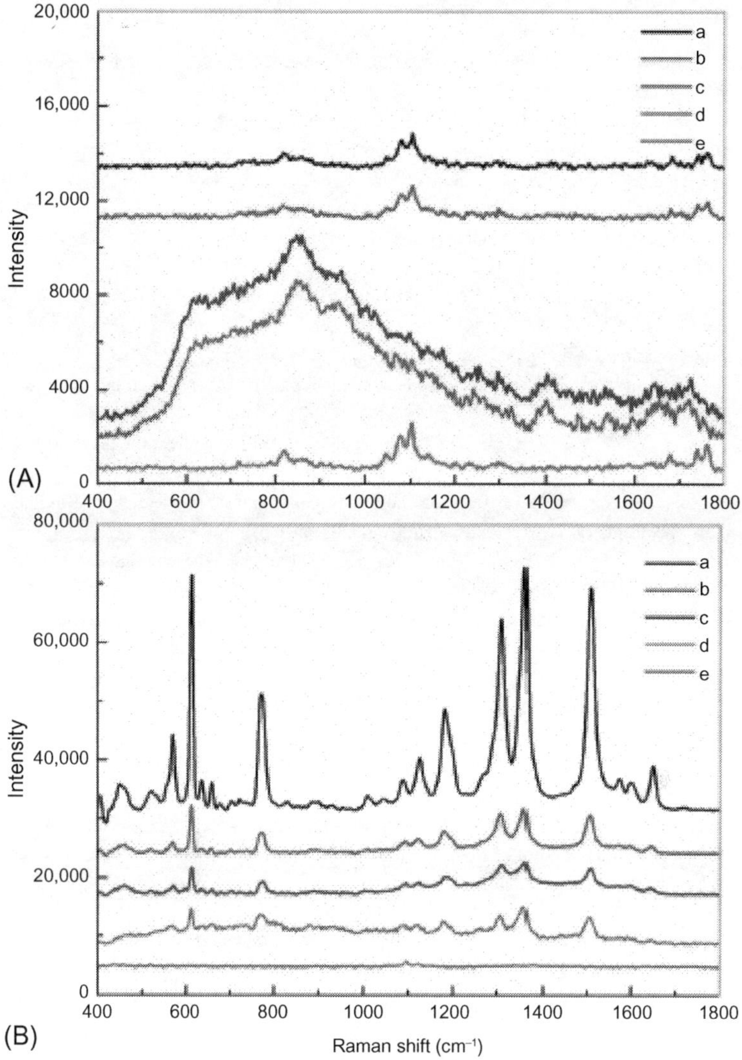

FIG. 6.25 (A) Background signals from the filter paper and wipers recorded by a portable Raman spectrometer. (B) SERS spectra of 10–4 g/mL Rhodamine 6G detected on (a) wiper 1, (b) wiper 2, (c) wiper 3, (d) wiper 4, and (e) bare filter paper. *(Reproduced with permission from D. Li, Q. Zhu, D. Lv, B. Zheng, Y. Liu, Y. Chai, Silver-nanoparticle-based surface-enhanced Raman scattering wiper for the detection of dye adulteration of medicinal herbs, Anal. Bioanal. Chem. 407 (2015) 6031–6039.)*

References

[1] A. Gurib-Fakim, Medicinal plants: traditions of yesterday and drugs of tomorrow, Mol. Asp. Med. 27 (2006) 1–93.

[2] S. Wachtel-Galor, I.F.F. Benzie, Herbal medicine an introduction to its history, usage, regulation, current trends, and research needs, in: I.F.F. Benzie, S. Wachtel-Galor (Eds.), Herbal Medicine: Biomolecular and Clinical Aspects, second ed., CRC Press/Taylor & Francis, Boca Raton, FL, 2011. (Chapter 1).

[3] N. Sahoo, P. Manchikanti, S. Dey, Herbal drugs: standards and regulation, Fitoterapia 81 (6) (2010) 462–471.

[4] World Health Organization, WHO Global Report on Traditional and Complementary Medicine, ISBN: 978-92-4-151543-6, 2019.

[5] https://medlineplus.gov/herbalmedicine.html, (Accessible on 25 December 2019).

[6] T. Hartmann, From waste products to ecochemicals: fifty years research of plant secondary metabolism, Phytochemistry 68 (22–24) (2007) 2831–2846.

[7] Y.Z. Liang, P.S. Xie, K. Chan, Quality control of herbal medicines, J. Chromatogr. B 812 (2004) 53–70.

[8] Y. Yang, J. Deng, Analysis of pharmaceutical products and herbal medicines using ambient mass spectrometry, TrAC Trends Anal. Chem. 82 (2016) 68–88.

[9] A. Zhang, H. Sun, G. Yan, X. Wang, Recent developments and emerging trends of mass spectrometry for herbal ingredients analysis, TrAC Trends Anal. Chem. 94 (2017) 70–76.

[10] C. Huck, Infrared spectroscopic technologies for the quality control of herbal medicines, in: Evidence-Based Validation of Herbal Medicine, Elsevier, 2015. (Chapter 22).

[11] S.Q. Sun, Q. Zhou, Z. Qin, Atlas of Two-Dimensional Correlation Infrared Spectroscopy for Traditional Chinese Medicine Identification, Chemical Industry Press, Beijing, 2003.

[12] D.D. Chen, X.F. Xie, H. Ao, J.L. Liu, C. Peng, Raman spectroscopy in quality control of Chinese herbal medicine, J. Chin. Med. Assoc. 80 (2017) 288–296.

[13] H.B. Zou, G.S. Yang, Z.Q. Qin, W.Q. Jiang, A.Q. Du, H.Y. Aboul-Enein, Progress in quality control of herbal medicine with IR fingerprint spectra, Anal. Lett. 38 (9) (2005) 1457–1475.

[14] A.A. Bunaciu, H.Y. Aboul-Enein, S. Fleschin, Recent applications of Fourier transform infrared spectrophotometry in herbal medicine analysis, Appl. Spectrosc. Rev. 46 (4) (2011) 251–260.

[15] L.K. Pei, S.Q. Sun, B.L. Guo, W.H. Huang, P.G. Xiao, Fast quality control of *Herba Epimedii* by using Fourier transform infrared spectroscopy, Spectrochim. Acta A 70 (2008) 258–264.

[16] H. Liu, S. Sun, G. Lv, X. Liang, Discrimination of extracted lipophilic constituents of Angelica with multi-steps infrared macro-fingerprinting method, Vib. Spectrosc. 40 (2006) 202–208.

[17] https://www.medicalnewstoday.com/articles/262982, (Accessible on 28 December 2019).

[18] https://www.fws.gov/international/permits/by-species/american-ginseng.html, (Accessible on 28 December 2019).

[19] H.G.M. Edwards, T. Munshi, K. Page, Analytical discrimination between sources of ginseng using Raman spectroscopy, Anal. Bioanal. Chem. 389 (2007) 2203–2215.

[20] K.Y.L. Yap, S.Y. Chan, C.S. Lim, Infrared-based protocol for the identification and categorization of ginseng and its products, Food Res. Int. 40 (5) (2007) 643–652.

[21] G.H. Lu, Q. Zhou, S.Q. Sun, K.S.Y. Leung, H. Zhang, Z.Z. Zhao, Differentiation of Asian ginseng, American ginseng and Notoginseng by Fourier transform infrared spectroscopy combined with two-dimensional correlation infrared spectroscopy, J. Mol. Struct. 883–884 (2008) 91–98.

[22] Editorial Committee of China Pharmacopoeia, China Pharmacopoeia, Part I, China Chemical Industry Press, Beijing, 2000253–254.

[23] S.C. Chao, An up-to-date review of phytochemicals and biological activities in *Chrysanthemum* Spp, Biosci. Biotechnol. Res. Asia 13 (2) (2016) 615–623.

[24] H.X. Liu, Q. Zhou, S.Q. Sun, H.J. Bao, Discrimination of different Chrysanthemums with Fourier transform infrared spectroscopy, J. Mol. Struct. 883–884 (2008) 38–47.

[25] M. Siwulski, K. Sobieralski, I. Golsk-Siwulska, S. Sokol, A. Sekara, *Ganoderma lucidum* (Curt.: Fr.) Karst.—health-promoting properties. A review, Herba Pol. 61 (3) (2015) 105–118.

[26] Y. Chen, M.Y. Xie, Y. Yan, S.B. Zhu, S.P. Nie, C. Li, Y.X. Wang, X.F. Gong, Discrimination of *Ganoderma lucidum* according to geographical origin with near infrared diffuse reflectance spectroscopy and pattern recognition techniques, Anal. Chim. Acta 618 (2008) 121–130.

[27] A.M. Mouazen, R. Karoui, J. De Baerdemaeker, H. Ramon, Presentation at the 2006 ASABE Annual International Meeting Sponsored by ASABE, Oregon Convention Center, Portland, Oregon, 2006. (Paper number 061067).

[28] Q.S. Chen, J.W. Zhao, H. Lin, Study on discrimination of roast green tea (*Camellia sinensis* L.) according to geographical origin by FT-NIR spectroscopy and supervised pattern recognition, Spectrochim. Acta A 72 (2009) 845–850.

[29] Q. Fan, C. Chen, Y. Lin, C. Zhang, B. Liu, S. Zhao, Fourier transform infrared (FT-IR) spectroscopy for discrimination of *Rhizoma gastrodiae* (Tianma) from different producing areas, J. Mol. Struct. 1051 (2013) 66–71.

[30] T. Li, C. Su, Authenticity identification and classification of *Rhodiola* species in traditional Tibetan medicine based on Fourier transform near-infrared spectroscopy and chemometrics analysis. Spectrochim. Acta A: Mol. Biomol. Spectrosc. 204 (2018) 131–140.

[31] G.H. Sung, N.L. Hywel-Jones, J.M. Sung, J.J. Luangsa-ard, B. Shrestha, J.W. Spatafora, Phylogenetic classification of *Cordyceps* and the clavicipitaceous fungi, Stud. Mycol. 57 (2007) 5–59.

[32] J.-H. Hsu, B.-Y. Jhou, S.-H. Yeh, Y.-L. Chen, C.-C. Chen, Healthcare functions of *Cordyceps cicadae*. J. Nutr. Food Sci. 5 (6) (2015) 432, https://doi.org/10.4172/2155-9600.1000432.

[33] Y.F. Sun, E. Kmonickova, R.L. Han, W. Zhou, K.B. Yang, H. Lu, Z.Q. Wang, H. Zhao, H. Wang, Comprehensive evaluation of wild *Cordyceps cicadae* from different geographical origins by TOPSIS method based on the macroscopic infrared spectroscopy (IR) fingerprint, Spectrochim. Acta A: Mol. Biomol. Spectrosc. 214 (2019) 252–260.

[34] K. Yoon, C.L. Hwang, Multiple Attribute Decision Making, Springer, Berlin Heidelberg, 1981, pp. 287–288.

[35] Pharmacopoeia of the People's Republic of China, Chinese Pharmacopoeia, vol. 1, China Medical Science Press, Beijing, China, 2015, p. 185.

[36] L. Qu, J.B. Chen, G.J. Zhang, S.Q. Sun, J. Zheng, Chemical profiling and adulteration screening of *Aquilariae Lignum Resinatum* by Fourier transform infrared (FT-IR) spectroscopy and two-dimensional correlation infrared (2D-IR) spectroscopy, Spectrochim. Acta A: Mol. Biomol. Spectrosc. 174 (2017) 177–182.

[37] J.F. Cheng, Z.Y.J. Lu, Y.C. Su, L.C. Chiang, R.Y. Wang, A traditional Chinese herbal medicine used to treat dysmenorrhoea among Taiwanese women, J. Clin. Nurs. 17 (2008) 2588–2595.

[38] C.K. Pezzei, M. Watschinger, V.A. Huck-Pezzei, C.B.S. Lau, Z. Zuo, P.C. Leung, C.W. Huck, Infrared spectroscopic techniques for the non-invasive and rapid quality control of Chinese traditional medicine Si-Wu-Tang, Spectrosc. Eur. 28 (3) (2016) 16–21.

[39] T. Wu, X. Yang, X. Zeng, P. Poole, Traditional Chinese medicine in the treatment of acute respiratory tract infections, Respir. Med. 102 (2008) 1093–1098.

[40] W. Li, L. Xing, L. Fang, J. Wang, H. Qu, Application of near infrared spectroscopy for rapid analysis of intermediates of Tanreqing injection, J. Pharm. Biomed. Anal. 53 (2010) 350–358.

[41] E. Deconinck, C.A.S. Djiogo, J.L. Bothy, P. Courselle, Detection of regulated herbs and plants in plant food supplements and traditional medicines using infrared spectroscopy, J. Pharm. Biomed. Anal. 152 (2018) 111–119.

[42] D. Custers, N. Van Praag, P. Courselle, S. Apers, E. Deconinck, Chromatographic fingerprinting as a strategy to identify regulated plants in illegal herbal supplements, Talanta 164 (2017) 490–502.

[43] G. Jin, Y. Su, M. Liu, Y. Xu, J. Yang, K. Liao, C. Yu, Medicinal plants of the genus *Gelsemium* (Gelsemiaceae, Gentianales)—a review of their phytochemistry, pharmacology, toxicology and traditional use, J. Ethnopharmacol. 152 (2014) 33–52.

[44] C.H. Ng, Y. Chen, Y.S. Ch'ng, C.S. Tan, Z.Q. Yeap, Y.C. Loh, S. Sheng Wu, M.F. Yam, Application of mid-infrared spectroscopy with multivariate analysis for the discrimination of toxic plant, *Gelsemium elegans*. Vib. Spectrosc. 99 (2018) 13–24.

[45] R. Cockburn, P.N. Newton, E.K. Agyarko, D. Akunyili, N.J. White, The global threat of counterfeit drugs: why industry and governments must communicate the dangers, PLoS Med. 2 (2005) 302–308.

[46] M. Kamil, M.A. Naji, Counterfeit herbal products—a global risk. Planta Med. 76 (2010) P115, https://doi.org/10.1055/s-0030-1251877.

[47] S.Y. Jiang, C.G. Ye, Experienced identification of adulterated slices of some common traditional Chinese medicines, Lishizhen Med. Mater. Med. Res. 17 (2006) 1509.

[48] P.C. Mahalanobis, On the generalised distance in statistics, Proc. Natl. Inst. Sci. India 2 (1936) 49–55.

[49] L. Zhang, L. Nie, Discrimination of geographical origin and adulteration of radix astragali using Fourier transform infrared spectroscopy and chemometric methods, Phytochem. Anal. 21 (6) (2010) 609–615.

[50] C.K. Ching, Y.H. Lam, A.Y.W. Chan, T.W.L. Mak, Adulteration of herbal antidiabetic products with undeclared pharmaceuticals: a case series in Hong Kong, Br. J. Clin. Pharmacol. 73 (5) (2012) 795–800.

[51] Y. Feng, D. Lei, C. Hu, Rapid identification of illegal synthetic adulterants in herbal antidiabetic medicines using near infrared spectroscopy, Spectrochim. Acta A: Mol. Biomol. Spectrosc. 125 (2014) 363–374.

[52] J. Najafian, M. Abdar-Esfahani, M. Arab-Momeni, A. Akhavan-Tabib, Safety of herbal medicine in treatment of weight loss, ARYA Atheroscler 10 (1) (2014) 55–58.

[53] C.A. Luque, J.A. Rey, The discovery and status of sibutramine as an antiobesity drug, Eur. J. Pharmacol. 440 (2002) 119–128.

[54] N.C. da Silva, R.S. Honorato, M.F. Pimentel, S. Garrigues, M.L. Cervera, M. de la Guardia, Near infrared spectroscopy detection and quantification of herbal medicines adulterated with Sibutramine, J. Forensic Sci. 60 (5) (2015) 1199–1205.

[55] E. Deconinck, C. Aouadi, J.L. Bothy, P. Courselle, Detection and identification of multiple adulterants in plant food supplements using attenuated total reflectance—infrared spectroscopy, J. Pharm. Biomed. Anal. 152 (2018) 111–119.

[56] T. Eteraf-Oskouei, M. Najafi, Traditional and modern uses of natural honey in human diseases: a review, Iran. J. Basic Med. Sci. 16 (6) (2013) 731–742.

[57] World Health Organization, Cough and Cold Remedies for the Treatment of Acute Respiratory Infections in Young Children, WHO, Geneva, 2001.

[58] S. Li, Y. Shan, X. Zhu, X. Zhang, G. Ling, Detection of honey adulteration by high fructose corn syrup and maltose syrup using Raman spectroscopy, J. Food Compos. Anal. 28 (2012) 69–74.

[59] Z.M. Zhang, S. Chen, Y.Z. Liang, Baseline correction using adaptive iteratively reweighted penalized least squares, Analyst 135 (5) (2010) 1138–1146.

[60] Y. Zhang, X.F. Huang, W.F. Liu, Z.N. Cheng, C.P. Chen, L.H. Yin, Analysis of drugs illegally added into Chinese traditional patent medicine using surface-enhanced Raman scattering, Anal. Sci. 29 (2013) 985–990.

[61] D. Li, Q. Zhu, D. Lv, B. Zheng, Y. Liu, Y. Chai, Silver-nanoparticle-based surface-enhanced Raman scattering wiper for the detection of dye adulteration of medicinal herbs, Anal. Bioanal. Chem. 407 (2015) 6031–6039.

Part IV

Food analysis applications

Chapter 7

Edible oil analysis

The term "edible oils" (also possibly referred to as "cooking oils") describes a group of fatty liquids physically extracted from several vegetables, some animal tissues, or microorganisms. They play an essential role in our healthy diet due to a high likelihood of being eaten as a food or food accessory [1]. They also serve as an ingredient for texture, flavor, and nutritional improvers or nondairy spreads and food additives in their hydrogenated form.

According to the processing, some of them are edible cold-pressed oils to preserve the content of biologically active micronutrients. But, most edible oils in the initial raw form are regularly refined, bleached, and deodorized before being commercially available as a colorless to golden yellow oily liquid. It is well known that the main components of edible oils are triglycerides (forming the bulk, 95–99%); whereas the minor components include mono- and diglycerides, free fatty acids, phosphatides, sterols, fat-soluble vitamins, tocopherols, pigments, waxes, and fatty alcohols. The fat molecules are basically not the same in different edible oils with respect to the type of fatty acid (saturated, monounsaturated, and polyunsaturated) attached to a specific position (sn-1, sn-2, and sn-3) on the glycerol backbone of the triglyceride molecule. Edible oils are, thus, different from each other in terms of fatty acid composition and physical properties (flavor, cloud point, etc.) [2]. To ensure the standard quality of edible oils, both their origin and classification need to be confirmed. Official methods, standardized by American Oil Chemists' Society (AOCS) [3] or International Union of Pure and Applied Chemistry (IUPAC) [4], are routinely used in quality control laboratories for edible oil analysis.

The exploitation of vibrational spectroscopy for determination of edible oils started with IR technique in the 1950s [5] and much later with Raman one in the 1990s [6]. So far, combining IR and Raman spectroscopy with chemometrics has strongly demonstrated the capacity of verifying the authenticity and quality of edible oils (as discussed later).

In the 1990s, FTIR spectroscopy (both transmission and ATR approaches) was developed as rapid, direct, and indirect methods for determination of free fatty acids (FFA) in the range of 0.2–8% in fats and oils [7]. For the direct method, calibration curves were prepared by spiking oleic acid to the oil under study and quantification was based on measuring the C=O absorption band at 1711 cm^{-1} after spectral ratioing for the sample against the same oil free of FFAs. For oxidized/thermally stressed oils, the indirect method was relied

Vibrational Spectroscopy Applications in Biomedical, Pharmaceutical and Food Sciences
https://doi.org/10.1016/B978-0-12-818827-9.00009-3

© 2020 Elsevier Inc. All rights reserved.

on the extraction and conversion of FFAs into their potassium salts in 1% KOH/methanol, and subsequent measurement of the carboxylate absorbance at 1570 cm^{-1}. In another study, Al-Alawi et al. realized the FTIR FFA determination by measuring the peak height of the ν(COO–) absorption of the FFA salt formed when mixing the oil sample with a suspension of potassium phthalimide in 1-propanol [8]. By using stoichiometric analytical idea, FTIR analysis was suitable for determination of FFAs at low levels (<0.005%) in refined edible oils [9]. It was simply done by vortexing the oil sample with methanol containing 2 g/L sodium carbodiimide and recording the FTIR spectrum of the methanol phase containing the FFA salts at 1573 cm^{-1} relative to an invariant baseline point at 1820 cm^{-1}. In addition to FFA analysis in edible oils, Al-Alawi's research group also developed an FTIR method for the determination of moisture with high accuracy (on the order of ± 10 ppm) by (i) mixing the oil sample with dry acetonitrile in a 1:1 (w/v) ratio and (ii) measuring water (extracted in the upper acetonitrile layer) in the resultant differential spectra by using either the OH stretching (3629 cm^{-1}) or bending (1631 cm^{-1}) bands (Fig. 7.1) [10].

Moreover, high-volume automated quantitative analysis of FFA or moisture in edible oils was workable (> 60 samples/h) when an FTIR spectrometer equipped with an autosampler [11]. It was also possible to simplify the direct transmission-based FTIR determination of FFAs in edible oils by employing the spectral reconstitution (SR) technique and 2D correlation analysis, i.e., conventional neat-oil and SR calibrations were constructed by adding hexanoic acid into

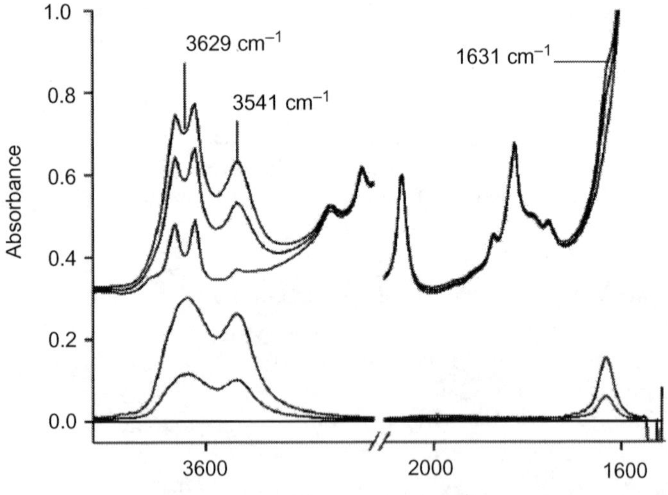

FIG. 7.1 Spectra of water/acetonitrile standards (0, 400, and 800 ppm added water) (upper series) and the corresponding spectra obtained after subtraction of the spectrum of the acetonitrile employed to prepare the standards (lower series). *(Reproduced with permission from A. Al-Alawi, F.R. van de Voort, J. Sedman, A new FTIR method for the analysis of low levels of moisture in edible oils, Appl. Spectrosc. 59 (2005) 1295–1299.)*

FFA-free canola oil and acquiring absorbance values ($1712\,cm^{-1}/1600\,cm^{-1}$); a correction equation was devised for blends of two FFA-free oils (canola and coconut) to compensate for underlying absorption due to a variation in oil saponification number, correlating the intensity of the triacylglycerol ester (C=O) absorption at the FFA measurement location with that of the first overtone of this vibration, measured at $3471\,cm^{-1}/3427\,cm^{-1}$; a generalized 2D correlation spectroscopy analysis led to the development of a second correction equation based on the absorbance at $4258\,cm^{-1}/4235\,cm^{-1}$ (Figs. 7.2 and 7.3) [12].

The acid values (AVs) of edible oils could be successfully FTIR determined by using the peak at $3535\,cm^{-1}$ corresponding to the O–H stretching band of

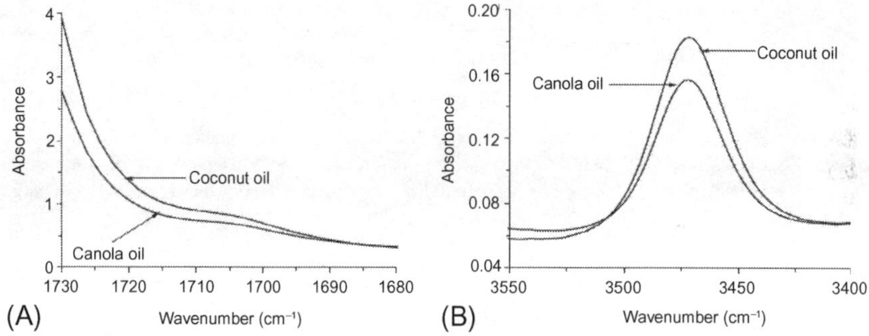

FIG. 7.2 SR spectra of FFA-free canola oil and coconut oil, illustrating their relative absorption profiles in the FFA measurement region (A) as well as the relative intensities of the first overtone of the triacylglycerol ester linkage ν(C=O) absorption (B). *(Reproduced with permission from X. Yu, F.R.V.D. Voort, J. Sedman, J.M. Gao, A new direct Fourier transform infrared analysis of free fatty acids in edible oils using spectral reconstitution, Anal. Bioanal. Chem. 401 (2011) 315–324.)*

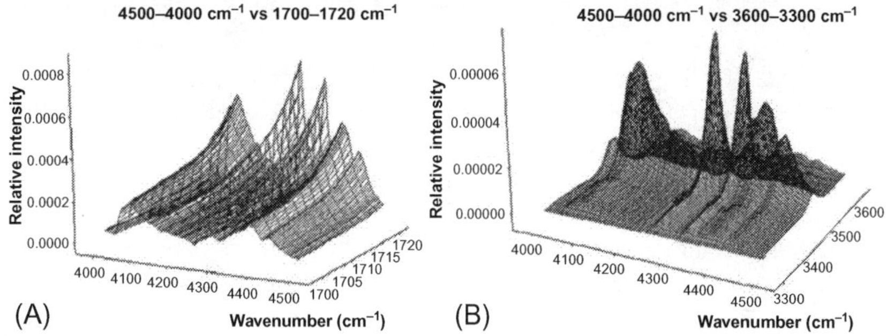

FIG. 7.3 (A) 3D contour map obtained by 2D correlation analysis of the FTIR spectra of FFA-free oil mixtures varying in saponification number, showing correlations between absorption in the FFA measurement region ($1700–1720\,cm^{-1}$) and the NIR combination band region ($4000–4500\,cm^{-1}$). (B) Corresponding contour map substituting the $3300–3600\,cm^{-1}$ region, containing the first overtone of the triacylglycerol ester linkage ν(C=O) absorption, for the $1700–1720\,cm^{-1}$ region. *(Reproduced with permission from X. Yu, F.R.V.D. Voort, J. Sedman, J.M. Gao, A new direct Fourier transform infrared analysis of free fatty acids in edible oils using spectral reconstitution, Anal. Bioanal. Chem. 401 (2011) 315–324.)*

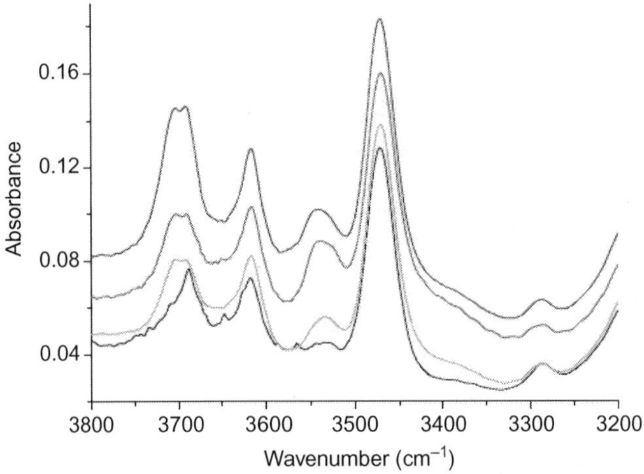

FIG. 7.4 FTIR spectra of edible oils (1% in CCl_4) with similar acid values. *(Reproduced with permission from X. Jiang, S. Li, G. Xiang, Q. Li, L. Fan, L. He, K. Gu, Determination of the acid values of edible oils via FTIR spectroscopy based on the O-H stretching band, Food Chem. 212 (2016) 585–589.)*

carboxyl groups in FFAs (as presented in Fig. 7.4, from bottom to top, the acid values of the edible oils are 0.33, 0.34, 0.36, and 0.35 mg g^{-1}, respectively) [13]. The developed method offered some advantages over other IR methods, including: (i) avoidance of the severe interference from other peaks that generally requires oil-specific calibration and/or spectral ratioing step (i.e., the peak at 1711 cm^{-1}, attributed to the C=O stretching in FFAs, occurs as a shoulder on the very strong peak at ca. 1746 cm^{-1} caused by the C=O stretching in triacylglycerol); (ii) observance of more spectral details (i.e., a long optical path of IR quartz cell allows further sample dilution to eliminate weak interactions, such as hydrogen bonding, between sample molecules); (iii) accurate control of the sample's optical path length (an IR quartz cell is more robust and convenient than a cavity cell composed of halide crystals).

With regard to peroxide value (PV) analytical methodology, the first FTIR method was developed by van de Voort et al. using a relatively simple calibration relied on rationed spectra with *t*-butyl hydroperoxide as a standard and measuring the characteristic O–H stretching absorption band of hydroxides in the mid-IR region [14]. Afterwards, a simpler, more accurate, and sensitive (~0.2 PV) FTIR method was proposed based on the stoichiometric reaction of triphenylphosphine with hydroperoxides to form triphenylphosphine oxide, measured at 542 cm^{-1} [15]. In a subsequent work, the determination of hydroperoxides in edible oils was investigated with disposable polyethylene infrared card [16]. With the use of spectral normalization to reduce sample-loading variability (based on the peak height of the ester linkage carbonyl overtone band at 3475 cm^{-1}) and PLS regression, the results of the 3M card-based FTIR

PV method were quantitatively comparable to those obtained with a flow cell. Originally, the FTIR method was devised for monitoring oxidative stress so that its measurable trends covered a broad range of PV (0–100). It could be further improved for a range of interest to edible oil industry, PV (0–10), by using convenient disposable 8-mm o.d. transparent glass vials for sample handling (Fig. 7.5) [17]. In comparison with the American Oil Chemists' Society primary reference method [18], the FTIR method was shown to be more accurate overall in tracking PV, but slightly less reproducible (0.9 PV).

Trans fatty acids (TFA) in edible fats and oils are believed to be etiologically implicated in arteriosclerosis and heart disease [19, 20]. They are commonly produced by isomerization of some of the *cis* double bonds of unsaturated fatty acids incompletely hydrogenated. As a consequence, monitoring TFA can help not only track and control the hydrogenation process but also access the quality and functionality of fats and oils.

According to the US FDA regulation promulgated in 2006 [21], the declaration of the total *trans* fat content on the Nutrition Fact label of foods was mandatory for any product containing ≥0.5 g of *trans* fatty acids per serving (i.e., approximately 2% of total fat). At that time, an IR spectroscopic method [22]

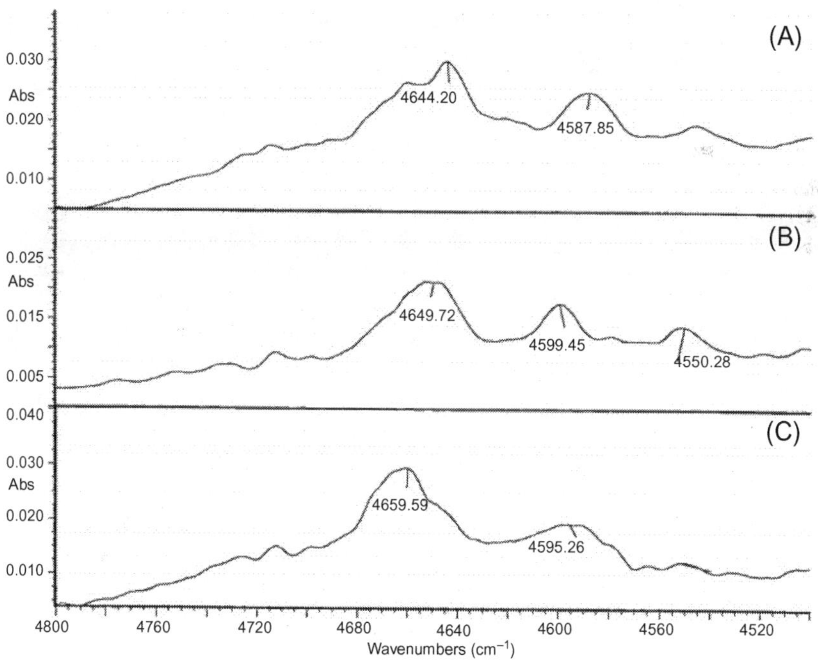

FIG. 7.5 Differential spectra of (A) triphenylphosphine (TPP), (B) triphenylphosphine oxide (TPPO), and (C) TPP+TPPO in canola oil produced by ratioing out the spectral contributions of canola oil. *(Reproduced with permission from H. Li, F.R. van de Voort, A.A. Ismail, R. Cox, Determination of peroxide value by Fourier transform near-infrared spectroscopy, J. Am. Oil Chem. Soc. 77 (2000) 137–42.)*

was inappropriate for determining <5% of total fat due to the sloping baseline near the *trans* absorption frequency leading to a great variability in measurement. To meet the US FDA labeling requirement, the negative second derivatives of ATR-FTIR spectra were generated to provide better sensitivity and accuracy as well as lower limit of quantitation from 5% to ca. 1% of total fat [23]. As illustrated in Figs. 7.6 and 7.7, the second-derivative transformation could eliminate spectral overlap: (i) the leading and tailing edges of the 966 cm^{-1} absorption band due to isolated *trans* double bonds, and (ii) the bands near 945, 985, and 990 cm^{-1} attributed to conjugated *cis/trans* or *trans*, *trans* double bonds.

It was shown that despite the lack of a discernible *trans* absorption band in the NIR region (Fig. 7.8), *trans* content of edible oils could be still FT-NIR spectroscopically determined by using 8-mm disposable glass vials for sample handling and measurement [24]. With the aid of a generalized PLS calibration constructed on a training set of a wide range of oils analyzed by the American Oil Chemists' Society single-bounce horizontal attenuated total reflectance (SB-HATR) mid-IR spectroscopic procedure [25], FT-NIR *trans* predictions could be achieved with an accuracy of ~±1.1% *trans*.

In another research, combining SB-HATR FTIR spectroscopy and PLS was investigated for simultaneous determination of iodine value and *trans* content of both neat fats and oils [26]. FTIR-predicted values for iodine value and *trans* content were in good agreement with those obtained by the two AOCS methods: gas chromatography [27] and FTIR/SBHATR method [28].

In a 12-laboratory international collaborative study, ATR-FTIR spectroscopic procedure (adopted as the official method AOCS Cd 14d-99 for determination of total *trans* fat content) exhibited high accuracy in the range of

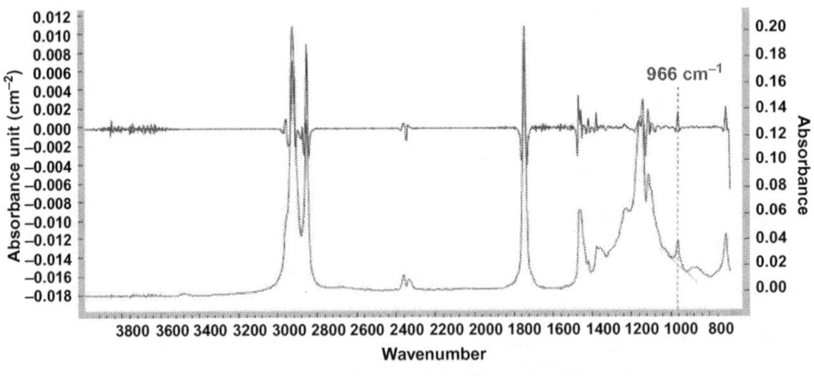

FIG. 7.6 Negative second-derivative (top) and absorption (bottom) spectra for a test sample consisting of trielaidin in tripalmitin with a *trans* level of 12.58% (as percent of total fat). A *vertical line* indicates the position of the unique band due to isolated *trans* double bonds at 966 cm^{-1}. The second derivative spectrum was multiplied by −1 only to have the bands point upwards for convenience. *(Reproduced with permission from M.M. Mossoba, A. Seiler, J.K.G. Kramer, V. Milosevic, M. Milosevic, H. Azizian, H. Steinhart, Nutrition labeling: rapid determination of total trans fats by using internal reflection infrared spectroscopy and a second derivative procedure, J. Am. Oil Chem. Soc. 86 (2009) 1037–1045.)*

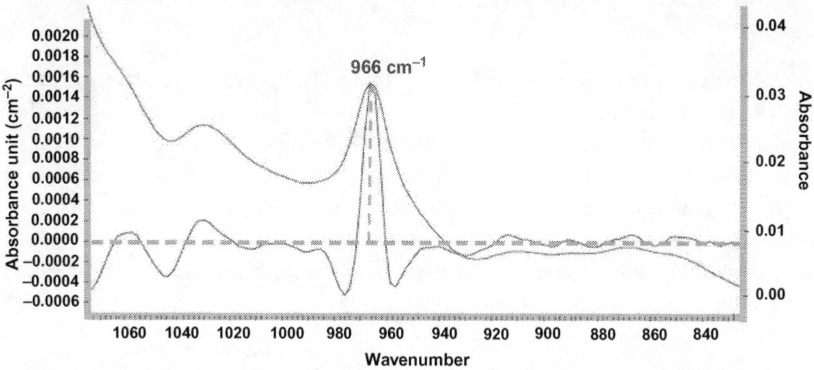

FIG. 7.7 The region of the spectra that exhibit the deformation band for isolated *trans* double bonds at 966 cm^{-1} is expanded for the negative second-derivative and absorption spectra for a test sample consisting of trielaidin in tripalmitin with a *trans* level of 12.58% (as percent of total fat). The height of the negative second-derivative band, as indicated by the *vertical arrow*, can be accurately measured for any test sample from the same horizontal baseline ($Y=0$). It is noted that several weak bands observed in the same spectral region are also more pronounced in the second-derivative spectrum (narrower bandwidths) than in the absorption spectrum. *(Reproduced with permission from M.M. Mossoba, A. Seiler, J.K.G. Kramer, V. Milosevic, M. Milosevic, H. Azizian, H. Steinhart, Nutrition labeling: rapid determination of total trans fats by using internal reflection infrared spectroscopy and a second derivative procedure, J. Am. Oil Chem. Soc. 86 (2009) 1037–1045.)*

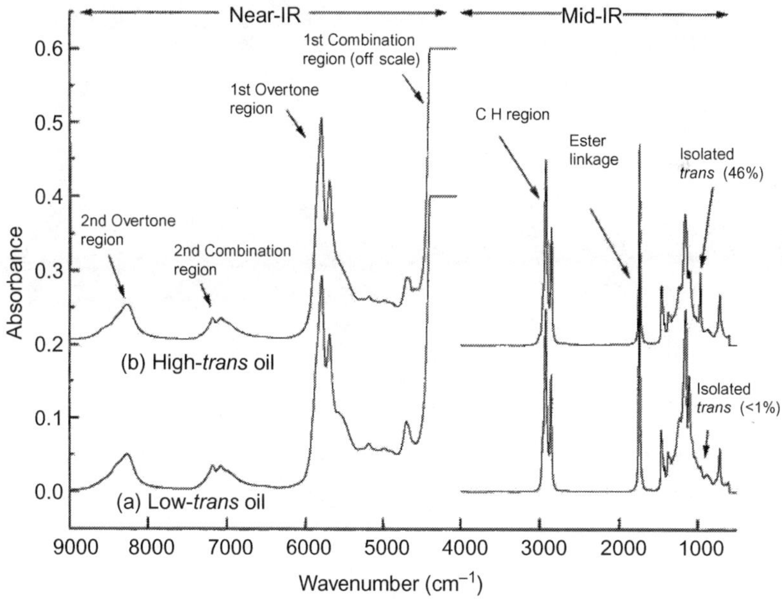

FIG. 7.8 Collated near/midinfrared (IR) spectra of (a) a low-*trans* oil and (b) a high-*trans* oil, illustrating the key spectral features associated with the two IR spectral regions. The fundamental *trans* absorption at 967 cm^{-1} is readily discernible in the mid-IR region. The absorbance values in the region from 10,000 to 4000 cm^{-1} have been magnified by a factor of 10 to provide greater detail. *(Reproduced with permission from H. Li, F.R. van de Voort, A.A. Ismail, J. Sedman, R. Cox, Trans determination of edible oils by Fourier transform near-infrared spectroscopy, J. Am. Oil Chem. Soc. 77 (2000) 1061–1067.)*

1–40% *trans* (as percent of total fat) and had <2% high bias relative to values gravimetrically determined [29]. This internal reflection method was judged to be more precise than other transmission IR official methods (Fig. 7.9).

Multivariate determination of FFA, moisture, peroxide index, iodine value in edible oils could be done by IR spectroscopy. It was stated that reflectance NIR spectroscopy was able to profitably monitor hydrolytic degradation of lipids in fish oil samples, which could critically influence their production and storage in a fishmeal factory [30]. To examine both outlier samples as well as spectra

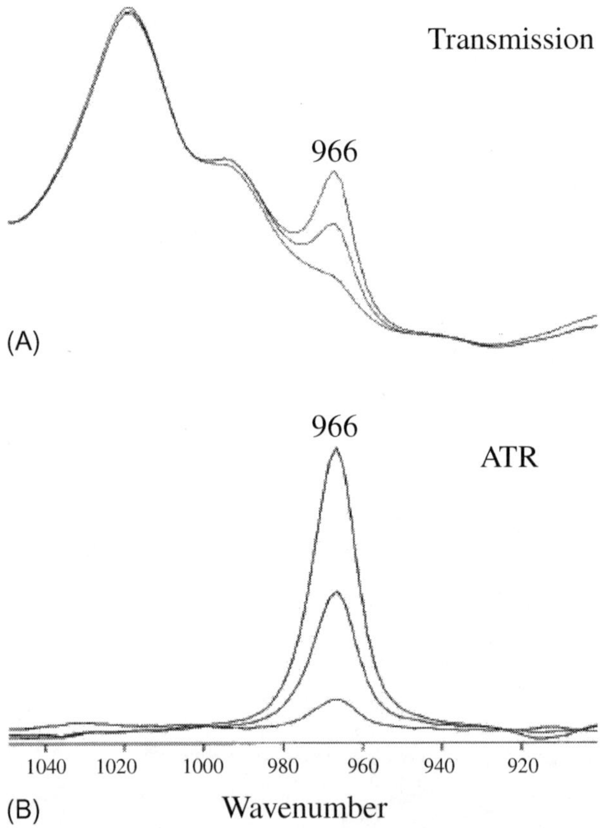

FIG. 7.9 (A) Transmission Fourier transform infrared (FTIR) spectra for oils containing approximately 1%, 4%, and 7% *trans* fat (as percentage of total fat). The spectra were obtained for carbon disulfide solutions by using official methods AOAC 965.34 and AOAC 994.14. The spectra were recorded in the range between 1050 and 900 cm^{-1} showing the 966 cm^{-1} band attributed to isolated *trans* double bonds. (B) Attenuated total reflection (ATR)-FTIR spectra for oils containing approximately 1%, 5%, and 10% *trans* fat (as percentage of total fat). Spectra were obtained by applying the "ratioing" and ATR-FTIR official method AOCS Cd 14d-99. Symmetric bands on a horizontal background were observed. *(Reproduced with permission from M. Adam, M.M. Mossoba, T. Lee, Rapid determination of total trans fat content by attenuated total reflection infrared spectroscopy: an international collaborative study, J. Am. Oil Chem. Soc. 77 (5) (2000) 457–462.)*

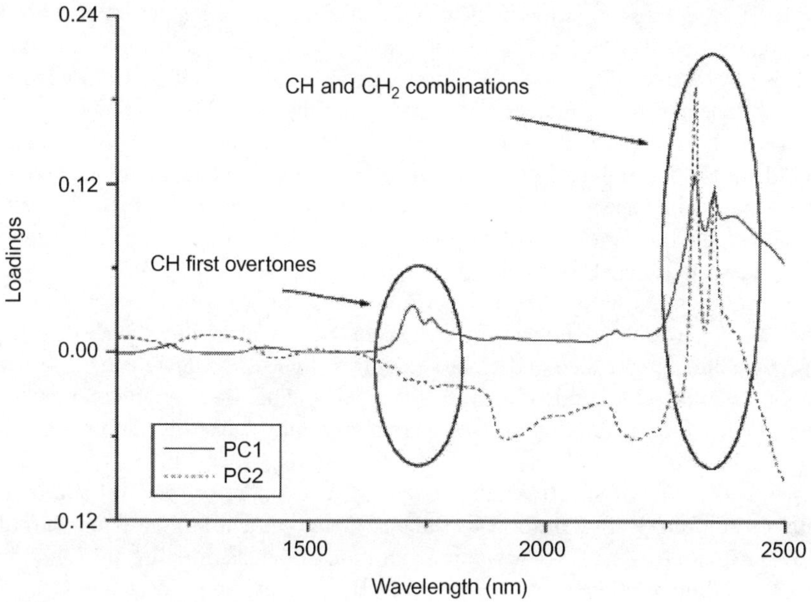

FIG. 7.10 Loadings for the first two PCs of fish oil samples (PC1=*line*; PC2=*dotted line*), accounting for >99% of spectral variation. *(Reproduced with permission from D. Cozzolino, I. Murray, A. Chree, J.R. Scaife, Multivariate determination of free fatty acids and moisture in fish oils by partial least squares regression and near-infrared spectroscopy, LWT- Food Sci. Technol. 38 (2005) 821–828.)*

discrimination between sampling periods, PCA was performed and presented in Fig. 7.10, before the development of PLS regression calibration models was realized for FFA and moisture ($R^2 \geq 0.94$). In other studies, the suitability was confirmed for estimation of iodine and acid values in red fruit oil by combining ATR-FTIR with PLS regression [31], for evaluation of acidity and peroxide index in various edible oils by coupling FT-NIR with hierarchical cluster analysis and PLS regression [32].

Olive oil has been reputedly used since 6000 BCE when it was exclusively extracted by the pressing of the olives for a food source in Galilee, Israel [33]. It is a main component of the Mediterranean diet, which experts consider "a heart-healthy eating plan." As the primary source of added fat, olive oil is rich in monounsaturated fats (of which oleic acid is the predominant fatty acid) making up 73% of the total oil content. The antioxidants in olive oil may help protect the body from cellular damage caused by free radicals. Many studies have looked at its health benefits (such as antibacterial and antiinflammatory properties; a lower risk of stroke and heart diseases, type 2 diabetes, and some cancers; favorable changes in weight loss; beneficial for treatment of Alzheimer's disease and rheumatoid arthritis) [34]. It is therefore unsurprising that olive oil has been the subject of numerous studies on physicochemical analysis over the last decade [35], especially IR spectroscopic application [36, 37]. As an example, fiber-based

ATR-FTIR spectroscopy (the probe either containing an exchangeable U-shaped section of the silver halide fiber or consisting of a fiber-coupled diamond crystal) was assessed for the analysis of extra virgin olive oil samples from various Mediterranean sites and their adulteration by sunflower oil [38]. It was shown that any olive oil sample contaminated by sunflower oil addition (>2% by weight) could be ambiguously quantified by using optimized PLS calibration models (with the use of a special variable selection strategy based on a pair wise consideration of significant respective minima and maxima of the optimum PLS regression vector, calculated for broad spectral intervals). Over a long time interval, a slight deterioration in the spectral performance did not perturb the analysis as long as proper background spectra were recorded for calculating the absorbance spectra. Although the diamond probe performed slightly less than the whole fiber probe with respect to random noise features, it could be overcome by accumulating more interferograms within transparent spectral regions (as can be seen in Fig. 7.11).

In another FTIR spectroscopic study, the assessment of olive oil adulteration by sunflower oil addition was reliably done by identifying a band shift at 3011 cm^{-1} (due to the C–H stretching vibration of the *cis*-double bond) and the degree of adulteration could be determined by measuring the absorbance of the wavenumbers close to 1303 or 1270 cm^{-1} or in the ranges 1100–1000 and 900–800 cm^{-1} [39].

FT-MIR spectroscopy, especially its first-derivative transformation, was also able to successfully monitor extra virgin olive oil (EVOO) adulterated with a wide range of concentration of old olive oil of lower quality (1.0–50.0%, w/w in EVOO) at frequencies of 3100–600 cm^{-1} with the aid of multivariate calibrations (PCR and PLS) as well as PLS-DA for classification analysis (as can be seen in Fig. 7.12) [40]. Similar results were also reported for FTIR spectroscopic analysis of EVOO adulterated with palm oil (1.0–50.0%, w/w in EVOO) [41]. To take advantage of the synergistic effect obtained from both NIR and MIR data, data fusion strategies (low-, mid-, and high-level) was investigated for olive oil adulteration with rapeseed oil [42]. To improve fruit harvesting and storage, an ATR-FTIR-PLS-based strategy was developed for the analysis of oil content directly performed on a section of the mesocarp from the olive trees grown in Morocco [43].

Besides IR analysis, a method of PCA-based Raman spectroscopy was developed to not only reliably authenticate genuine olive oil of different brands and grades collected in the United States but also accurately identify its adulteration with other edible oils collected in the Mainland China with volume percentage $\geq 5\%$ (such as soybean oil, rapeseed oil, sunflower seed oil, and corn oil) [44]. Spectra obtained from a portable Raman spectrometer showed that there was a marked difference in the corresponding band intensities among the edible oils under study due to the fact that they contained different percentages of monounsaturated, polyunsaturated, and saturated fatty acids (Fig. 7.13).

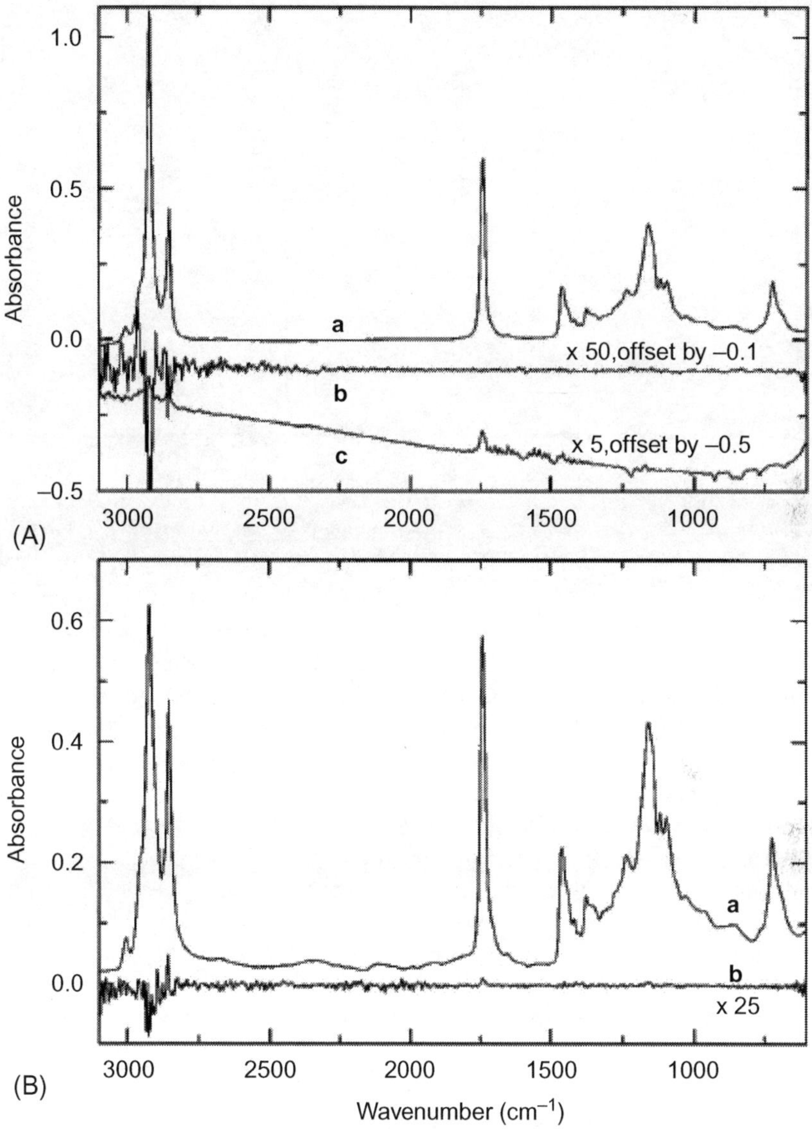

FIG. 7.11 (A) Absorbance spectrum of an olive oil sample measured by the silver halide fiber probe (a), absorbance noise from two consecutive single-beam spectra with olive oil sample (128 scans, 4 cm^{-1} spectral resolution) (b), and degradation of fiber sensor transmission after measurement of 100 olive oil spectra (c). (B) Absorbance spectrum of an olive oil sample measured by the fiber-coupled diamond probe (a); absorbance noise under same conditions as given above (b). *(Reproduced with permission from L. Kupper, H.M. Heise, P. Lampen, A.N. Davies, P. McIntyre, Authentication and quantification of extra virgin olive oils by attenuated total reflectance infrared spectroscopy using silver halide fiber probes and partial least-squares calibration, Appl. Spectrosc. 55 (2001) 563–570.)*

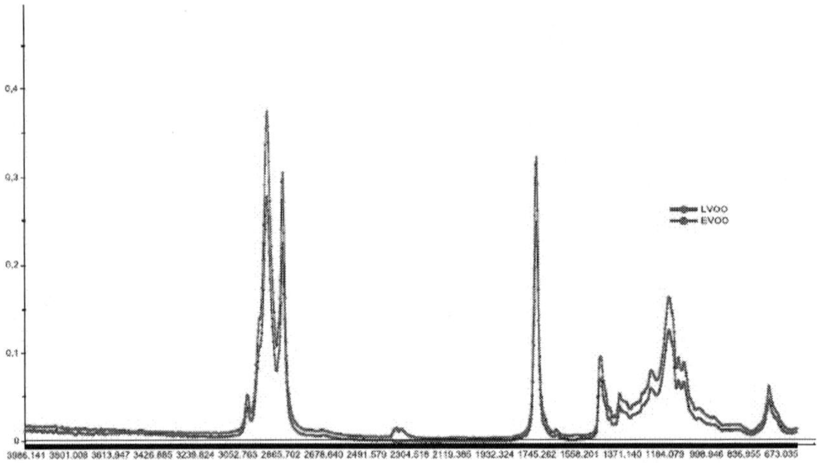

FIG. 7.12 FTIR spectra of extra and lampante virgin olive oil in the spectral range 4000–600 cm^{-1}. *(Reproduced with permission from A. Hirri, M. Gammouh, A. Gorfti, F. Kzaiber, M. Bassbasi, S. Souhassou, A. Balouki, A. Oussama, The use of Fourier transform mid infrared (FT-MIR) spectroscopy for detection and estimation of extra virgin olive oil adulteration with old olive oil, Sky J. Food Sci. 4(5) (2015) 060–066.)*

Using PLS-based FT-Raman spectroscopy, authentication of EVOO was additionally evaluated in terms of quantitative information (prediction of fatty acid contents) and qualitative information (classification of diverse EVOOs as a function of the harvest year, olive variety, geographical origin, and Andalusian Protected Designations of Origin) [45]. Raman spectroscopic identification of EVOO adulterated by soybean oil (5%, w/w) could be further improved by probing an optimal sample temperature [46]. This experiment was done by separately freezing oleic and linoleic acids (the main components in EVOO and soybean oil) in liquid nitrogen and consecutively collecting Raman spectra at regular intervals as samples simply exposed to ambient condition to increase temperature. At –36.4°C, the lowest discrimination error of 8.0% was obtained for principal component analysis (PCA) and linear discriminant analysis (LDA) of temperature-induced spectral variations of pure and adulterated EVOOs (Fig. 7.14).

To better authenticate olive oil, a transesterification step was used prior to applying Raman and FTIR spectroscopic analysis [47]. FTIR fingerprints were shown to be more discriminative than Raman's in both classification and regression scenarios. They had excellent classification ability, especially when coupling with partial least squared-discriminant analysis (PLS-DA) or support vector machine (SVM) algorithms, i.e., 100% of olive oil samples and 92% of other oil samples (soybean, sunflower, rapeseed, corn) were correctly classified (as presented in Figs. 7.15 and 7.16).

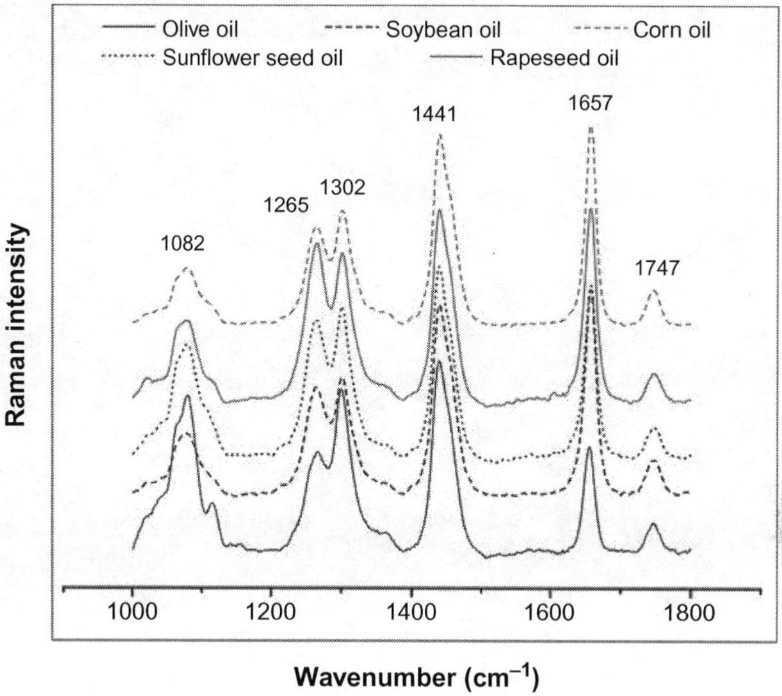

FIG. 7.13 Raman spectra of olive oil, soybean oil, sunflower seed oil, rapeseed oil, and corn oil. *(Reproduced with permission from Z. Xiaofang, Q. Xiaohua, Z. Mingqiang, L. Feng, Rapid authentication of olive oil by Raman spectroscopy using principal component analysis, Anal. Lett. 44 (2011) 2209–2220.)*

In a different study, the adulteration of EVOO with hazelnut oil was Raman and FTIR spectroscopically addressed by the use of a continuous locality preserving projections (CLPP) technique [48]. This dimensionality reduction technique allows the modeling of the continuous nature of the produced in-house admixtures as data series instead of discrete points. The highest outperformance was achieved by accompanying CLPP with *k*-nearest neighbors algorithm as compared to any other state-of-the-art pattern recognition techniques such as SIMCA, PLS-DA. For the classification problem with four classes (pure EVOO, low adulterated EVOO, high adulterated EVOO, and mostly pure hazelnut oil), it was found that ca. 80% and 75% overall mean classification rate was obtained with $\geq$82% and 69% in low percentages (1–12%) for Raman and FT-IR data, respectively.

With advanced chemometric tools, vibrational spectroscopic techniques were again successfully exercised for analysis of edible oils of different origins [49], such as ostrich oil [50], adulterated chia and sesame oils [51], cold pressed grape seed oil in blends with refined soybean oils [52], butter oil adulteration with soybean oil [53], adulterated neem and flaxseed oil [54], hazelnut oil adulteration [55], cod liver oil from beef fat [56].

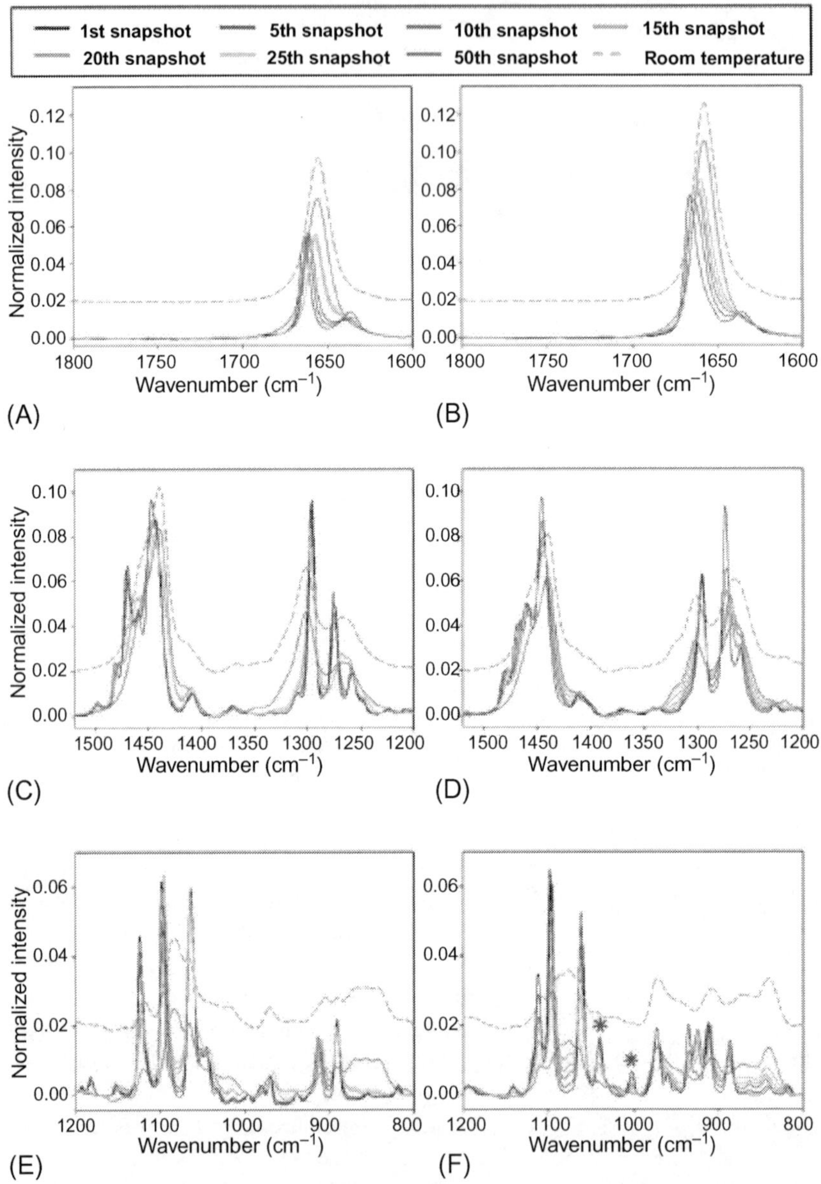

FIG. 7.14 Raman spectra of EVOO (left column) and soybean oil (right column) collected during temperature elevation. The 1800–1600 (A,B), 1520–1200 (C,D), and 1200–800 cm^{-1} (E,F) ranges are highlighted. *(Reproduced with permission from D. Ryoo, J. Hwang, H. Chung, Probing temperature able to improve Raman spectroscopic discrimination of adulterated olive oils, Microchem. J. 134 (2017) 224–229.)*

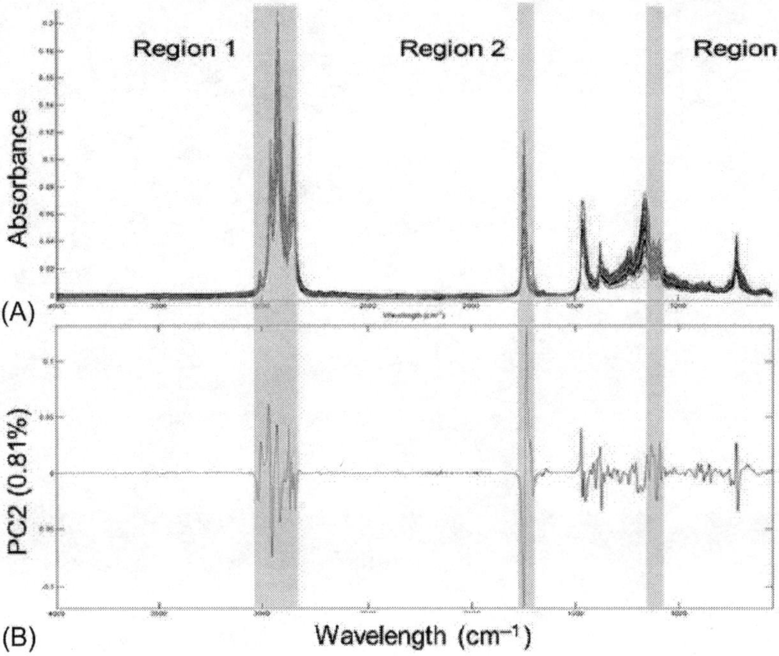

FIG. 7.15 (A) Superposed FTIR spectra and (B) loading plot of the 146 vegetable oil samples showing the three regions selected. *(Reproduced with permission from A.M. Jiménez-Carvelo, M.T. Osorio, A. Koidis, A. González-Casado, L. Cuadros-Rodríguez, Chemometric classification and quantification of olive oil in blends with any edible vegetable oils using FTIR-ATR and Raman spectroscopy, LWT 86 (2017) 174–184.)*

One of the key points in assessing safety of food-contact materials is to determine the diffusivity of a migrant in a plastic packaging material, particularly in the worst-case food-simulating liquid, i.e., the olive oil. In a vibrational spectroscopic study [57], FTIR spectroscopy was used to investigate simultaneous desorption of Uvitex OB (an additive) and sorption of olive oil (a food constituent) in a nondestructive and "in situ" mode. The diffusivity of Uvitex OB and olive oil was found to be $(8.0 \pm 2.5 \times 10^{-14} m^2 s^{-1})$ and $(6.9 \pm 3.4 \times 10^{-14} m^2 s^{-1})$ in linear low-density polyethylene (LLDPE), respectively. On the other hand, Raman microspectroscopy gave the concentration profile of Uvitex OB in the depth of the polymer sample, confirming that mass transfer in the system roughly followed a Fickian kinetics (as are presented in Figs. 7.17 and 7.18).

Frying is a versatile and ubiquitous way of cooking that improves sensory appeal of food. Nonetheless, it can modify both the food and frying medium, in particular deep frying or repeated frying cycles (i.e., a deterioration of the frying medium leads to changes in the performance of oils and fats). PLS-based mid-FTIR data showed that palm oil was more resistive than soybean

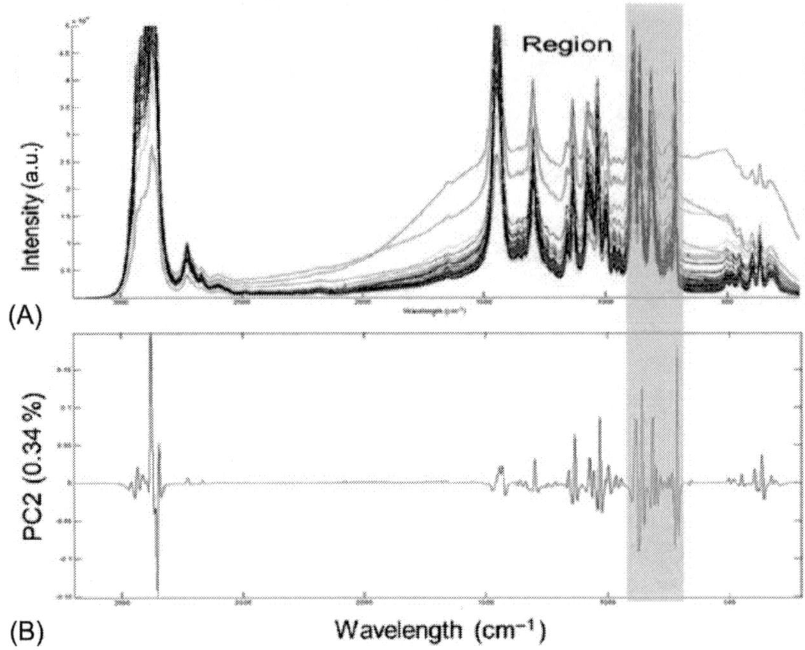

FIG. 7.16 Plot of (A) superposed Raman spectra and (B) loading plot of all the vegetable edible oil samples showing the region selected. *(Reproduced with permission from A.M. Jiménez-Carvelo, M.T. Osorio, A. Koidis, A. González-Casado, L. Cuadros-Rodríguez, Chemometric classification and quantification of olive oil in blends with any edible vegetable oils using FTIR-ATR and Raman spectroscopy, LWT 86 (2017) 174–184.)*

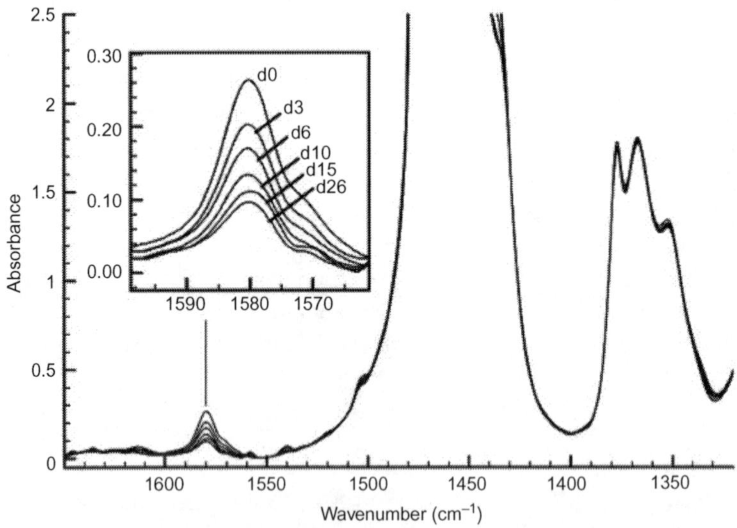

FIG. 7.17 FTIR spectral fingerprint of LLDPE + Uvitex OB. The focused region shows the band selected to follow Uvitex OB desorption and its decrease throughout the test. *(Reproduced with permission from M. Mauricio-Iglesias, V. Guillard, N. Gontard, S. Peyron, Application of FTIR and Raman microspectroscopy to the study of food/packaging interactions, Food Addit. Contam. 26(11) (2009) 1515–1523.)*

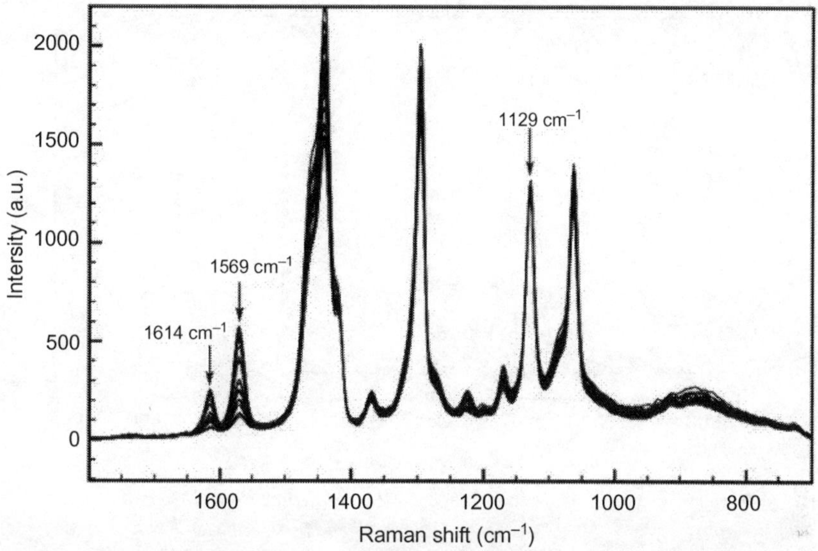

FIG. 7.18 Raman spectral fingerprint of LLDPE + Uvitex OB. The focused region shows the doublet selected to follow Uvitex OB (C=N, C=C) desorption and its variation in the depth of the film. *(Reproduced with permission from M. Mauricio-Iglesias, V. Guillard, N. Gontard, S. Peyron, Application of FTIR and Raman microspectroscopy to the study of food/packaging interactions, Food Addit. Contam. 26(11) (2009) 1515–1523.)*

oil to physicochemical changes due to a difference in the degree of unsaturation (soybean oil, 61.9%; palm oil, 13.8%) [58]. After being in service for 48 h, however, both oils still had a good frying power because their level of total polar compounds had not exceeded the recommended limit. In another research, ATR-FTIR spectroscopy was suggested for a rapid determination of the degree of degradation of 156 rapeseed oil samples used for frying potatoes [59]. Although the entire range of spectra looked very similar for both fresh and used oils, the weak absorption at 966 cm^{-1} was obviously seen for used oil samples (probably ascribed to the C–H out-of-plane deformation of isolated *trans* double bonds or some *trans* conjugated unsaturated fatty acids) (Fig. 7.19). The prediction of acid value and total polar compounds was done with PLS regression with full cross-validation, based on the raw, first-, and second-derivative FTIR spectra (4000–650 cm^{-1}). Using second-derivative spectral data, a precise calibration model was obtained clearly indicating that the degradation of frying oils could be accurately measured by using FTIR spectroscopy.

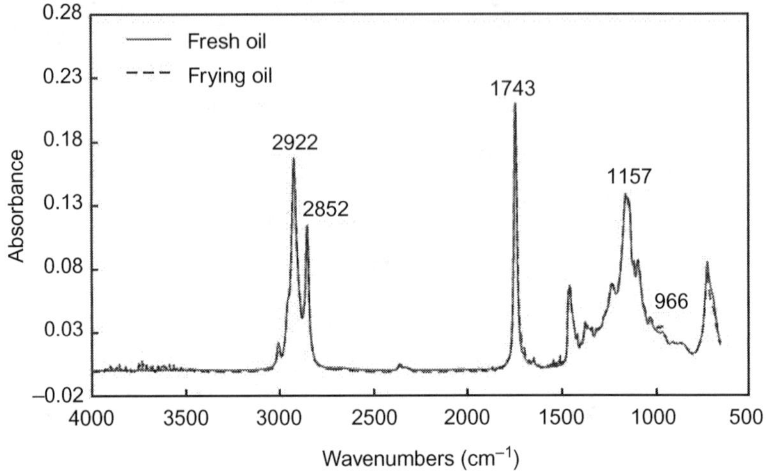

FIG. 7.19 FTIR spectra of the fresh and used canola oil. *(Reproduced with permission from J.Y. Chen, H. Zhang, J. Ma, T. Tuchiya, Y. Miao, Determination of the degree of degradation of frying rapeseed oil using fourier-transform infrared spectroscopy combined with partial least-squares regression, Int. J. Anal. Chem. 2015 (2015), 185367.)*

References

[1] https://encyclopedia2.thefreedictionary.com/edible+oil, (Accessible on December 27 2019).

[2] T.K.W. Ng, M. Appukutty, S. Shyam, P.T. Voon, K.R. Selvaduray, Cooking oils in health and sports, in: Nutrition and Enhanced Sports Performance—Muscle Building, Endurance, and Strength, second ed., Academic Press, 2019, pp. 751–756.

[3] AOCS, Official Methods and Recommended Practices of the American Oil Chemists' Society, fifth ed., AOCS Press, Champaign, IL, 1998.

[4] IUPAC, Standard Methods for the Analysis of Oils, Fats and Derivatives, seventh ed., Blackwell Scientific, London, 1992.

[5] J.G. Bartlet, Identification of edible oils and the detection of oil adulteration by differential infrared spectroscopy, Nature 180 (1957) 1071–1072.

[6] V. Baeten, M. Meurens, M.T. Morales, R. Aparicio, Detection of virgin olive oil adulteration by Fourier transform Raman spectroscopy, J. Agric. Food Chem. 44 (1996) 2225–2230.

[7] A.A. Ismail, F.R. van de Voort, J. Sedman, Rapid quantitative determination of free fatty acids in fats and oils by FTIR spectroscopy, J. Am. Oil Chem. Soc. 70 (1993) 335–341.

[8] A. Al-Alawi, F.R. van de Voort, J. Sedman, New method for the quantitative determination of free fatty acids in oil by FTIR spectroscopy, J. Am. Oil Chem. Soc. 81 (2004) 441–446.

[9] A. Al-Alawi, F.R. van de Voort, J. Sedman, A new FTIR method for the analysis of low levels of FFA in refined edible oils, Spectrosc. Lett. 38 (2005) 389–403.

[10] A. Al-Alawi, F.R. van de Voort, J. Sedman, A new FTIR method for the analysis of low levels of moisture in edible oils, Appl. Spectrosc. 59 (2005) 1295–1299.

[11] A. Al-Alawi, F.R. van de Voort, J. Sedman, A. Ghetler, Automated FTIR analysis of free fatty acids or moisture in edible oils, J. Assoc. Lab. Autom. 11 (2006) 23–29.

[12] X. Yu, F.R.V.D. Voort, J. Sedman, J.M. Gao, A new direct Fourier transform infrared analysis of free fatty acids in edible oils using spectral reconstitution, Anal. Bioanal. Chem. 401 (2011) 315–324.

[13] X. Jiang, S. Li, G. Xiang, Q. Li, L. Fan, L. He, K. Gu, Determination of the acid values of edible oils via FTIR spectroscopy based on the O-H stretching band, Food Chem. 212 (2016) 585–589.

[14] F.R. van de Voort, A.A. Ismail, J. Sedman, J. Dubois, T. Nicodemo, The determination of peroxide value by Fourier transform infrared (FTIR) spectroscopy, J. Am. Oil Chem. Soc. 71 (1994) 921–926.

[15] K. Ma, F.R. van de Voort, J. Sedman, A.A. Ismail, Stoichiometric determination of hydroperoxides in fats and oils by Fourier transform infrared spectroscopy, J. Am. Oil Chem. Soc. 74 (8) (1997) 897–906.

[16] K. Ma, F.R. van de Voort, A.A. Ismail, J. Sedman, Quantitative determination of hydroperoxide formation by Fourier transform infrared spectroscopy using a disposable infrared card, J. Am. Oil Chem. Soc. 71 (1998) 1095–1101.

[17] H. Li, F.R. van de Voort, A.A. Ismail, R. Cox, Determination of peroxide value by Fourier transform near-infrared spectroscopy, J. Am. Oil Chem. Soc. 77 (2000) 137–142.

[18] American Oil Chemists' Society (Eds.), Official Methods and Recommended Practices of the American Oil Chemists' Society, fourth ed., American Oil Chemists' Society, Champaign, IL, 1989.

[19] B.F. Haumann, Dietary fats still spark controversy, Inform 5 (1994) 358.

[20] A. Aro, Epidemiological studies of trans fatty acids and coronary heart disease, in: J.L. Sebedio, W.W. Christie (Eds.), Trans Fatty Acids in Human Nutrition, The Oily Press, Dundee, 1998, pp. 235–260.

[21] DHHS/FDA, Food labeling: *trans* fatty acids in nutrition labeling, nutrient content claims, and health claims, Fed. Reg. 68 (2003) 41433.

[22] M.M. Mossoba, J.K.G. Kramer, P. Delmonte, M.P. Yurawecz, J.I. Rader, Determination of trans fats by gas chromatography and infrared methods, in: D.K. Kodali, G.R. List (Eds.), Trans Fats Alternatives, AOCS Press, Champaign, IL, 2005, pp. 47–70.

[23] M.M. Mossoba, A. Seiler, J.K.G. Kramer, V. Milosevic, M. Milosevic, H. Azizian, H. Steinhart, Nutrition labeling: rapid determination of total *trans* fats by using internal reflection infrared spectroscopy and a second derivative procedure, J. Am. Oil Chem. Soc. 86 (2009) 1037–1045.

[24] H. Li, F.R. van de Voort, A.A. Ismail, J. Sedman, R. Cox, Trans determination of edible oils by Fourier transform near-infrared spectroscopy, J. Am. Oil Chem. Soc. 77 (2000) 1061–1067.

[25] D. Firestone (Ed.), Official Methods and Recommended Practices of the American Oil Chemists' Society, fifth ed., AOCS Press, Champaign, 1998. (Methods Ce 1g-96, Cd 14d-97, and Ce 1f-97).

[26] J. Sedman, F.R. van De Voort, A.A. Ismail, Simultaneous determination of iodine value and trans content of fats and oils by single-bounce horizontal attenuated total reflectance Fourier transform infrared spectroscopy, J. Am. Oil Chem. Soc. 77 (2000) 399–403.

[27] American Oil Chemists' Society (Eds.), Official Methods and Recommended Practices of the American Oil Chemists' Society, fifth ed., American Oil Chemists' Society, Champaign, 1998. (Method Ce 1f-96).

[28] American Oil Chemists' Society (Eds.), Official Methods and Recommended Practices of the American Oil Chemists' Society, fifth ed., American Oil Chemists' Society, Champaign, 1998. (Method Cd 14d-96).

[29] M. Adam, M.M. Mossoba, T. Lee, Rapid determination of total trans fat content by attenuated total reflection infrared spectroscopy: an international collaborative study, J. Am. Oil Chem. Soc. 77 (5) (2000) 457–462.

[30] D. Cozzolino, I. Murray, A. Chree, J.R. Scaife, Multivariate determination of free fatty acids and moisture in fish oils by partial least squares regression and near-infrared spectroscopy, LWT- Food Sci. Technol. 38 (2005) 821–828.

[31] L. Triyasmono, S. Riyanto, A. Rohman, Determination of iodine value and acid value of red fruit oil by infrared spectroscopy and multivariate calibration, Int. Food Res. J. 20 (6) (2013) 3259–3263.

[32] S. Armenta, S. Garrigues, M. de la Guardia, Determination of edible oil parameters by near infrared spectrometry, Anal. Chim. Acta 596 (2007) 330–337.

[33] J. McMathis, Ancient pottery unearthed in Israel contains 8000 year old olive oil, Isr. J. Plant Sci. 62 (2015) 65–74.

[34] D. Boskou (Ed.), Olive and Olive Oil Bioactive Constituents, first ed., 2015. Hardcover ISBN: 9781630670412. eBook ISBN: 9781630670429. Imprint: Academic Press and AOCS Press.

[35] M. Meenu, Q. Cai, B. Xu, A critical review on analytical techniques to detect adulteration of extra virgin olive oil, Trends Food Sci. Technol. 91 (2019) 391–408.

[36] J. Delgadillo, A. Barros, A. Nunes, Quality evaluation of olives, olive pomace and olive oil by infrared spectroscopy. in: Boskou Dimitrios, Olive Oil—Constituents, Quality, Health Properties and Bioconversions, IntechOpen, 2012, https://doi.org/10.5772/30679.

[37] A. Rohman, Infrared spectroscopy for quantitative analysis and oil parameters of olive oil and virgin coconut oil: a review, Int. J. Food Prop. 20 (7) (2016) 1447–1456.

[38] L. Kupper, H.M. Heise, P. Lampen, A.N. Davies, P. McIntyre, Authentication and quantification of extra virgin olive oils by attenuated total reflectance infrared spectroscopy using silver halide fiber probes and partial least-squares calibration, Appl. Spectrosc. 55 (2001) 563–570.

[39] M.A. Poiana, G. Mousdis, E. Alexa, D. Moigradean, M. Negrea, C. Mateescu, Application of FT-IR spectroscopy to assess the olive oil adulteration, J. Agroaliment. Process. Technol. 18 (4) (2012) 277–282.

[40] A. Hirri, M. Gammouh, A. Gorfti, F. Kzaiber, M. Bassbasi, S. Souhassou, A. Balouki, A. Oussama, The use of Fourier transform mid infrared (FT-MIR) spectroscopy for detection and estimation of extra virgin olive oil adulteration with old olive oil, Sky J. Food Sci. 4 (5) (2015) 060–066.

[41] A. Rohman, Y.B.C. Man, Fourier transform infrared (FTIR) spectroscopy for analysis of extra virgin olive oil adulterated with palm oil, Food Res. Int. 43 (3) (2010) 886–892.

[42] Y. Li, Y. Xiong, S. Min, Data fusion strategy in quantitative analysis of spectroscopy relevant to olive oil adulteration, Vib. Spectrosc. 101 (2019) 20–27.

[43] A. Hirri, M. Bassbasi, F. Kzaiber, A. Oussama, Prediction of oil content in olive fruit using Fourier transformed infrared spectroscopy FT-IR coupled with partial least squares regression, Int. Food Res. J. 21 (2) (2014) 723–727.

[44] Z. Xiaofang, Q. Xiaohua, Z. Mingqiang, L. Feng, Rapid authentication of olive oil by Raman spectroscopy using principal component analysis, Anal. Lett. 44 (2011) 2209–2220.

[45] E. Sánchez-López, M.I. Sánchez-Rodríguez, A. Marinas, J.M. Marinas, M. Moalem, Chemometric study of andalusian extra virgin olive oils Raman spectra: qualitative and quantitative information, Talanta 156–157 (2016) 180–190.

[46] D. Ryoo, J. Hwang, H. Chung, Probing temperature able to improve Raman spectroscopic discrimination of adulterated olive oils, Microchem. J. 134 (2017) 224–229.

[47] A.M. Jiménez-Carvelo, M.T. Osorio, A. Koidis, A. González-Casado, L. Cuadros-Rodríguez, Chemometric classification and quantification of olive oil in blends with any edible vegetable oils using FTIR-ATR and Raman spectroscopy, LWT 86 (2017) 174–184.

[48] K. Georgouli, J. Martinez-del-Rincon, A. Koidis, Continuous statistical modelling for rapid detection of adulteration of extra virgin olive oil using mid infrared and Raman spectroscopic data, Food Chem. 217 (2017) 735–742.

[49] O. Jović, Durbin-Watson partial least-squares regression applied to MIR data on adulteration with edible oils of different origins, Food Chem. 213 (2016) 791–798.

[50] Q.S. Martins, P.V. Almeida, Q.S. Ferreira, A. Oliveira, J.L.B. Faria, Investigation of ostrich oil via Raman and infrared spectroscopy and predictions using the DFT method, Vib. Spectrosc. 104 (2019) 102945.

[51] S.D. Rodríguez, M. Gagneten, A.E. Farroni, N.M. Percibaldi, M. Pilar Buera, FT-IR and untargeted chemometric analysis for adulterant detection in chia and sesame oils, Food Control 105 (2019) 78–85.

[52] G. Akin, Ş. Nihan, K. Elmas, F. Nur Arslan, İ. Yılmaz, A. Kenar, Chemometric classification and quantification of cold pressed grape seed oil in blends with refined soybean oils using, attenuated total reflectance–mid infrared (ATR–MIR) spectroscopy, LWT 100 (2019) 126–137.

[53] C.G. Pereira, A.I.N. Leite, J. Andrade, M.J.V. Bell, V. Anjos, Evaluation of butter oil adulteration with soybean oil by FT-MIR and FT-NIR spectroscopies and multivariate analyses, LWT 107 (2019) 1–8.

[54] B. Elzey, D. Pollard, S.O. Fakayode, Determination of adulterated neem and flaxseed oil compositions by FTIR spectroscopy and multivariate regression analysis, Food Control 68 (2016) 303–309.

[55] B.F. Ozen, L.J. Mauer, Detection of hazelnut oil adulterated spectroscopy, J. Agric. Food Chem. 50 (2002) 3898–3901.

[56] A. Rohman, Y.B. Che Man, Authentication analysis of cod liver oil from beef fat using fatty acid composition and FTIR spectra, Food Addit. Contam. 28 (11) (2011) 1469–1474.

[57] M. Mauricio-Iglesias, V. Guillard, N. Gontard, S. Peyron, Application of FTIR and Raman microspectroscopy to the study of food/packaging interactions, Food Addit. Contam. 26 (11) (2009) 1515–1523.

[58] Y.S. Al-Degs, M. Al-Ghouti, N. Salem, Determination of frying quality of vegetable oils used for preparing falafel using infrared spectroscopy and multivariate calibration, Food Anal. Methods 4 (4) (2011) 540–549.

[59] J.Y. Chen, H. Zhang, J. Ma, T. Tuchiya, Y. Miao, Determination of the degree of degradation of frying rapeseed oil using fourier-transform infrared spectroscopy combined with partial least-squares regression, Int. J. Anal. Chem. 2015 (2015) 185367.

Chapter 8

Milk analysis

Analysis of foods is especially intricate due to the complex food system, in which coexist not only physicochemical and biological characteristics of nutrient materials but also possible microorganisms, contaminants, and additives. Aligned with legislation and consumer demand, this branch of analytical chemistry is frequently appealing for analytical methodologies that are more robust, efficient, sensitive, time saving, and cost effective to assure the safety, quality, and traceability of foods. In food analysis, vibrational spectroscopy has been applied, with good results, for a great number of dairy products, such as milk [1,2] and milk products [3–6].

Chemically speaking, milk is a natural biological emulsion composed of liquid fat globules heterogeneously dispersed within a water-based solution. Its composition (fat, lactose, casein, whey proteins, minerals, etc.) varies with species, genetic breeding programs, feeding schemes, stages of lactation, and climatic conditions [7]. For routine analysis of the main constituents of milk, the traditional methods (such as Kjeldahl [8] and Rose-Gottlieb [9]) were replaced by spectrophotometric assays [10], in order to obtain better results in a shorter time and save the environmental medium from the reagents used before. The application of IR spectroscopy was evolved into a standard AOAC (Association of Official Analytical Chemists) method [11] for quantitative determination of milk. In line with this trend, vibrational spectroscopy has been numerously exploited for milk analysis, of which some of the most important applications are presented here.

Payment for milk may not be necessarily made on the basis of its protein content. Nevertheless, from a nutritional point of view, milk is a great protein-rich source because it contains two types of high-quality protein: whey (20%) and casein (80%), providing all essential amino acids in amounts adequate to maintain the multiple roles of protein in the body. It makes the determination of protein content become essential in milk analysis. To perform nondestructive measurement of protein in milk powder, an IR spectroscopic method (using mid-IR and near-IR), coupled with least-squares support vector machine (LS-SVM) was proposed [12]. In Fig. 8.1 are presented comparatively the transmission spectra of different samples of milk powder obtained from several suppliers. It was shown that the frequencies and intensities of MIRS bands are richer and stronger than NIRS ones because MIRS covers fundamental bands of NIRS. As compared with MIRS, the main disadvantage of NIRS is related to

Vibrational Spectroscopy Applications in Biomedical, Pharmaceutical and Food Sciences
https://doi.org/10.1016/B978-0-12-818827-9.00008-1
© 2020 Elsevier Inc. All rights reserved.

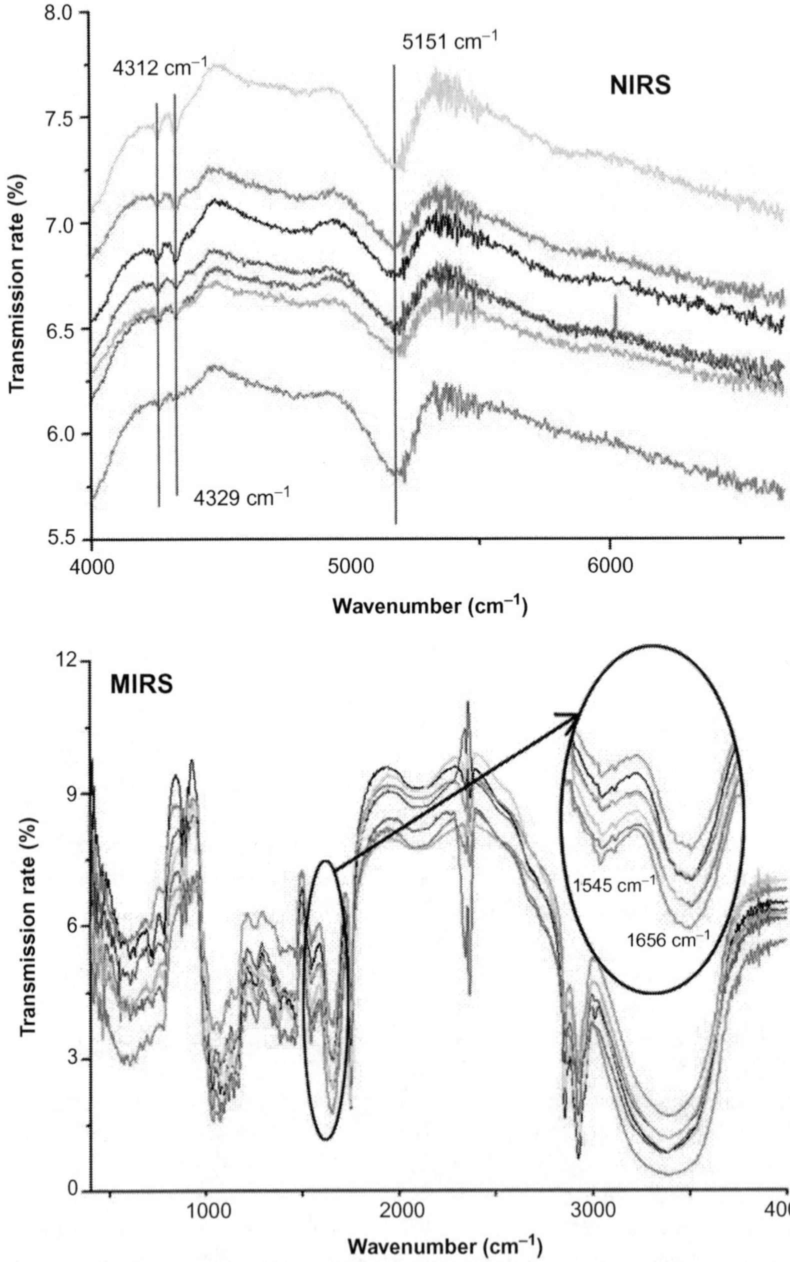

FIG. 8.1 NIRS and MIRS transmission rate of different varieties of milk powder. *(Reproduced with permission from Y. He, D.W. Sun, Study on infrared spectroscopy technique for fast measurement of protein content in milk powder based on LS-SVM, J. Food Eng. 84 (1) (2008) 124–131.)*

not only its low signal sensitivity but also the superposition of various overtone and combination bands in the NIR region.

One of the most important aspects in milk analysis is the quantitative determination of fatty acids. The usual methods for this assay are gas chromatography (GC) [13,14] and Fourier Transform Infrared (FTIR) spectroscopy [15]. Based on a comparison study [16], it was established that although there was no equivalency in the measurements of fatty acid from milk, a similar pattern of milk composition could be obtained by these two methods.

For milk authentication, the combination of different types of spectroscopic techniques is advisable. This point of view was illustrated by a study, in which ewes milk samples were discriminated [17] by using MIR and fluorescence spectroscopies coupled with factorial discriminant analysis. MIR spectra recorded for different types of ewes milk showing bands characteristic of hydrocarbon chains are presented in Fig. 8.2. This study proved that concatenation of MIR and fluorescence data sets is an efficient tool to authenticate ewes milk samples according to their feeding systems (controls, ewes fed scotch bean, and ewes fed soybean), regardless of the lactation stage. In another research, the potential capacity of MIR and NIR spectroscopic data was tested in order to discriminate bulk milk samples according to the feeding system, cow breed, and altitude of the dairy farm [18].

Breast milk is the most important food for a baby and its composition relies upon a mother's metabolism [19]. From existing data sources, human milk contains cholesterol at a level lower than that of cows [20] ($20\,mg/100\,cm^3$) and higher than that of infant formulas [21] (0.93–$5.45\,mg/100\,cm^3$). This component could be regularly regarded as a main cause of atherosclerosis, but its consumption should not be restricted in infants due to multiple positive functions. Taking into account this consideration, an inexpensive and rapid method using ATR-FTIR spectroscopic technique was proposed for evaluating the cholesterol concentration in human milk at different stages of lactation [22]. This study showed no significant differences in cholesterol concentration of the intrasubject samples (early (1–3 months), medium (4–6 months), and late (> 6 months) lactation stages), but a broad range of cholesterol concentrations for the intersubject samples. In contrast, the effect of seasonal variation of bovine milk fat composition on modeling was quantified [23]. It was concluded that the IR spectroscopic predictability of milk fat composition (except for fatty acids with low concentrations) could be improved by using more observations. For fatty acids showing large differences (in level and standard deviation) between summer and winter, it was critical to use a representative sample (including observations collected in various seasons) for unbiased prediction. In Fig. 8.3 are presented the wavenumber regions used for fat determination, with regard to the period of the year.

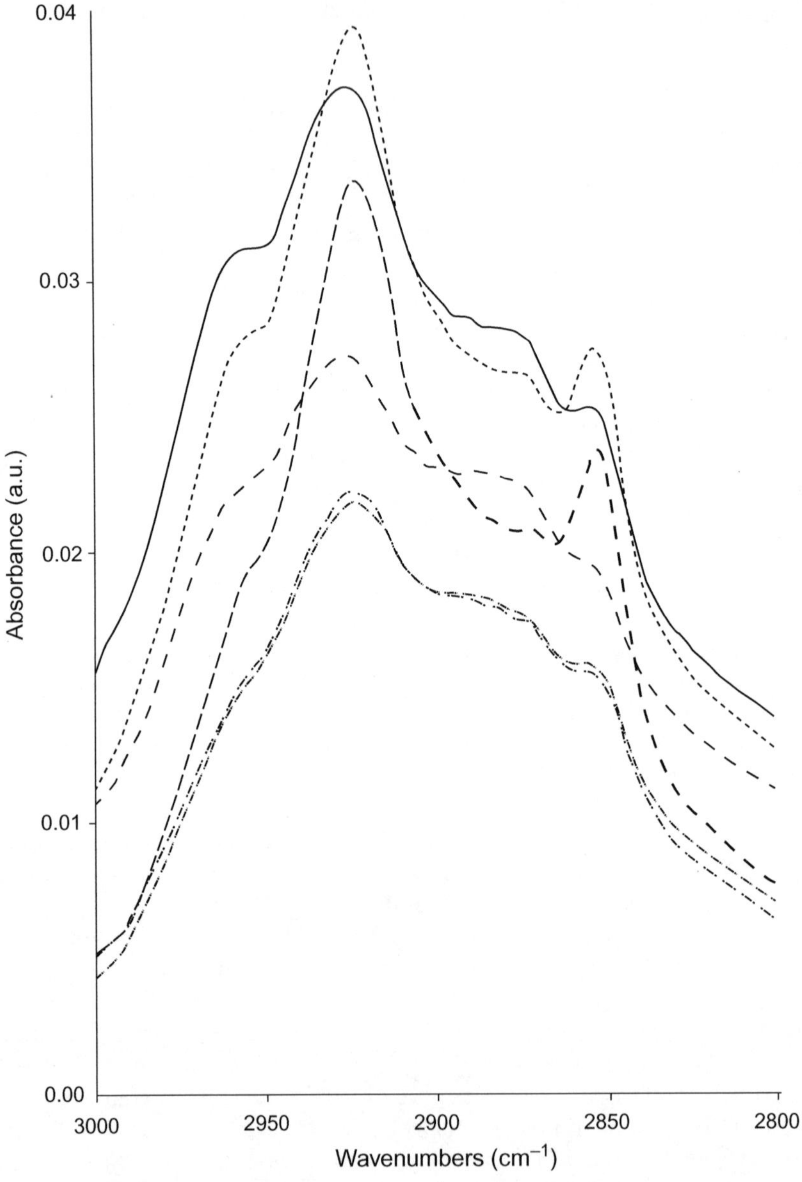

FIG. 8.2 MIR spectra recorded in 3000–2800 cm^{-1} for milk samples of 1-week lactation (control (—), scotch bean (...), and soybean (– – –) groups) and 11-week lactation (control (-..-..), scotch bean (——), and soybean (–.–.–) groups). *(Reproduced with permission from R. Karoui, M. Hammami, H. Rouissi, C. Blecker, Mid infrared and fluorescence spectroscopies coupled with factorial discriminant analysis technique to identify sheep milk from different feeding systems, Food Chem. 127 (2011) 743–748.)*

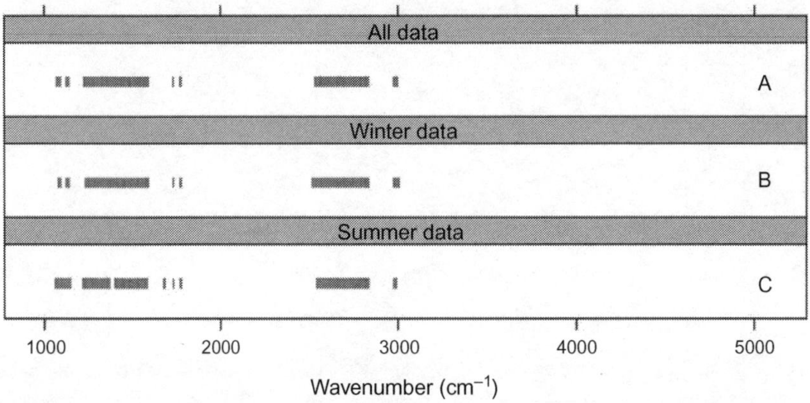

FIG. 8.3 Wavenumbers selected for modeling of milk fat composition for different period of time. *(Reproduced with permission from J.M. Rutten, H. Bovenhuis, K.A. Hettinga, H.J.F. van Valenberg, J.A.M. van Arendonk, Predicting bovine milk fat composition using infrared spectroscopy based on milk samples collected in winter and summer, J. Diary Sci. 92 (12) (2009) 6202–6209.)*

In addition to IR spectroscopy, a rapid determination of milk content could be realized by using Raman spectroscopy coupled with chemometrics. This coupling aims at overcoming a poor signal-to-noise (S/N) ratio inherently characterized by Raman spectra of milk samples. The low S/N ratio of Raman spectra could affect the characteristic of fat quantification, especially for greatly defatted milk samples [24].

Thus, in combination with PLS, fat content in milk was directly determined by using Raman spectra (employing the 514.5-nm emission line of an argon ion laser for excitation) recorded from samples contained in an open Al dish, as presented in Fig. 8.4 [25]. Raman spectral characteristics could be associated with proteins, but mainly milk's fat content (i.e., the highest and lowest spectral intensities correspond to the highest (4.0%) and lowest (0.3%) fat content in the milk samples, respectively). The Raman bands for protein contribution were observed to be much weaker than expected, and this variation could be due to the coexistence of fat and carbohydrates in milk [26].

In another study, FT-Raman spectra obtained in the 4000–200 cm^{-1} wavenumber range when excited by a 1064-nm Nd:YAG laser, demonstrated no annoying fluorescence effects for a wide assortment of infant food formula and powdered milk, as displayed in Fig. 8.5 [27]. This study showed that the calibration data set could be properly selected for corrected Raman spectra of infant formula and powdered milk samples by the use of hierarchical cluster analysis.

Adulteration of milk does not only decrease its quality but also make it even hazardous because most of the chemicals used as adulterants are poisonous. Normally, adulterants can be ranged from 20% to 25% of the milk powder

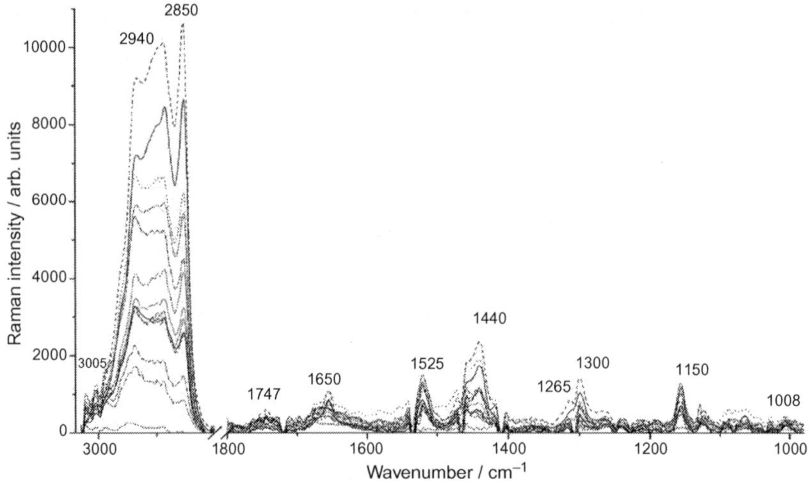

FIG. 8.4 Raman spectra of milk with different fat concentrations in Al dish. *(Reproduced with permission from R.M. El-Abassy, P.J. Eravuchira, P. Donfack, B. von der Kammer, A. Materny, Fast determination of milk fat content using Raman spectroscopy, Vib. Spectrosc. 56 (2011) 3–8.)*

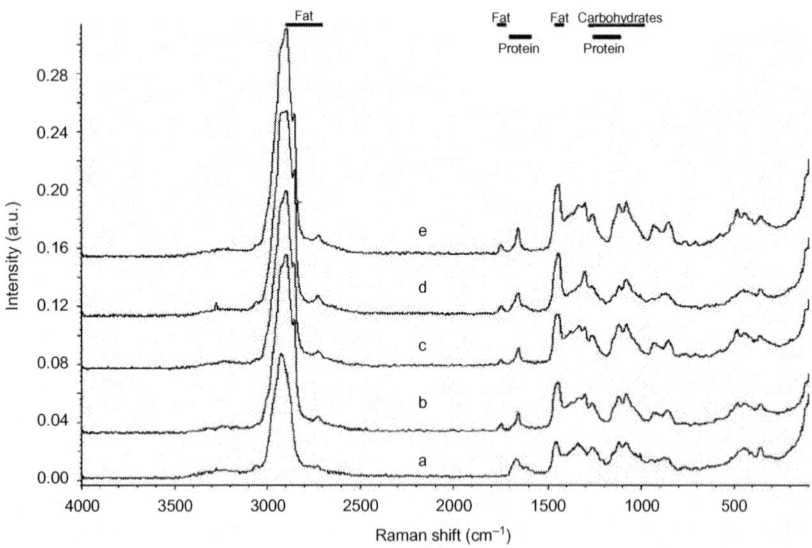

FIG. 8.5 Raw FT-Raman spectra of skimmed powdered milk (a), breast-feeding milk enhanced with iron (b), continuation milk enhanced with minerals and vitamins (c), whole instantaneous powdered milk (d), and breast-feeding powdered milk enhanced with iron (e). Experimental conditions: 700 mW laser power, scanning on back scattering mode (180°) averaging 50 scans per spectrum using a nominal resolution of 4 cm^{-1} without background correction. *Note*: Spectra were shift on the y-axis to clearly show their bands. *(Reproduced with permission from J. Moros, S. Garrigues, M. de la Guardia, Evaluation of nutritional parameters in infant formulas and powdered milk by Raman spectroscopy, Anal. Chim. Acta, 593 (2007) 30–38.)*

content to increase the shelf-life of milk products while avoiding detectable flavor changes; but in most cases, adulteration levels may climb to 60% [28]. To assure adulterant-free milk for consumption, FT-Raman analysis was investigated as a potential method for a fast and reliable quality screening of several kinds of milk powder commercially available in Argentinean, Brazilian, and Uruguayan markets [29]. The classification of adulterated samples (i.e., whole milk powder, skimmed milk powder, low-fat milk powder, and modified milk powder) was absolutely done (100%) by using principal component analysis (PCA) and partial least squares discriminate analysis (PLS-DA). Fig. 8.6 displays the Raman spectra of plain whey powder (composed of lactose, 70% w/w), plain whole milk powder, and milk powder with addition of whey powder (5, 10, and 20% w/w). In the existence of at least 10% (*w*/w) whey powder in milk powders, an increase was clearly seen in the intensities of the Raman bands, specifically 2978 and 2888 cm^{-1}, and in the range 1120–850 cm^{-1} (i.e., changes in the spectral regions that have vibrational modes primarily characteristic of whey).

Melamine (1,3,5-triazine-2,4,6-triamine), a small nitrogen-rich (67% per mass unit) molecule (Fig. 8.7), is usually used in the production of dinnerware sets and durable building materials. This organic compound has been

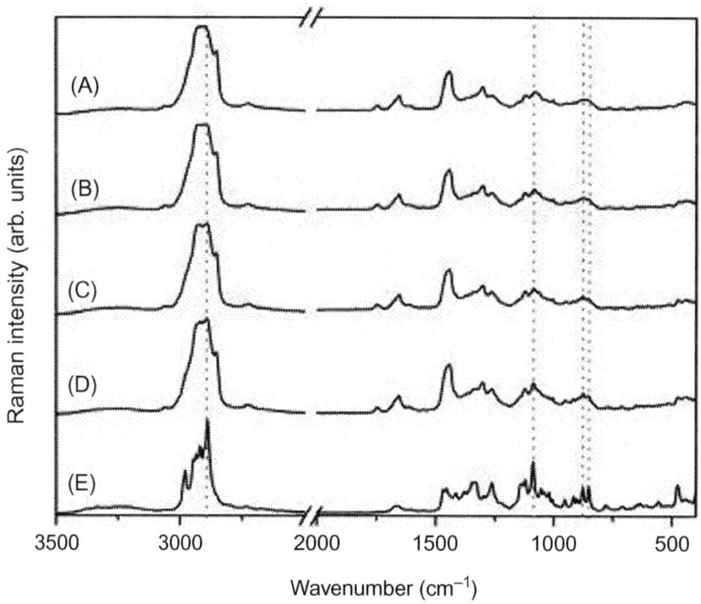

FIG. 8.6 FT-Raman spectra for (A) whole milk powder, whole milk powder with (B) 5% (w/w), (C) 10% (w/w), and (D) 20% (w/w) whey, (E) whey powder. *(Reproduced with permission from M.R. Almeida, K.d.S. Oliveira, R. Stephani, L.F.C. de Oliveira, Fourier-transform Raman analysis of milk powder: a potential method for rapid quality screening, J. Raman Spectrosc. 42 (2011) 1548–1552.)*

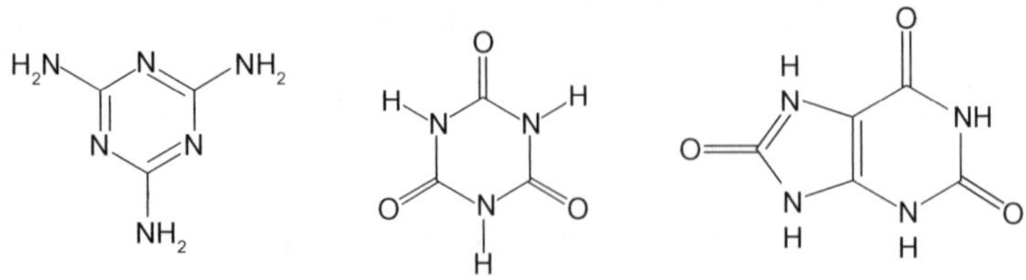

FIG. 8.7 Melamine and its structural analogues. *(Reproduced with permission from C.G. Skinner, J.D. Thomas, J.D. Osterloh, Melamine toxicity, J. Med. Toxicol. 6 (2010) 50–55.)*

also known to be fraudulently added to milk to fool protein content tests, as the conversion of nitrogen content in the samples into protein content constitutes the basis of food protein determination for some of the most frequently used methods. Although melamine is not considered acutely toxic and rapidly eliminated in unmetabolized form in the urine [30], it may lead to the formation of urinary stones in infants when being ingested alone with a large amount or coingested with cyanuric acid (Fig. 8.8) [31].

In 2008, the world biggest milk powder scandal was deplorable in China involving mainly milk infant formula being tainted with melamine. This contamination incident prompted some regulations (e.g., US Food and Drug Administration, European Community) to be approved by allowing tolerable

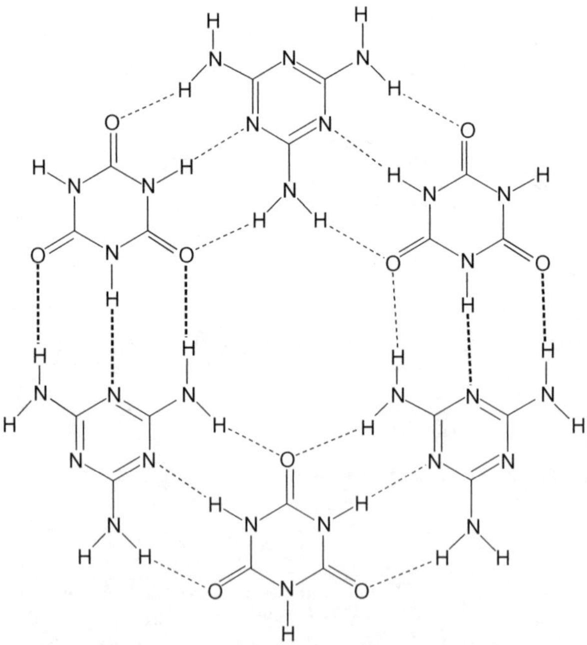

FIG. 8.8 Melamine and acid cyanuric complex structure. *(Reproduced with permission from C.G. Skinner, J.D. Thomas, J.D. Osterloh, Melamine toxicity, J. Med. Toxicol. 6 (2010) 50–55.)*

daily intake (TDI) and Maximum Residue Limits (MRLs) for melamine in various everyday products. This also triggered the quality control departments to get involved in the fabrication of an efficient sensor for detecting small amounts of melamine present in milk samples without any crossreactivity. As an illustration, the determination of melamine in whole milk was studied with a biosensor combining molecularly imprinted polymers and surface-enhanced Raman spectroscopy (MIPs-SERS) [32] (as schematically shown in Fig. 8.9).

During industrial production, a fast identification of excessive additives is vitally important for powdered milk. With regard to this, four NIR imaging methods (i.e., relationship imaging, RI; chemical imaging, CI; principal component analysis, PCA imaging; and classical least square, CLS imaging) were evaluated to acquire chemical information on the spatial distribution and cluster side of two allowed additives (namely, $ZnSO_4$ and lactose) and one banned chemical (melamine) in simulated milk powder (Fig. 8.10) [33]. Clear distribution was observed in the RI image (Fig. 8.11B); wherein the red and green regions represent powdered milk and $ZnSO_4$, respectively. Whereas the distribution image of $ZnSO_4$ could be detected by using RI, the identification of lactose was able by assigning a proper wavenumber region in applying PCA coupled with correlation coefficient imaging approach. Melamine, in this

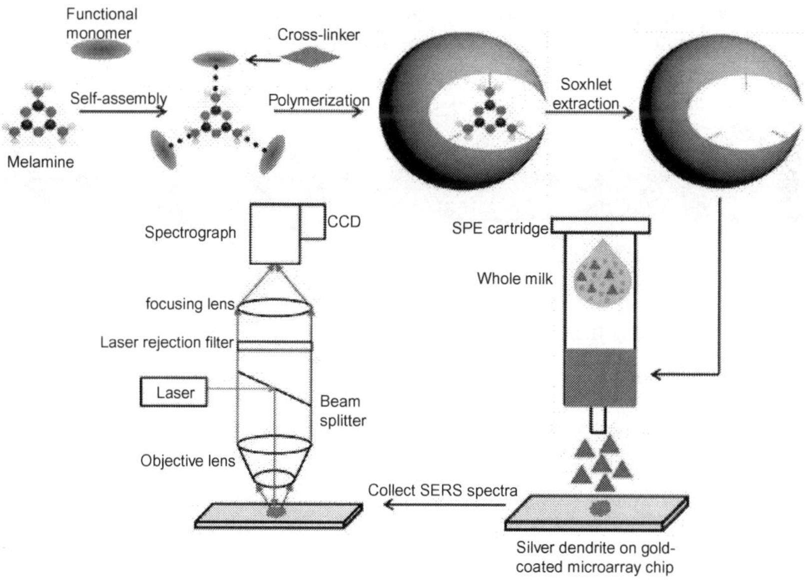

FIG. 8.9 Schematic illustration of MIPs-SERS biosensor for detecting melamine in whole milk. *CCD*, charge-coupled device; *MIPs*, molecularly imprinted polymers; *SERS*, surface enhanced-Raman spectroscopy. *(Reproduced with permission from Y. Hu, S. Feng, F. Gao, E.C.Y. Li-Chan, E. Grant, X. Lu, Detection of melamine in milk using molecularly imprinted polymers–surface enhanced Raman spectroscopy, Food Chem. 176 (2015) 123–129.)*

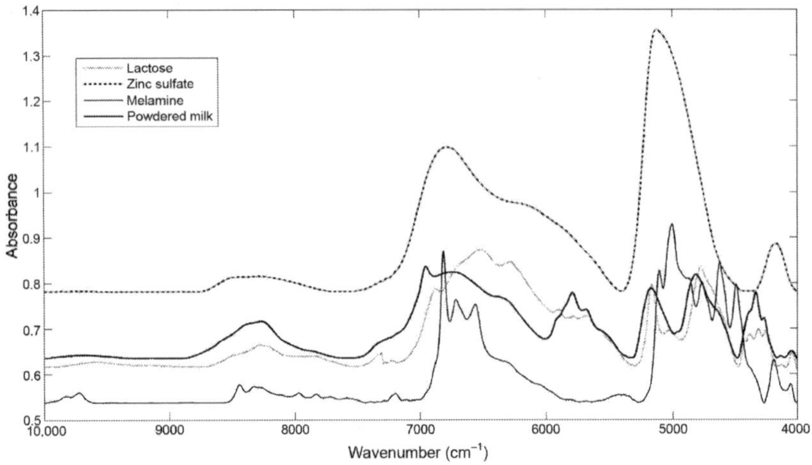

FIG. 8.10 NIR spectra of lactose, $ZnSO_4$, melamine, and powdered milk. *(Reproduced with permission from Y. Huang, S. Min, J. Duan, L. Wu, Q. Li, Identification of additive components in powdered milk by NIR imaging methods, Food Chem. 145 (2014) 278–283.)*

study, was possibly semiquantified in the powdered milk by CLS imaging. In Fig. 8.12, the red speckles and light blue region represent melamine and the matrix, respectively (the left image); the deep blue spots present melamine (the right image); whereas, except for melamine, a high correlation of the other components in the matrix is indicated by a profile of yellow and red spots (the rest of the image). Despite being less sensitive than LC-MS and ELISA [34] in detecting melamine at trace concentration levels, the NIR imaging method may be adopted as a simple and intuitive detection approach for analysis of powdered milk products whose melamine content surpasses its maximum permissible limit.

To quantify common adulterants in powdered milk (such as starch, whey, and sucrose as well as their binary mixtures), the use of least-squares support vector machine (LS-SVM) and NIR spectroscopy with diffuse reflectance measurements was proposed as an alternative methodology [35]. Taking into account the spectral differences of these adulterants, a nonlinear behavior was seen when all groups of adulterants were in the same data set (Fig. 8.13), making the use of linear algorithms such as PLS difficult. LS-SVM, a nonlinear multivariate calibration procedure, demonstrated a better prediction ability than PLS, i.e., the presence of adulterants could be correctly predicted by using LS-SVM, while PLS models might give false-positive results.

Camel milk possesses low cholesterol content, high mineral contents (sodium, potassium, iron, copper, zinc, and magnesium) and high vitamin C levels in comparison with other ruminant milk. It is a good source of nutrients

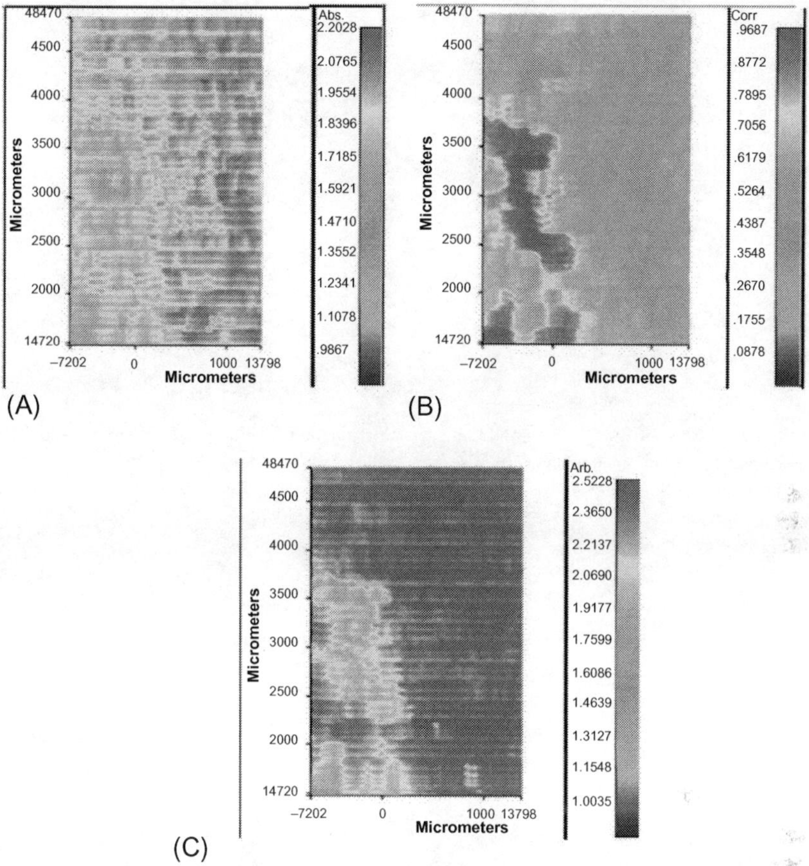

FIG. 8.11 Scanned region of the visible image of the mixed sample of powdered milk and zinc sulfate: (A) total absorbance image; (B) relationship image; (C) chemical image. *(Reproduced with permission from Y. Huang, S. Min, J. Duan, L. Wu, Q. Li, Identification of additive components in powdered milk by NIR imaging methods, Food Chem. 145 (2014) 278–283.)*

for people living especially in the arid and urban areas, but it is rarely available in the market and usually sold at a high price. It was shown that NIR spectroscopy combined with multivariate analysis could be reasonably employed for identification (PLS-DA) and quantification (PLS) of camel milk adulterated with goat milk [36].

In an effort to record high-quality spectra of adulterated liquid milk at the point of collection, reflective focusing wells simply fabricated in aluminum were combined with a portable mini-Raman and a focusing fiber optic probe (as shown in Fig. 8.14) [37]. Spectra were recorded directly from 1-μL milk in hemispherical Al wells, which were specially designed for optimization

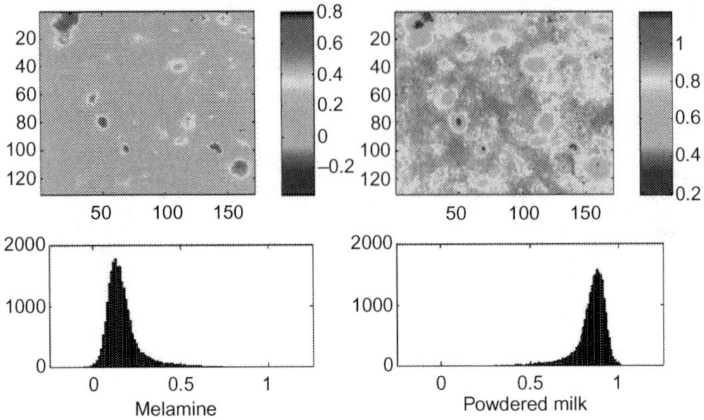

FIG. 8.12 CLS images (after conducting calculation and spatial pixel reconstruction; the color band represents components' relative content) and concentration histograms of melamine and powdered milk. *(Reproduced with permission from Y. Huang, S. Min, J. Duan, L. Wu, Q. Li, Identification of additive components in powdered milk by NIR imaging methods, Food Chem. 145 (2014) 278–283.)*

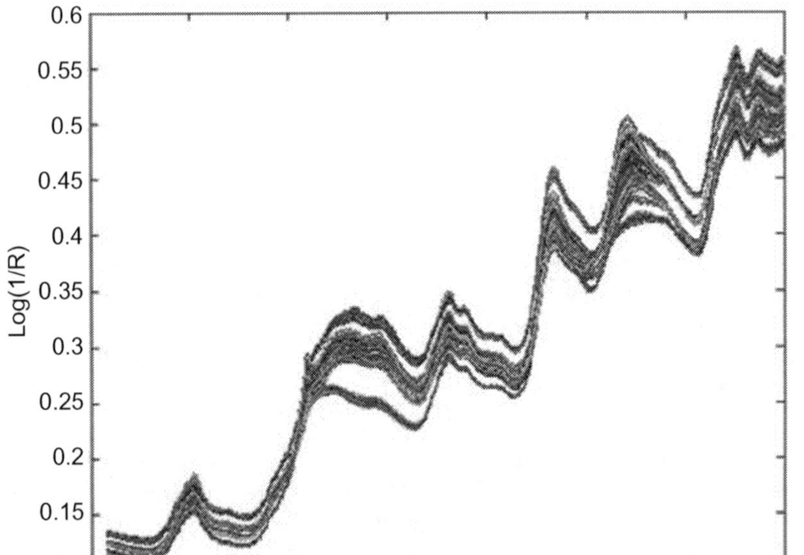

FIG. 8.13 NIR spectra of adulterated powdered milk after being preprocessed with a multiplicative scattered correction (MSC) to correct spectral baseline deviation. *(Reproduced with permission from A. Borin, M.F. Ferrão, C. Mello, D.A. Maretto, R.J. Poppi, Least-squares support vector machines and near infrared spectroscopy for quantification of common adulterants in powdered milk, Anal. Chim. Acta, 579 (2006) 25–32.)*

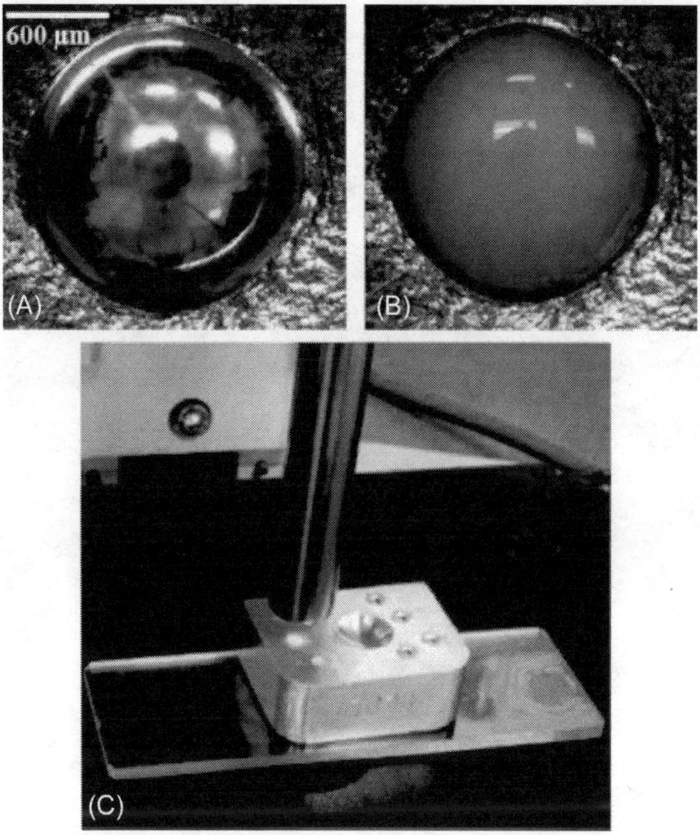

FIG. 8.14 (A) Optical image of a hemispherical shaped well indented in an aluminum block using a 2.38-mm SS316 stainless steel ball bearing to a depth of around 530 µm. (B) Hemispherical aluminum well made by indentation with a 2.38-mm SS316 ball, filled with 1 µL of liquid milk. (C) Fiber optic probe focused on the surface of milk in a hemispherical well. Focal distance was 5.7 mm. *(Reproduced with permission from M.K. Nieuwoudt, S.E. Holroyd, M. McGoverin, M.C. Simpson, D.E. Williams, Rapid, sensitive, and reproducible screening of liquid milk for adulterants using a portable Raman spectrometer and a simple, optimized sample well, J. Dairy Sci. 99 (2016) 7821–7831.)*

of internal reflection and sampling volume by matching the focal length of the mirror to the depth of focus of the laser probe. This method was viable for rapidly screening liquid milk for adulterants such as 4 different nitrogen-rich compounds (melamine, urea, dicyandiamide, and ammonium sulfate as shown in Fig. 8.15) and sucrose. The recorded spectra were reproducible (RSD values of 8% for N-rich compounds and 10% for sucrose) and the limit of detection interval calculated from PLS calibrations for 4 N-rich compounds and sucrose to be in the ranges of 140–520 mg/L and 7000–36,000 mg/L, respectively.

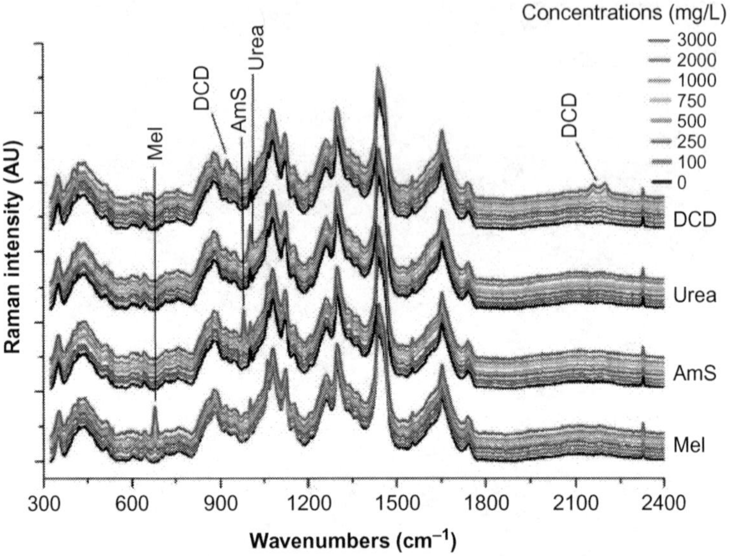

FIG. 8.15 Spectra (arbitrary units, AU) of milk solutions containing different concentrations of melamine (Mel), urea, dicyandiamide (DCD), and ammonium sulfate (AmS). The spectra have been offset for clarity. *Arrows* indicate the strongest bands characteristic for each component. Spectra have been baseline-corrected and normalized to the 1300 cm^{-1} lipid ethyl deformation peak. For each adulterant, spectra are shown in order from the highest to lowest concentration, matching the key. *(Reproduced with permission from M.K. Nieuwoudt, S.E. Holroyd, M. McGoverin, M.C. Simpson and D.E. Williams, Rapid, sensitive, and reproducible screening of liquid milk for adulterants using a portable Raman spectrometer and a simple, optimized sample well, J. Dairy Sci. 99 (2016) 7821–7831.)*

References

[1] B. Aernouts, E. Polshin, W. Saeys, J. Lammertyn, Mid-infrared spectrometry of milk for dairy metabolomics: a comparison of two sampling techniques and effect of homogenization, Anal. Chim. Acta 705 (2011) 88–97.

[2] G. Leitner, Y. Lavi, U. Merin, L. Lemberskiy-Kuzin, G. Katz, Online evaluation of milk quality according to coagulation properties for its optimal distribution for industrial applications, J. Dairy Sci. 94 (2011) 2923–2932.

[3] N. Koca, N.A. Kocaoglu-Vurma, W.J. Harper, L.E. Rodriguez-Saona, Application of temperature-controlled attenuated total reflectance-mid-infrared (ATR-MIR) spectroscopy for rapid estimation of butter adulteration, Food Chem. 121 (2010) 778–782.

[4] M.L. Oca, M.C. Ortiz, L.A. Sarabia, A.E. Gredilla, D. Delgado, Prediction of Zamorano cheese quality by near-infrared spectroscopy assessing false non-compliance and false compliance at minimum permitted limits stated by designation of origin regulations, Talanta 99 (2012) 558–565.

[5] A.P. Pax, L. Ong, J. Vongsvivut, M.J. Tobin, S.E. Kentish, S.L. Gras, The characterisation of mozzarella cheese microstructure using high resolution synchrotron transmission and ATR-FTIR microspectroscopy, Food Chem. 291 (2019) 214–222.

[6] J. Moros, F.A. Iñón, M. Khanmohammadi, S. Garrigues, M. de la Guardia, Evaluation of the application of attenuated total reflectance–Fourier transform infrared spectrometry (ATR–FTIR) and chemometrics to the determination of nutritional parameters of yogurt samples, Anal. Bioanal. Chem. 385 (2006) 708–715.

[7] M.R. Cavalcanti Inacio, M. de Fátima Vitória de Moura, K.M.G. de Lima, Classification and determination of total protein in milk powder using near infrared reflectance spectrometry and the successive projections algorithm for variable selection, Vib. Spectrosc. 57 (2011) 342–345.

[8] J. Kjeldahl, New method for the determination of nitrogen in organic substances, Z. Anal. Chem. 22 (1) (1883) 366–383.

[9] H.D. Richmond, The Röse-Gottlieb method of milk analysis, Lancet 209 (5412) (1927) 1107.

[10] N.K.K. Kamizake, M.M. Goncalves, C.T.B.V. Zaia, D.A.M. Zaia, Determination of total proteins in cow milk powder samples: a comparative study between the Kjeldahl method and spectrophotometric methods, J. Food Compos. Anal. 16 (2003) 507–516.

[11] D.A. Biggs, Performance specifications for infrared milk analysis, J. AOAC Int. 62 (1979) 1211–1214.

[12] Y. He, D.W. Sun, Study on infrared spectroscopy technique for fast measurement of protein content in milk powder based on LS-SVM, J. Food Eng. 84 (1) (2008) 124–131.

[13] M. Collomb, T. Buhler, Analyse de la composition en acides gras de la graisse de lait, Mitteilungen aus Lebensmitteluntersuchung und Hygiene 91 (2000) 306–332.

[14] F. Dorey, D. Brodin, J.F. Le Querler, S. Kuzdzalsavoie, Analyse des acides gras du beurre par chromatographie en phase gazeuse couplée avec la spectrometrie de masse, Ind. Aliment. Agric. 10 (1988) 437–442.

[15] H. Soyeurt, P. Dardenne, F. Dehareng, G. Lognay, D. Veselko, M. Marlier, C. Bertozzi, P. Mayeres, N. Gengler, Estimating fatty acid content in cow milk using mid-infrared spectrometry, J. Dairy Sci. 89 (2006) 3690–3695.

[16] M.A. Petersen Rodriguez, J. Petrini, E.M. Ferreira, L.R.M.B. Mourão, M. Salvian, L.D. Cassoli, A.V. Pires, P.F. Machado, G.B. Mourão, Concordance analysis between estimation methods of milk fatty acid content, Food Chem. 156 (2014) 170–175.

[17] R. Karoui, M. Hammami, H. Rouissi, C. Blecker, Mid infrared and fluorescence spectroscopies coupled with factorial discriminant analysis technique to identify sheep milk from different feeding systems, Food Chem. 127 (2011) 743–748.

[18] B. Valenti, B. Martin, D. Andueza, C. Leroux, C. Labonne, F. Lahalle, H. Larroque, P. Brunschwig, C. Lecomte, M. Brochard, A. Ferlay, Infrared spectroscopic methods for the discrimination of cows' milk according to the feeding system, cow breed and altitude of the dairy farm, Int. Dairy J. 32 (2013) 26–32.

[19] B. Kowalewska-Kantecka, Breastfeeding – the gold standard of infant nutrition the gold standard, Contemp. Pediatr. Gastroenterol. Hepatol. Child Feed. 9 (2007) 65–68.

[20] E. Viturro, H.H. Meyer, C. Gissel, M. Kaske, Rapid method for cholesterol analysis in bovine milk and options for applications, J. Dairy Res. 77 (2010) 85–89.

[21] A.M. Kamelska, R. Pietrzak-Fiećko, K. Bryl, Poznań Univ, Econ. Rev. 196 (2011) 38–45. (in Polish).

[22] A.M. Kamelska, R. Pietrzak-Fiećko, K. Bryl, Determination of cholesterol concentration in human milk samples using attenuated total reflectance Fourier transform infrared spectroscopy, J. Appl. Spectrosc. 80 (2013) 148–152.

[23] J.M. Rutten, H. Bovenhuis, K.A. Hettinga, H.J.F. van Valenberg, J.A.M. van Arendonk, Predicting bovine milk fat composition using infrared spectroscopy based on milk samples collected in winter and summer, J. Diary Sci. 92 (12) (2009) 6202–6209.

[24] S. Mazurek, R. Szostak, T. Czaja, A. Zachwieja, Analysis of milk by FT-Raman spectroscopy, Talanta 138 (2015) 285–289.

[25] R.M. El-Abassy, P.J. Eravuchira, P. Donfack, B. von der Kammer, A. Materny, Fast determination of milk fat content using Raman spectroscopy, Vib. Spectrosc. 56 (2011) 3–8.

[26] P.F. Fox, P.L.H. McSweeney, Dairy Chemistry and Biochemistry, Blackie Academic & Professional, 1998.

[27] J. Moros, S. Garrigues, M. de la Guardia, Evaluation of nutritional parameters in infant formulas and powdered milk by Raman spectroscopy, Anal. Chim. Acta 593 (2007) 30–38.

[28] M.A.M. Salih, Y. Shuming, Common milk adulteration in developing countries cases study in China and Sudan: a review. J. Adv. Diary Res. 5 (2017) 192.

[29] M.R. Almeida, K.d.S. Oliveira, R. Stephani, L.F.C. de Oliveira, Fourier-transform Raman analysis of milk powder: a potential method for rapid quality screening, J. Raman Spectrosc. 42 (2011) 1548–1552.

[30] https://www.who.int/foodsafety/areas_work/chemical-risks/melamine/en/index1.html, (Accessed January 20, 2020).

[31] C.G. Skinner, J.D. Thomas, J.D. Osterloh, Melamine toxicity, J. Med. Toxicol. 6 (2010) 50–55.

[32] Y. Hu, S. Feng, F. Gao, E.C.Y. Li-Chan, E. Grant, X. Lu, Detection of melamine in milk using molecularly imprinted polymers–surface enhanced Raman spectroscopy, Food Chem. 176 (2015) 123–129.

[33] Y. Huang, S. Min, J. Duan, L. Wu, Q. Li, Identification of additive components in powdered milk by NIR imaging methods, Food Chem. 145 (2014) 278–283.

[34] P. Lutter, M.C. Savoy-Perroud, E. Campos-Gimenez, L. Meyer, T. Goldmann, M.C. Bertholet, P. Mottier, A. Desmarchelier, F. Monard, C. Perrin, F. Robert, T. Delatour, Screening and confirmatort methods for the determination of melamine in cow's milk and milk-based powdered infant formula: Validation and proficiency-test of ELISA, HPLC-UV, GC-MS and LC-MS/MS, Food Control 22 (2011) 903–913.

[35] A. Borin, M.F. Ferrão, C. Mello, D.A. Maretto, R.J. Poppi, Least-squares support vector machines and near infrared spectroscopy for quantification of common adulterants in powdered milk, Anal. Chim. Acta 579 (2006) 25–32.

[36] F. Mabood, F. Jabeen, M. Ahmed, J. Hussain, S.A.A. Al Mashaykhi, Z.M.A. Al Rubaiey, S. Farooq, R. Boqué, L. Ali, Z. Hussain, A. Al-Harrasi, A. Latif Khan, Z. Naureen, M. Idrees, S. Manzoor, Development of new NIR-spectroscopy method combined with multivariate analysis for detection of adulteration in camel milk with goat milk, Food Chem. 221 (2017) 746–750.

[37] M.K. Nieuwoudt, S.E. Holroyd, M. McGoverin, M.C. Simpson, D.E. Williams, Rapid, sensitive, and reproducible screening of liquid milk for adulterants using a portable Raman spectrometer and a simple, optimized sample well, J. Dairy Sci. 99 (2016) 7821–7831.

Chapter 9

Alcoholic drink analysis

In a modern society, food quality and safety are always consumers' expectations and concerns. It has been globally evidenced by an interconnected system for the production and distribution of food, and proliferation of enormously higher levels of food standards in the 2000s [1]. A food product as a whole is not simply the sum of its nutrients. In the last decade, the ability of IR spectroscopy was experimentally assessed for food and beverage composition in such a complex matrix [2] as well as for control of food spoilage and pathogenic microorganisms [3].

An alcoholic beverage (or alcoholic drink) is defined as any liquor (such as wine, beer, or distilled spirit) fermented from the sugars in fruits, berries, grains, and such other ingredients (e.g., plant saps, tubers, honey, and milk) [4]. Wine is one of the most consumed beverages per capita, which is composed mainly of water (70–90%), ethanol, sugars, glycerol and polysaccharides, acids, and volatile compounds. NIR spectral data were first recorded by Kaffka and Noms, who prepared the samples by adding some main components of interest (ethanol, fructose, tartaric acid) to a red or white wine [5]. During the 1980s, NIR continued to be the only vibrational spectroscopic technique applied to wine analysis [6]. At that time, the measurement of ethanol was based on the use of infrared filter instruments with two or three wavelengths [7]. Wine composition was later analyzed by a fully automated sequential injection system with FTIR detection [8].

Although FTIR spectroscopy offered good precision and accuracy for a great number of wine quality parameters (e.g., ethanol, total amount of acidic substances, total amount of sugars in sweet wines, sulfate), spectral interferences could exist in the analysis of compounds (such as volatile acids and sugars in dry wines) having strong IR absorption bands differed insignificantly from other abundant compounds such as ethanol and water [9]. To diminish this problem, it is advisable to minimize intersample variation by performing specific calibrations with different sample sets grouped by wine types (red/white or dry/sweet).

It is noted that dissimilarities in IR calibration data are mainly related to cuvette pathlength or penetration depth into a sample, not the measurement technique itself. Using an identical sample set of 166 international wines, Friedel and coworkers [10] showed that results for different FTIR instruments (single bounce attenuated total reflection (SB-ATR), variable- and defined-pathlength transmission) depend strongly on the parameters analyzed, i.e., for substances in low concentration (e.g., organic acids), a variable-pathlength transmission instrument yielded better results than an SB-ATR instrument; for relative density, ethanol and sugars, an SB-ATR instrument offered comparable to or better results than

Vibrational Spectroscopy Applications in Biomedical, Pharmaceutical and Food Sciences
https://doi.org/10.1016/B978-0-12-818827-9.00011-1
© 2020 Elsevier Inc. All rights reserved.

long-pathlength transmission instruments; whereas a defined-pathlength transmission instrument was well suited for analysis of all parameters under study.

With red wine, tannins are a crucial descriptor for tasting as they basically refer to the dryness, bitterness, and astringency. FT-MIR spectroscopy combined with chemometric techniques [multivariate PLS regression and spectral interval selection procedures (interval PLS and changeable size moving window PLS)] could allow an accurate determination of tannin concentration and the average molecular weight of tannin components in Chilean red wines [11].

In another study, the prediction of MIR spectra was further improved when being combined with visible spectra for quantification of total polyphenol compounds, *o*-coumaric acid, glycerol and glycerol/ethanol ratio, especially anthocyanin compounds, in monovarietal red and white wines with the aid of orthogonal PLS regression technique [12]. Coupling FT-MIR with PLS regression also manifested its feasibility in routinely screening for 12 anthocyanins (3-*O*-glucosides of delphinidin, cyanidin, petunidin, peonidin, and malvidin, as well as acetic acid esters and *p*-coumaric acid esters of petunidin, peonidin, and malvidin and caffeic acid ester of malvidin) and three sums (sum of nonacylated anthocyanins, sum of acetylated anthocyanins, and sum of coumaroylated anthocyanins), in red wines differently matured using a sample set inclusive of wines from the Protected Designation of Origin Rioja (Spain) [13]. This method was judged to be suitable only for a quick determination of anthocyanins and their sums in young wines newly bottled, i.e., much less than 1 year near to the date of analysis (standard error of prediction of 15–30%). Fig. 9.1 displays FT-IR spectra of wine samples acquired in the region 5012–926 cm^{-1}; some specific regions were eliminated from analysis because they contained strong water absorption bands and very little useful information.

Chinese rice wine or Mijiu is a traditional wine brewed from glutinous rice and wheat. Originally produced in Shaoxing, a city in Eastern China's Zhejiang province, it is well known for sweet taste, low alcoholicity, and rich nutrients (amino acids, proteins, oligosaccharides, vitamins, and mineral elements), especially for beneficial effects on the prevention of cancer and cardiovascular disease that may be related to its antioxidant properties [14]. Raman or ATR-IR spectroscopy and the combination of these two techniques together with chemometrics (i.e., synergy interval PLS and support vector machine) demonstrated a great potential in the concurrent prediction of total antioxidant capacity and total phenolic content of Chinese rice wine [15]. Fig. 9.2 displays the average ATR-IR and Raman spectra of 111 commercial rice wine samples collected from the 5 most well-known rice wine wineries, i.e., Shikumen, Hewine, Xianheng, Pagoda, and Minzuhong.

In order to obtain a high-quality wine, it is necessary to monitor the concentration of organic acids during the winemaking process. A simple distinction can be made between acids originally found in grapes (i.e., tartaric, malic, and citric acids) and those fundamentally produced in the fermentation processes (such as succinic, lactic, and acetic acids). So far, the analysis of these substances has been really considered as challenges to the application of vibrational

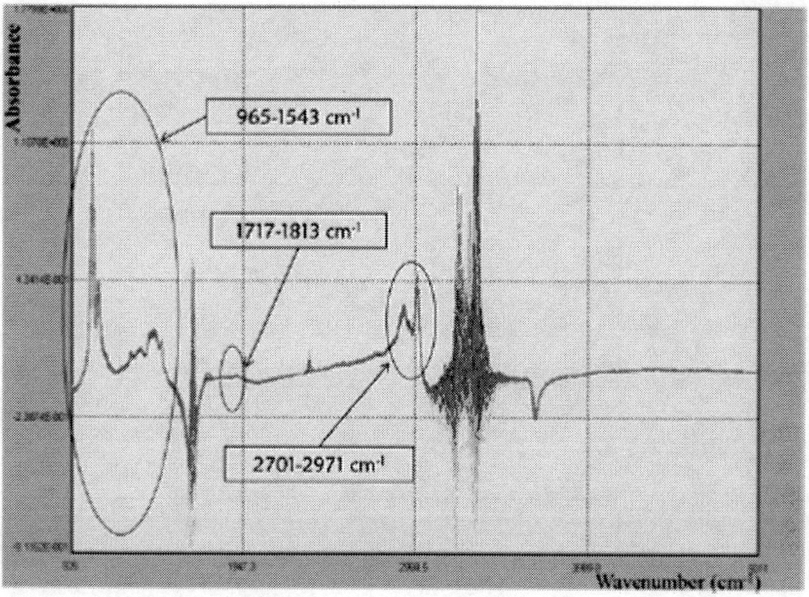

FIG. 9.1 Spectral ranges used to select the useful frequencies for wine analysis. *(Reproduced with permission from M. Romera-Fernández, L.A. Berrueta, S. Garmón-Lobato, B. Gallo, F. Vicente, J.M. Moreda, Feasibility study of FT-MIR spectroscopy and PLS-R for the fast determination of anthocyanins in wine, Talanta 88 (2012) 303–310.)*

spectroscopic techniques [16]. For instance, Regmin and coworkers showed that direct determination of organic acids (tartaric, malic, lactic, succinic, citric, and acetic) in wine and wine-derived products (vinegars and spirits) could be realized by using FTIR and chemometric technique (PLS) [17]. Although a strong correlation with reference values (acquired by Ion Exclusion Chromatography with conductimetric detection) was demonstrated for high levels of concentration ($>0.6\,g\,L^{-1}$) in all acids, poor results were obtained for low levels of concentration ($<0.6\,g\,L^{-1}$) (Fig. 9.3) proving that the calibration and validation of FTIR data highly depended on not only the composition of the sample set but also the predictive ability of the reference method.

Element analysis in wine plays a vital role in determining legal limits as well as inspecting potential changes in color, oxidation, and health benefits. Cozzolino and coworkers proposed the use of VIS-NIR spectroscopy as simple measurement technique for a range of practically relevant elements in Australian wines [18]. All samples were analyzed for elements (Ca, K, Mg, P, Na, S, Fe, B, Mn) by inductively coupled plasma mass spectrometry (ICP-MS). Calibration models between elements and VIS-NIR spectra (pretreated by second-derivative transformation and standard normal variate) were constructed by using PLS regressions with full cross-validation. As shown in Fig. 9.4, loadings in the NIR region between 1600 and 1700 nm (associated with O–H and C–H tones, e.g., sugars,

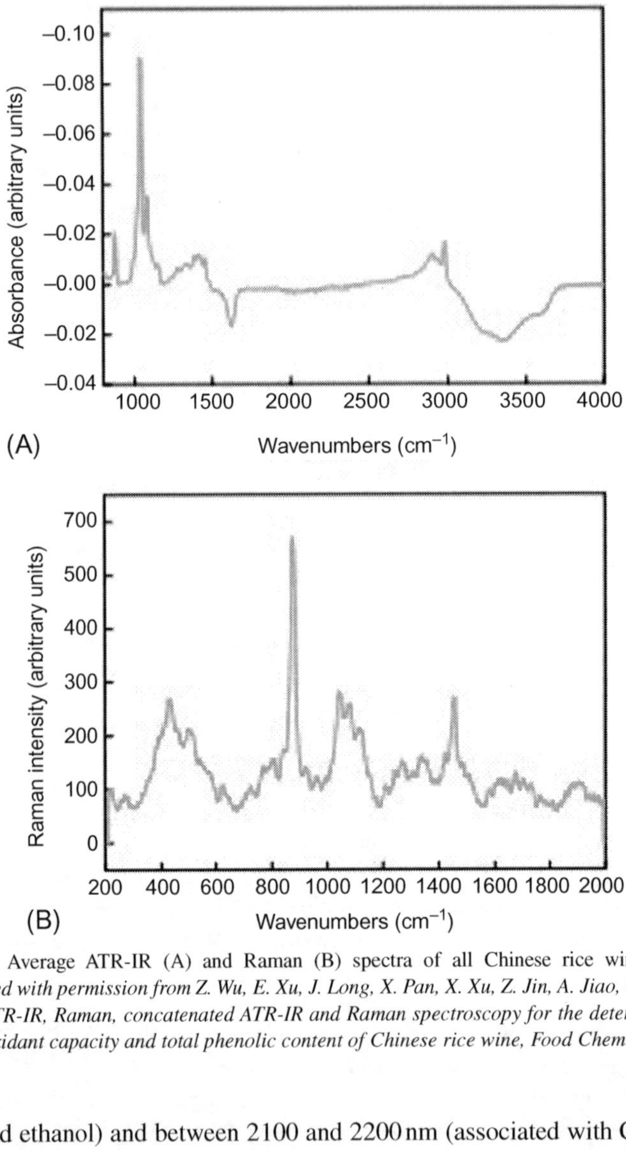

FIG. 9.2 Average ATR-IR (A) and Raman (B) spectra of all Chinese rice wine samples. *(Reproduced with permission from Z. Wu, E. Xu, J. Long, X. Pan, X. Xu, Z. Jin, A. Jiao, Comparison between ATR-IR, Raman, concatenated ATR-IR and Raman spectroscopy for the determination of total antioxidant capacity and total phenolic content of Chinese rice wine, Food Chem. 194 (2016) 671–679.)*

water, and ethanol) and between 2100 and 2200 nm (associated with C—H combinations tones) had a large influence on the different calibration models, in particular for Ca and Fe, although no distinct NIR region dominated. Inverse loadings were seen mainly between 2100 and 2200 nm (C—H combination tones) for Ca and K, as compared with Fe. In contrast, for other elements it was difficult to assign to any specific bond as the loadings were spread along the NIR region.

Together with polyphenols and proteins, polysaccharides are one of the major groups of macromolecules in wines. They are released from either grape cell walls after pectic enzyme-catalyzed degradation (during berry maturation and vinification treatments) or yeast walls (during alcoholic fermentation) [19].

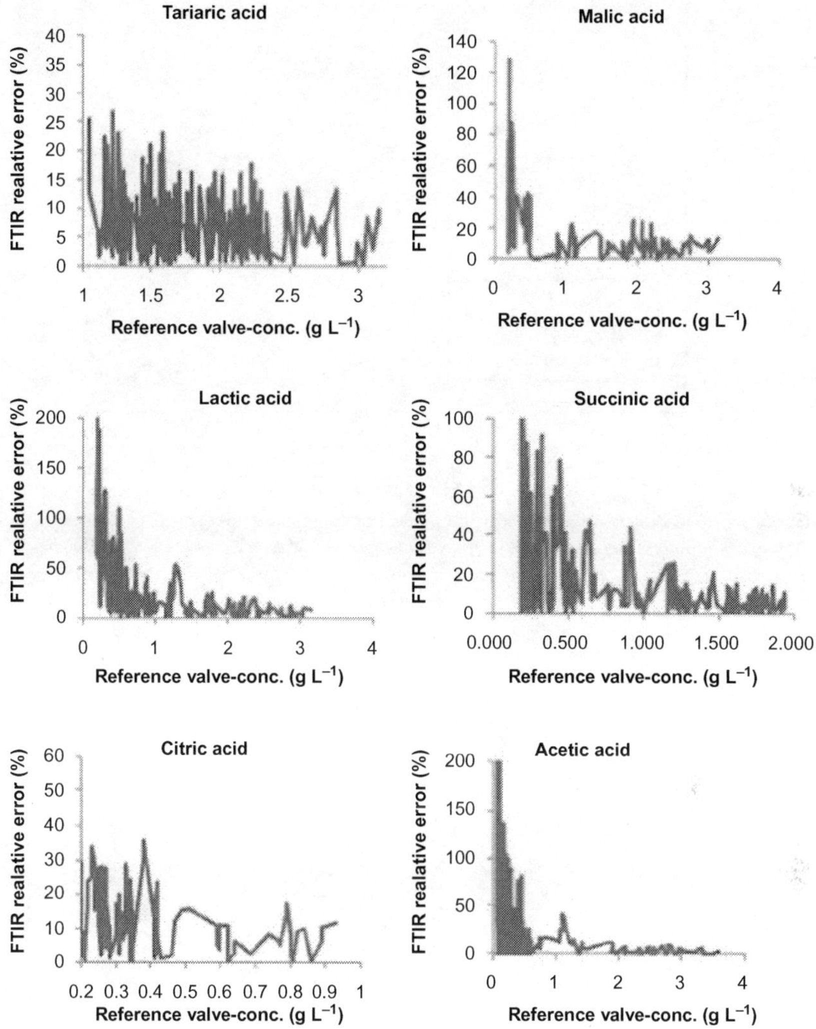

FIG. 9.3 Validation result (FT-IR relative error vs. reference value). *(Reproduced with permission from U. Regmi, M. Palma, C.G. Barroso, Direct determination of organic acids in wine and wine-derived products by Fourier transform infrared (FT-IR) spectroscopy and chemometric techniques, Anal. Chim. Acta 732 (2012) 137–144.)*

In the literature, MIR spectra were reported for the main polysaccharides in wines from the degradation of the grape berry cell walls (i.e., rhamnogalacturonans (RG-Is, RG-IIs) and polysaccharide rich in arabinose and galactose) and from yeast walls (i.e., mannoproteins) [20]. In Fig. 9.5, a greater absorbance between 1200 and 1800 cm^{-1} was assigned to the presence of more or less esterified uronic acids. Although each polysaccharide family had its own characteristic profile, there existed non-negligible intrafamily variability and a strong colinearity between polysaccharide reference spectra. It is thus quite difficult

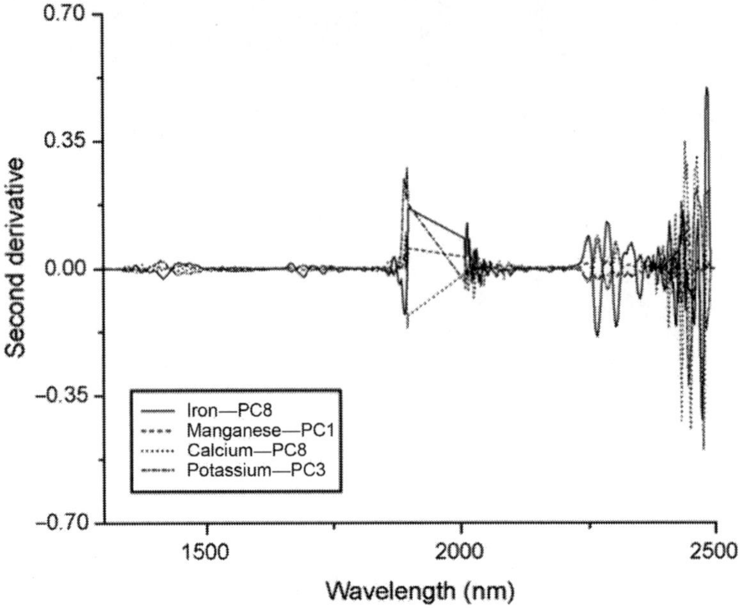

FIG. 9.4 PLS loadings for optimal near-infrared calibration models for elements in wine (1400–2500 nm). *(Reproduced with permission from D. Cozzolino, M.J. Kwiatkowski, R.G. Dambergs, W.U. Cynkar, L.J. Janik, G. Skouroumounis, M. Gishen, Analysis of elements in wine using near infrared spectroscopy and partial least squares regression, Talanta 74 (2008) 711–716.)*

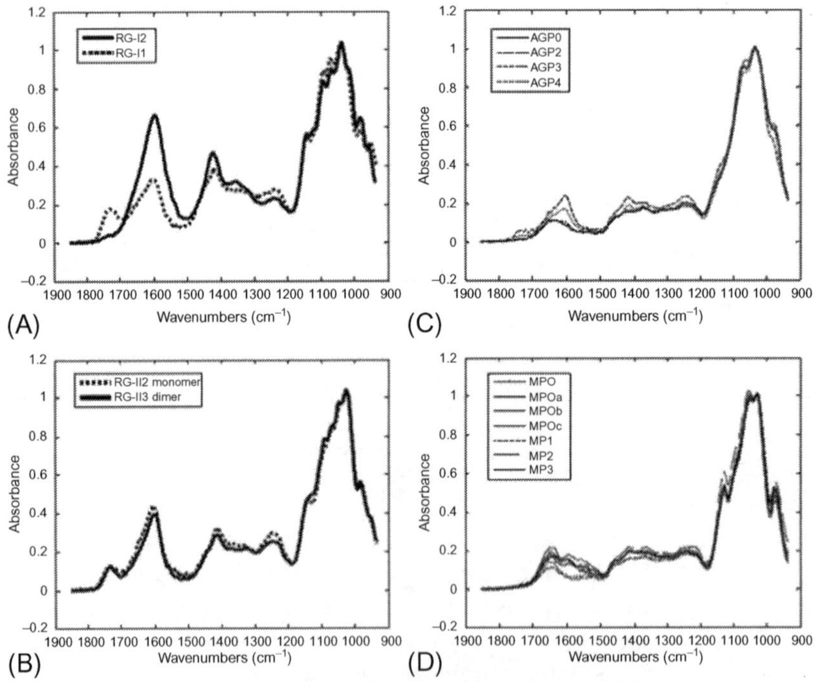

FIG. 9.5 FTIR spectra of some polysaccharide in red wine [spectra of RG-Is (A), RG-II monomer and dimer (B), AGPs (arabinogalactan proteins) (C), and MPs (mannoproteins) (D), extracted from a red wine]. *(Reproduced with permission from J.C. Boulet, P. Williams, T. Doco, A Fourier transform infrared spectroscopy study of wine polysaccharides, Carbohydr. Polym. 69 (2007) 79–85.)*

to directly quantify these polysaccharides that display great structural diversity and are often found in the concentration range of 0.2–2.0 $g L^{-1}$ in wines.

The determination of alcoholic content is tedious and time-consuming because it is principally based on either volumetric measurement of distillate sample [21, 22] or redox-based titrimetric and spectrophotometric methods [23]. To determine ethanol in all types of alcoholic beverages, a simple strategy was proposed by employing FTIR detection [24]. This methodological proposal was effortlessly adaptable to any FTIR spectrometer equipped with a conventional and standard transmission IR cell. It included two steps: (i) online liquid-liquid ethanolic extraction with chloroform through a sandwich-type cell equipped with a polytetrafluoroethylene (PTFE) membrane, using a double channel manifold; and (ii) direct quantification of ethanol by measuring FTIR signal at 877 cm^{-1} with baseline correction established between 844 and 929 cm^{-1}. In the spectral region 1200–800 cm^{-1} chloroform presents a wide transparency interval, while ethanol shows an intense band at 1048 cm^{-1} and two other bands of medium intensity centered at 1086 and 877 cm^{-1} due to vibrational transitions of the –C–O–H system, C–O stretching vibration (p-OH), and –O–H bending vibration out of plane. In a comparative manner, the FTIR spectra obtained from real samples were displayed in Figs. 9.6 and 9.7. Except for simple degassing of beer and bubby wine samples or dilution of spirits with water, this procedure did not require any sample pretreatment and gave results highly comparable to those obtained by head space-gas chromatography with flame-ionization detection (HS-GC-FID).

In the context of screening alcoholic beverages (e.g., spirit drinks and beer), FTIR spectroscopy in combination with PLS regression experimentally showed superiority over conventional methods as an efficient high-throughput tool (2 min per sample) [25]. For doing so, a task-specific FTIR interferometer was used and equipped with an injection nozzle plunged directly into the sample and an automatic thermostat. Before diving into any kind of data analysis, FTIR spectra were visually checked for abnormalcy (maybe due to incorrect sampling with air bubbles as illustrated in Fig. 9.8). PLS-processed FTIR spectra were found to be well correlated with reference values for parameters (such as density, ethanol, methanol, ethyl acetate, propanol-1, isobutanol and 2-/3-methyl-1-butanol, original gravity, and lactic acid). Besides, a differentiation was possible for deteriorated fruit spirits distilled from microbiologically spoiled mashes when applying PCA to spectral data.

In an effort to develop equations independent of the type of wine, the applicability of both NIRS and FT-MIRS techniques was investigated for multidetermination of several enological parameters (i.e., alcoholic degree, volumic mass, total acidity, glycerol, total polyphenol index, lactic acid, and total sulfur dioxide) in Spanish wines (red, rosé, and white; young and aged) from different appellations d'origine and grape varieties [26]. Data showed that NIRS was better than FT-MIR in terms of signal/noise ratio; nonetheless, the combination of these two regions could improve the determination of glycerol and total sulfur dioxide because it generated a larger number of variables (wavelengths) than when separately used, helping surpass the noise

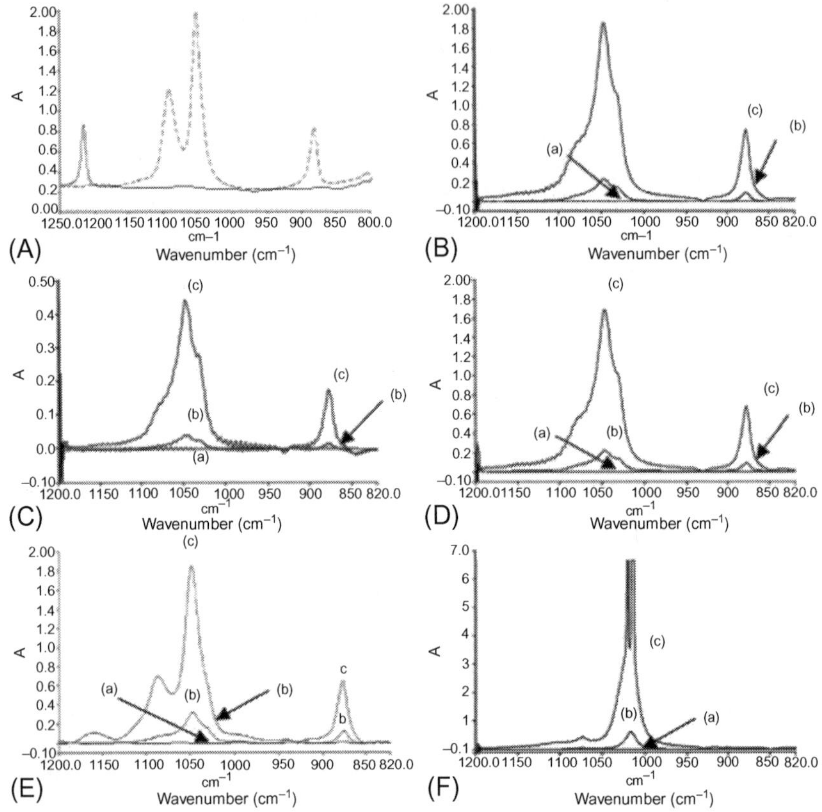

FIG. 9.6 FTIR spectra of pure chloroform, absolute ethanol, and ethanol standard solutions. (A) Spectra obtained in film corresponding to: (—) pure chloroform, (---) absolute ethanol. (B) Spectra obtained in batch from solutions of ethanol in chloroform. (C) Spectra (in batch) corresponding to the organic extract obtained by offline extraction (1:1) of ethanol with chloroform from aqueous solutions of ethanol. (D) Spectra obtained from aqueous standards of ethanol in the proposed system (injection volume, V_i=1.0 mL). (E) Spectra obtained from aqueous standards of ethanol in the proposed system (pathlength (b)=0.25 mm, [Ethanol]: (a) 0% (v/v) (DI), (b) 2% (v/v), and (c) 20% (v/v); V_i=1 mL). (F) Spectra obtained from aqueous standards of methanol in the proposed system (V_i=1.0 mL). For (B), (C), (D), and (F): b=0.5 mm, [Alcohol]=(a) 0% (v/v), (b) 1% (v/v), and (c) 10% (v/v). In all cases, the spectra were obtained by accumulating three scans and using pure and dried chloroform as reference. *(Reproduced with permission from M. Gallignani, C. Ayala, M. del Rosario Brunetto, J.L. Burguera, M. Burguera, A simple strategy for determining ethanol in all types of alcoholic beverages based on its on-line liquid–liquid extraction with chloroform, using a flow injection system and Fourier transform infrared spectrometric detection in the mid-IR, Talanta 68(2) (2005) 470–479.)*

of MIR spectra. Fig. 9.9 presents the FTIR spectra in the whole measurement range (400–2500 nm) (inclusive of MIR region used, 800–3500 cm^{-1}); but the zones, 400–800 and 3000–4000 cm^{-1}, were not used by reason of the high irreproducibility related to high absorbance values. In another study, a slightly better predictive performance was also gained for NIR-MIR combined spectra for determination of real and original extract, and ethanol content in beers (Fig. 9.10) [27].

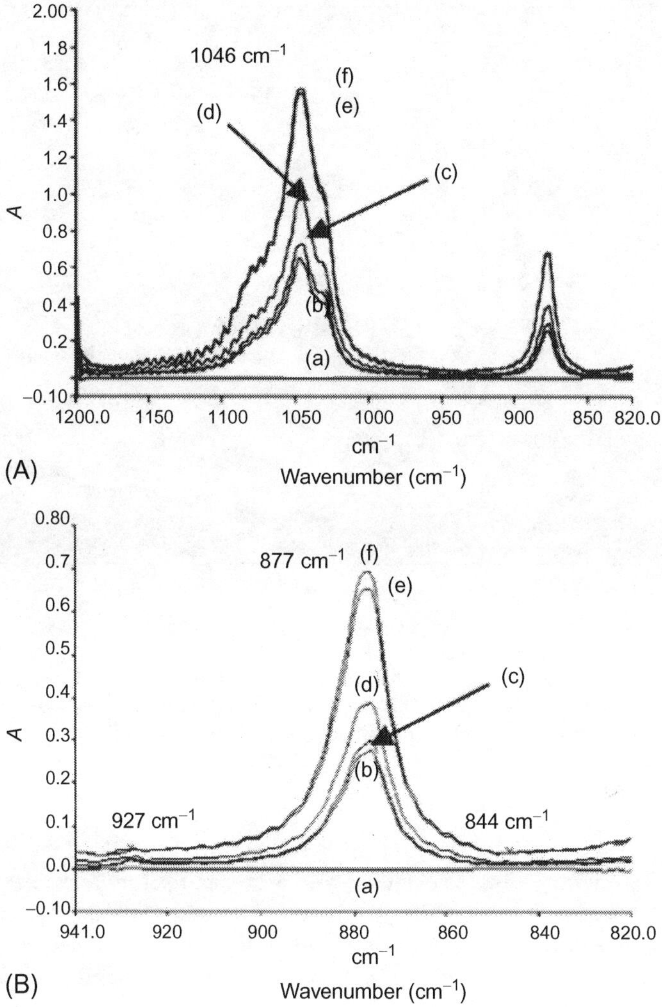

FIG. 9.7 FTIR spectra of different commercial alcoholic beverages. (A) Amplification of the $1046\,cm^{-1}$ band. (B) Amplification of the $877\,cm^{-1}$ band. Samples: (a) DI (double deionized water of $18\,M\Omega\,cm^{-1}$ specific resistivity), (b) beer (4.5%, v/v), (c) aqueous standard of ethanol (5.0%, v/v), (d) red wine (12.5%, v/v) diluted 50% (v/v) with DI, (e) white rhum (40%, v/v) diluted 25% (v/v) with DI, (f) whisky (43%, v/v) diluted 25% (v/v) with DI. *(Reproduced with permission from M. Gallignani, C. Ayala, M. del Rosario Brunetto, J.L. Burguera, M. Burguera, A simple strategy for determining ethanol in all types of alcoholic beverages based on its on-line liquid–liquid extraction with chloroform, using a flow injection system and Fourier transform infrared spectrometric detection in the mid-IR, Talanta 68(2) (2005) 470–479.)*

In a comparative study, the competency of different vibrational techniques (Raman, NIR, and MIR spectroscopy) was rated for routine analysis of several parameters often considered as indicators of wine quality [28]. PLS regression was used to construct calibration models correlating between reference results (determined according to the Compendium of International Methods of Wine

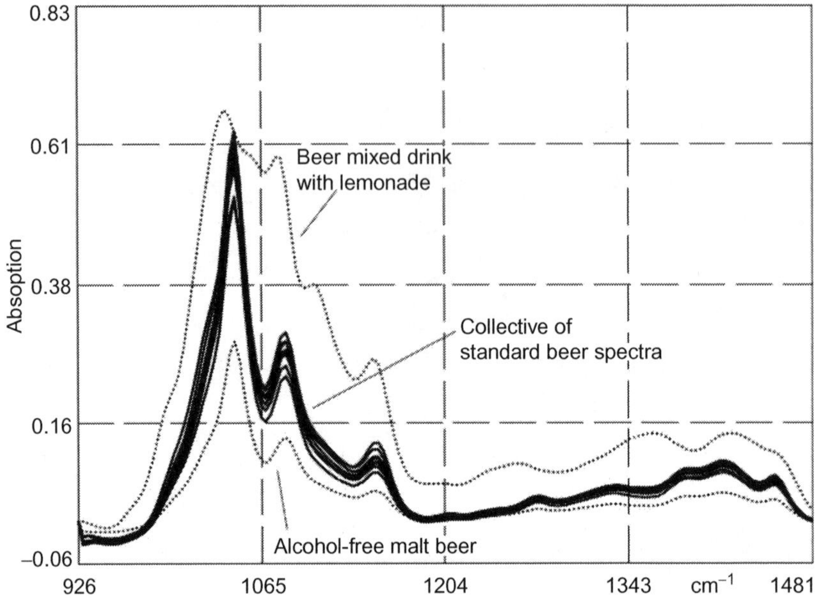

FIG. 9.8 Spectral approach for outlier detection. The FTIR spectra of beer samples in the characteristic range 926–1481 cm^{-1} clearly include two abnormal spectra, which could be identified as a beer mixed with lemonade and an alcohol-free beer erroneously integrated into the sample set. *(Reproduced with permission from D.W. Lachenmeier, Rapid quality control of spirit drinks and beer using multivariate data analysis of Fourier transform infrared spectra, Food Chem. 101 (2007) 825–832.)*

and Must Analysis [29]) and spectral information about alcoholic strength, density, total acidity, volatile acidity, total sugars, and pH in Portuguese white wines. It was revealed that MIR spectroscopy offered a better performance for the measurement of alcoholic strength and total acidity; Raman spectroscopy was the most suitable for the determination of total sugars; similar results were obtained by all techniques for density; either technique was qualified for the accurate determination of volatile acidity.

In another development, beer quality screening by FTIR spectrometry was intensively scrutinized in the matter of measurement strategies (i.e., horizontal- and microattenuated total reflection, HATR and μATR; high-throughput transmission (HTT), as schematically shown in Fig. 9.11) [30]. It was found that the prediction performance of PLS models was nearly the same for both HATR and μATR, except for apparent and real extracts, apparent and real degrees of fermentations of 50 Belgian and Dutch beers (dark and blonde ales, lager beers, white (wheat), lambic fruit and trappist beers). The evidence that 3-reflections μATR accessory slightly outperformed 9-reflections HATR, was possibly related to a higher reproducibility of the signal and a lower cross-contamination of the crystal's surface by samples and their replicas. In contrast, good results were achieved by HTT fast measurements

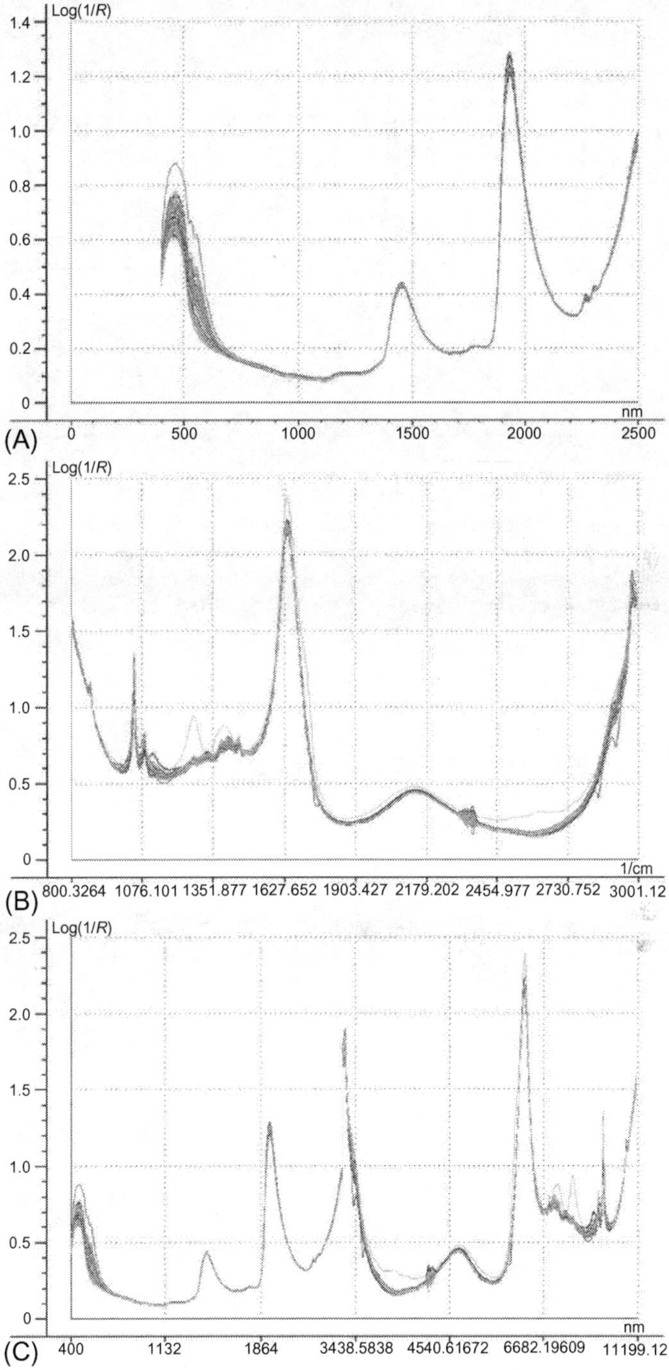

FIG. 9.9 (A) NIR spectra, (B) MIR spectra, and (C) combination of NIR and MIR spectra of different wines. *(Reproduced with permission from F.A. Iñón, S. Garrigues, M. de la Guardia, Combination of mid- and near-infrared spectroscopy for the determination of the quality properties of beers, Anal. Chim. Acta 571 (2008) 167–174.)*

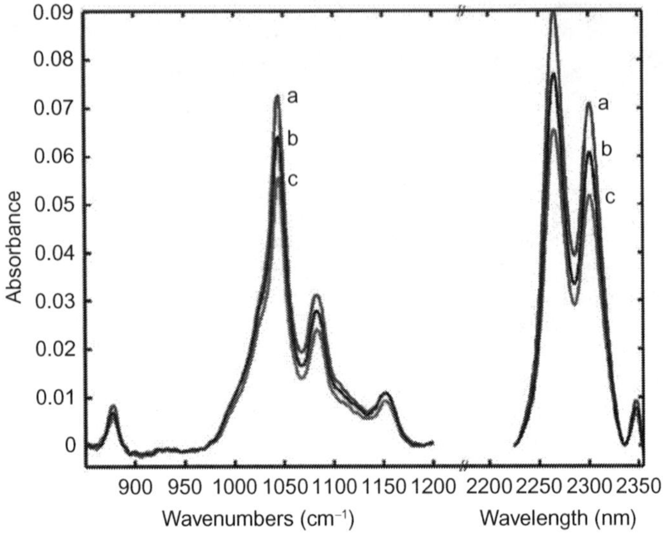

FIG. 9.10 Combined NIR-MIR spectra of main types of beer samples. (a) 100% malt beer; (b) normal beer; and (c) beer without alcohol. *(Reproduced with permission from F.A. Iñón, S. Garrigues, M. de la Guardia, Combination of mid- and near-infrared spectroscopy for the determination of the quality properties of beers, Anal. Chim. Acta 571 (2008) 167–174.)*

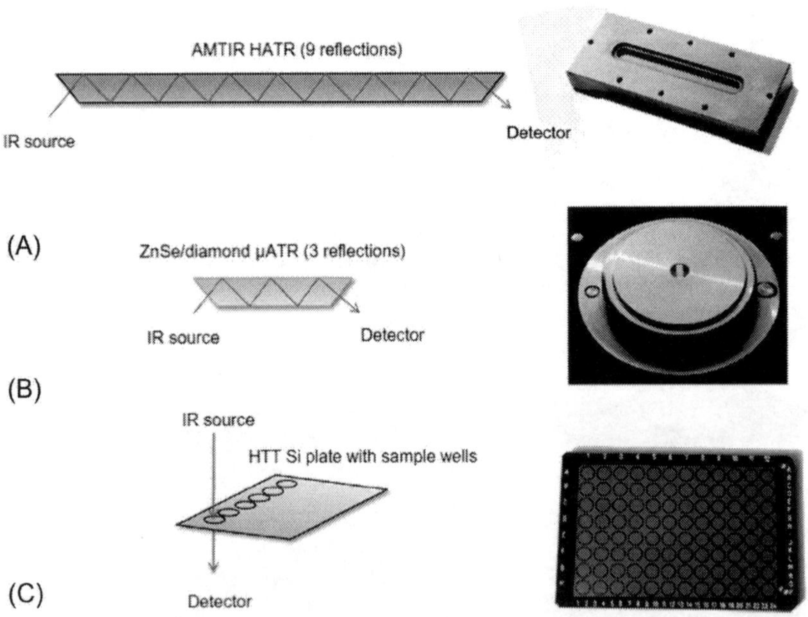

FIG. 9.11 A scheme showing differences in measurement strategies of FT-IR together with their corresponding photo. In (A) and (B), the IR light is directed into the crystal, where it reflects from its internal surface, and creates an evanescent wave, which projects orthogonally into the sample and decays exponentially with distance from the surface of the crystal. In (C), the IR radiation passes through the wells of the HTT Si plate with the samples on it. *(Reproduced with permission from E. Polshin, B. Aernouts, W. Saeys, F. Delvaux, F.R. Delvaux, D. Saison, M. Hertog, B.M. Nicolaï, J. Lammertyn, Beer quality screening by FT-IR spectrometry: impact of measurement strategies, data pre-processings and variable selection algorithms, J. Food Eng. 106 (2011) 188–198.)*

(1 sample/min) for prediction of real extract (Fig. 9.12). This accessory may be further modified for a potential increase of sample throughput and automation.

In the wine industry, the oenologist can achieve an optimal vinification process if problematic fermentations are detected soon. In this regard, MIR technology was explored for detecting early indications of abnormalities by comparing three processes of fermentation with artificial musts: a given normal behavior and two problems frequently encountered (i.e., assimilable nitrogen deficiency and a steep temperature gradient) [31]. Each process was tracked through changes in both MIR spectra and must composition. Fig. 9.13 shows that the evolution of MIR spectra is rather complicated as a typical wine fermentation progresses, making direct analysis difficult. It was concluded that the MIR estimation of metabolite concentration errors was sufficiently small to discern between ongoing normal and problematic fermentations, but it required calibrations to be performed with fermenting synthetic must samples. Hence, it was easier and cheaper to detect potential problems clearly in the progress of fermentation by monitoring wine metabolite concentrations than through direct spectra analysis. These findings, however, were promising enough to warrant further MIR research in monitoring commercial fermentation runs of a specific grape must.

Many wines improve in quality during storage that may take place in barrels and/or bottles. This ageing period is very valuable in the production process because diverse physicochemical properties change (e.g., reduction in acidity, better clarification and stabilization, enhancement in flavor and aroma).

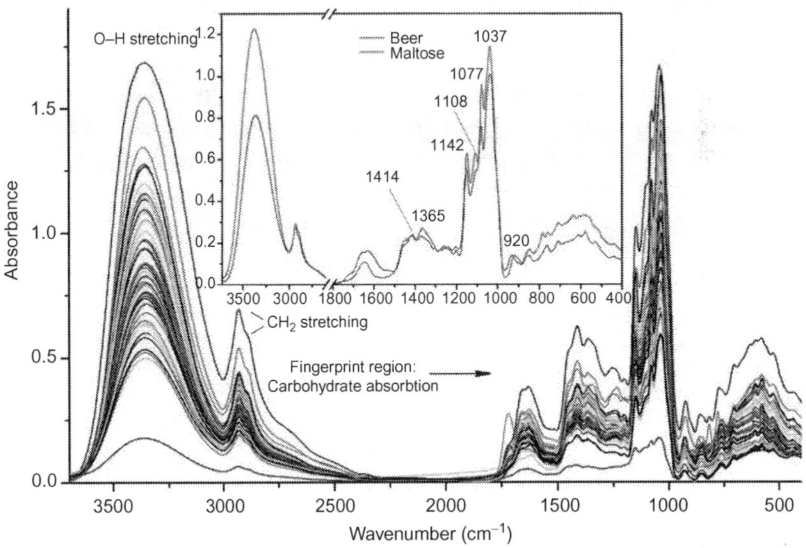

FIG. 9.12 (A) Mid-IR absorption spectra of all beer samples measured with HTT accessory for 10-μL sample volume; (B) mid-IR absorption spectra of beer and maltose measured with HTT accessory. *(Reproduced with permission from E. Polshin, B. Aernouts, W. Saeys, F. Delvaux, F.R. Delvaux, D. Saison, M. Hertog, B.M. Nicolaï, J. Lammertyn, Beer quality screening by FT-IR spectrometry: impact of measurement strategies, data pre-processings and variable selection algorithms, J. Food Eng. 106 (2011) 188–198.)*

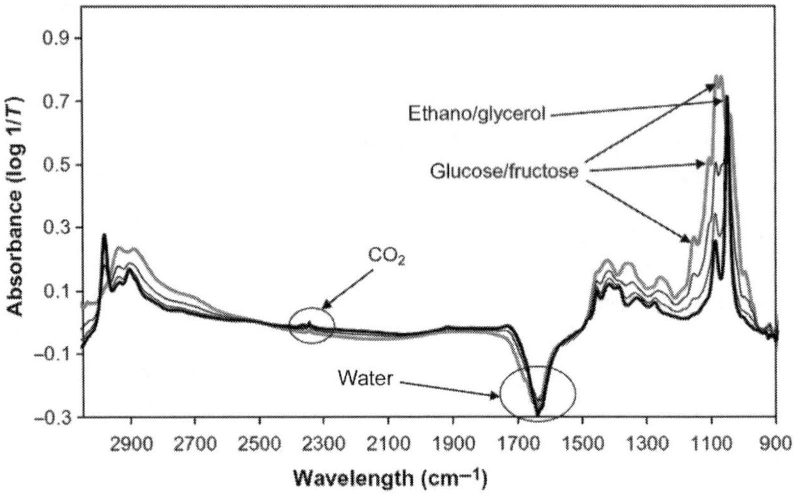

FIG. 9.13 Spectrum evolution in a typical normal wine fermentation. *(Reproduced with permission from A. Urtubia, J.R. Pérez-Correa, F. Pizarro, E. Agosin, Exploring the applicability of MIR spectroscopy to detect early indications of wine fermentation problems, Food Control 19 (2008) 382–388.)*

To optimize aged wines in a simple and fast way, the employment of FTIR and Vis spectroscopic techniques and PLS as a screening method was suggested for correlation of spectral and chromatic parameters with the aging level of a particular Sherry wine [32]. It was shown that the spectral data in IR (1100–2000 and 2300–2999 cm^{-1}) and Vis (380–450 nm) regions could be well developed into three regression models ($R^2 > 0.95$) for a full differentiation between the seven levels of aging in the wine under study.

Nowadays, food authentication has been in the phase of exponential growth and has been attracting a high level of attention from authorities and media. There is a growing demand for reliable analytical methods to ascertain the authenticity of foodstuffs for protection of bona fide products from any counterfeit and illegal substitutes around the world [33]. As an example of this, the verification of claimed brand identify among Trappist beers (only brewed by monks in a monastery, according to tradition and extremely high-quality standards) would help brewers and regulatory authorities uncover fraudulent labeling. FTIR-ATR spectroscopic analysis was applied, in this case, to a set of 267 bottles of both Trappist and non-Trappist beers (53 different brands) for confirmation of the claimed identity for samples of a single beer brand (Rochefort 8°) [34]. It is obviously illustrated in Fig. 9.14 that visual inspection of FTIR spectra could not differentiate between the sample types, indicating the requirement of a more extensive multivariate data analysis. As a result, data classification (by various algorithms such as linear discriminant analysis, quadratic discriminant analysis, *k*-nearest neighbors) was extensively preceded by chemometrics in the following order: (i) outlier detection by skewness-adjusted robust principal component analysis, (ii) data dimensionality reduction by extended canonical

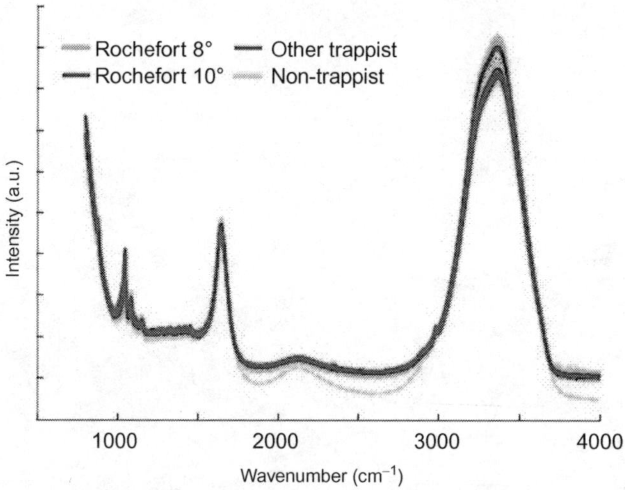

FIG. 9.14 Raw FT-IR spectra of different samples of beer. *(Reproduced with permission from J. Engel, L. Blanchet, L.M.C. Buydens, G. Downey, Confirmation of brand identity of a Trappist beer by mid-infrared spectroscopy coupled with multivariate data analysis, Talanta 99 (2012) 426–432.)*

variates analysis. It was shown that spectral regions (1000–1200 cm^{-1}) were identified important for the authentication of Rochefort beers. This approach could be served as an effective tool for identification of Rochefort 81 beers using Rochefort vs. non-Rochefort and Rochefort 8° vs. Rochefort 6°, 10° models, with overall correct prediction abilities ≥93.3%. Unfortunately, MIR identification of Trappist from non-Trappist beers did not seem to be a viable option (prediction ability of only 76.8%), suggesting GC-MS to be more suitable for this classification problem [35].

The potential of combining different types of spectroscopy (UV, Vis, NIR, and MIR) and chemometric techniques [principal component analysis (PCA), soft independent modeling of class analogy (SIMCA) and partial least squares discriminant analysis (PLS-DA)] was demonstrated as a rapid method for classification of Sauvignon Blanc wines from Australia and New Zealand, according to their geographical origin [36].

In parallel to IR analysis, the possibilities of discriminating wines could be given by FT-Raman spectroscopy (using an excitation laser at 1064 nm) in conjunction with supervised chemometric technique (i.e., stepwise linear discriminant analysis) [37]. More specifically, nonresonant excitation generated by FT-Raman technique, which may reflect wine differences, could be probed either in the Stokes (50–850 and 1600–1750 cm^{-1}) or anti-Stokes range as shown in Fig. 9.15. In this study, the classification of 30 Romanian wines was greatly performed by testing a control sample set of 4 French samples, i.e., 100% obtained in both initial and cross-validation percentages for discrimination of variety and geographical origin; for vintage differentiation, 100% and 94.1% were reached in initial and cross-validation percentages, respectively.

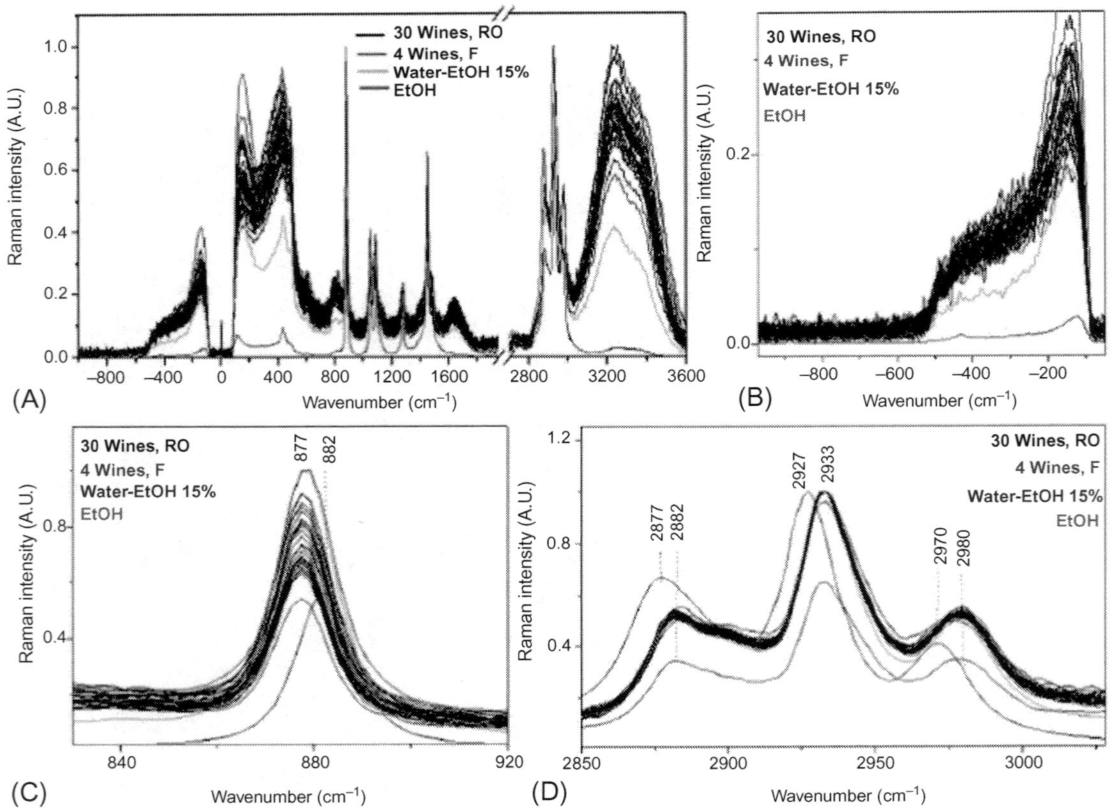

FIG. 9.15 (A) FT-Raman spectra of 34 wine samples (30 Romanian samples, showed in black, 4 French wines, showed in *red*) displayed both in Stokes (0–3600 cm^{-1}) and anti-Stokes (– 1000 to 0 cm^{-1}) spectral range. FT-Raman spectra of analytical grade ethanol *(blue)* and water-ethanol solution (15%) *(green)* measured in a similar quartz cuvette are comparatively showed. (B) Detail of the anti-Stokes range highlighting the large spectral variety in wine signal; (C, D) zoom of the main band shifts of ethanol due to hydrogen-bonding interactions in the water-ethanol solutions. Excitation: 1064 nm, 350 mW. *(Reproduced with permission from D.A. Magdas, F. Guyon, I. Feher, S. Cinta-Pinzaru, Wine discrimination based on chemometric analysis of untargeted markers using FT-Raman spectroscopy, Food Control 85 (2018) 385–391.)*

Using label-free SERS spectroscopy, the discrimination between wines (180 samples of three different wine types) and wineries (three different winemakers) from northeastern Italy could be also feasible (overall efficiencies, 87–93%) with the help of PCA and a robust data-driven version of SIMCA [38]. Samples were analyzed by using Ag citrate-reduced colloids and a portable Raman instrument with a 785-nm laser. Spectral pattern was dominated by three main metabolites: purines (adenine, adenosine), carboxylic acids, and glutathione, whose intensity ratios changed from wine to wine (Figs. 9.16 and 9.17). Moreover, stepwise linear discriminant analysis-based SERS technique was profitably studied for differentiation of French and Romanian white wine samples (with respect to variety and geographical origin) by employing silver nanoparticles as substrates and a compact DeltaNU532 Raman 152 system equipped with a frequency-doubled Nd:YAG (532 nm) laser [39].

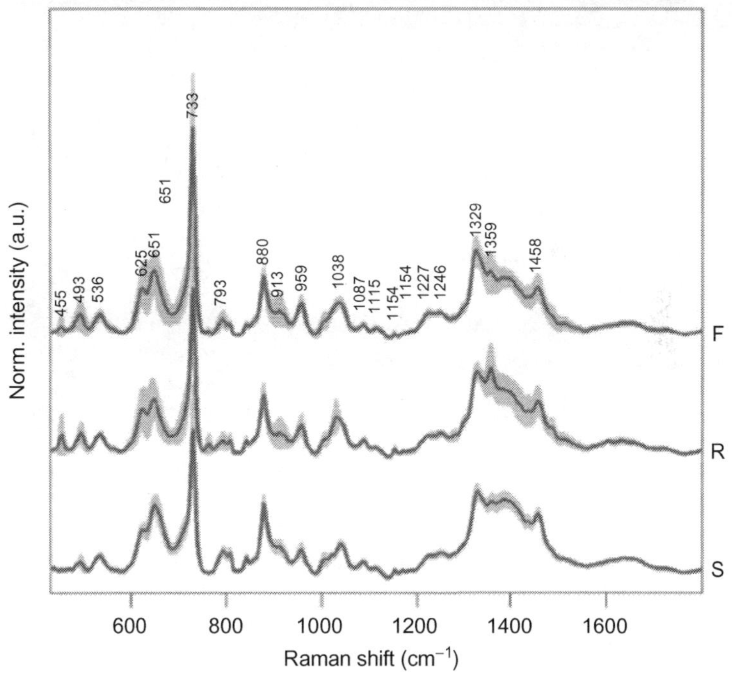

FIG. 9.16 Average SERS spectra of the wines grouped according to the Friulano (F), Ribolla (R), and Sauvignon (S) wine type (calculated from the normalized spectra of wines from all producers A, B, and C), together with ± 1 intensity standard deviation, indicated as *gray*. Spectra were stacked for better clarity. *(Reproduced with permission from F. Zanuttin, E. Gurian, I. Ignat, S. Fornasaro, A. Calabretti, G. Bigot, A. Bonifacio, Characterization of white wines from north-eastern Italy with surface-enhanced Raman spectroscopy, Talanta 203 (2019) 99–105.)*

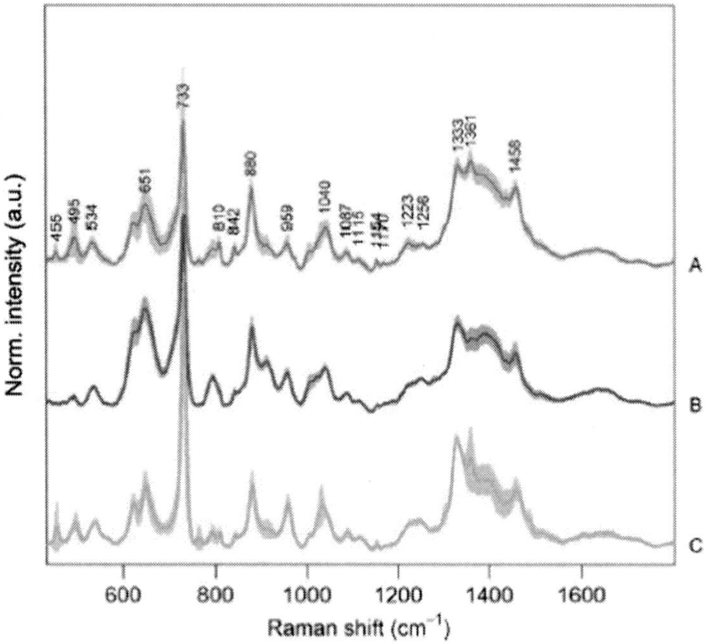

FIG. 9.17 Average SERS spectra of the wines grouped according to the producer A *(blue)*, B *(red)*, and C *(green)* (calculated from the normalized spectra of all the F, R, and S wine types), together with ±1 intensity standard deviation, indicated in lighter colors. Spectra were stacked for better clarity. *(Reproduced with permission from F. Zanuttin, E. Gurian, I. Ignat, S. Fornasaro, A. Calabretti, G. Bigot, A. Bonifacio, Characterization of white wines from north-eastern Italy with surface-enhanced Raman spectroscopy, Talanta 203 (2019) 99–105.)*

References

[1] J. Trienekens, P. Zuurbier, Quality and safety standards in the food industry, developments and challenges, Int. J. Prod. Econ. 113 (2008) 107–122.

[2] T. Woodcock, C. O'Donnell, G. Downey, Better quality food and beverages: the role of near infrared spectroscopy, J. Near Infrared Spectrosc. 16 (2008) 1–29.

[3] X. Lu, B. Rasco, Investigating food spoilage and pathogenic microorganisms by midinfrared spectroscopy, in: Handbook of Vibrational Spectroscopy, John Wiley & Sons, Ltd., NJ, USA, 2010.

[4] https://www.britannica.com/topic/alcoholic-beverage. (Accessible on 31 January 2020).

[5] K.J. Kaffka, K.H. Norris, Rapid instrumental analysis of composition of wine, Acta Aliment. 5 (1976) 267–279.

[6] G.F. Baumgarten, The determination of alcohol in wines by means of near infrared technology, S. Afr. J. Enol. Vitic. 8 (1987) 75–77.

[7] B.G. Osborne, T. Fearn, P.H. Hindle, Practical NIR Spectroscopy With Applications in Food and Beverage Analysis, Longman Scientific and Technical, Harlow, UK, 1993.

[8] R. Schindler, R. Vonach, B. Lendl, R. Kellner, A rapid automated method for wine analysis based upon sequential (SI)-FTIR spectrometry, Fresen. J. Anal. Chem. 362 (1998) 130–136.

[9] J.L.S. Moreira, L. Santos, Spectroscopic interferences in Fourier transform infrared wine analysis, Anal. Chim. Acta 513 (2004) 263–268.

[10] M. Friedel, C.D. Patz, H. Dietrich, Comparison of different measurement techniques and variable selection methods for FT-MIR in wine analysis, Food Chem. 141 (4) (2013) 4200–4207.

[11] K. Fernandez, E. Agosin, Quantitative analysis of red wine tannins using Fourier-transform mid-infrared spectrometry, J. Agric. Food Chem. 55 (2007) 7294–7300.

[12] I. Sen, B. Ozturk, F. Tokatli, B. Ozen, Combination of visible and mid-infrared spectra for the prediction of chemical parameters of wines, Talanta 161 (2016) 130–137.

[13] M. Romera-Fernández, L.A. Berrueta, S. Garmón-Lobato, B. Gallo, F. Vicente, J.M. Moreda, Feasibility study of FT-MIR spectroscopy and PLS-R for the fast determination of anthocyanins in wine, Talanta 88 (2012) 303–310.

[14] F. Que, L. Mao, C. Zhu, G. Xie, Antioxidant properties of Chinese yellow wine, its concentrate and volatiles, LWT- Food Sci. Technol. 39 (2) (2006) 111–117.

[15] Z. Wu, E. Xu, J. Long, X. Pan, X. Xu, Z. Jin, A. Jiao, Comparison between ATR-IR, Raman, concatenated ATR-IR and Raman spectroscopy for the determination of total antioxidant capacity and total phenolic content of Chinese rice wine, Food Chem. 194 (2016) 671–679.

[16] D. Cozzolino, Sample presentation, sources of error and future perspectives on the application of vibrational spectroscopy in the wine industry, J. Sci. Food Agric. 95 (5) (2015) 861–868.

[17] U. Regmi, M. Palma, C.G. Barroso, Direct determination of organic acids in wine and wine-derived products by Fourier transform infrared (FT-IR) spectroscopy and chemometric techniques, Anal. Chim. Acta 732 (2012) 137–144.

[18] D. Cozzolino, M.J. Kwiatkowski, R.G. Dambergs, W.U. Cynkar, L.J. Janik, G. Skouroumounis, M. Gishen, Analysis of elements in wine using near infrared spectroscopy and partial least squares regression, Talanta 74 (2008) 711–716.

[19] S. Vidal, P. Williams, T. Doco, M. Moutounet, P. Pellerin, The polysaccharides of red wine: total fractionation and characterisation, Carbohydr. Polym. 54 (2003) 439–447.

[20] J.C. Boulet, P. Williams, T. Doco, A Fourier transform infrared spectroscopy study of wine polysaccharides, Carbohydr. Polym. 69 (2007) 79–85.

[21] M.A. Amerine, C.S. Ought, Análisis de vinos y mostos, Acribia, Zaragoza, Espãna, 1976, p. 57.

[22] K. Helrich (Ed.), Official Methods of Analysis of the Association of Official Analytical Chemist (AOAC), fifteenth ed., Association of Official Analytical Chemist, Arlington, VA, 1990.

[23] G.J. Pilone, Determination of ethanol in wine by titrimetric and spectrophotometric dichromate methods: collaborative study, J. Assoc. Off. Anal. Chem. 68 (1985) 188–190.

[24] M. Gallignani, C. Ayala, M. del Rosario Brunetto, J.L. Burguera, M. Burguera, A simple strategy for determining ethanol in all types of alcoholic beverages based on its on-line liquid–liquid extraction with chloroform, using a flow injection system and Fourier transform infrared spectrometric detection in the mid-IR, Talanta 68 (2) (2005) 470–479.

[25] D.W. Lachenmeier, Rapid quality control of spirit drinks and beer using multivariate data analysis of Fourier transform infrared spectra, Food Chem. 101 (2007) 825–832.

[26] M. Urbano Cuadrado, M.D. Luque de Castro, P.M. Pérez Juan, M.A. Gómez-Nieto, Comparison and joint use of near infrared spectroscopy and Fourier transform mid infrared spectroscopy for the determination of wine parameters, Talanta 66 (2005) 218–224.

[27] F.A. Iñón, S. Garrigues, M. de la Guardia, Combination of mid- and near-infrared spectroscopy for the determination of the quality properties of beers, Anal. Chim. Acta 571 (2008) 167–174.

[28] C.A. Teixeira dos Santos, R.N.M.J. Páscoa, P.A.L.S. Porto, A.L. Cerdeira, J.M. González-Sáiz, C. Pizarro, J.A. Lopes, Raman spectroscopy for wine analyses: a comparison with near and mid infrared spectroscopy, Talanta 186 (2018) 306–314.

[29] International Organisation of Vine and Wine (OIV), Compendium of International Methods of Wine and Must Analysis, 2016, Available online at: http://www.oiv.int/public/medias/4231/compendium-2016-en-vol1.pdf. (Accessible on 31 December 2019).

[30] E. Polshin, B. Aernouts, W. Saeys, F. Delvaux, F.R. Delvaux, D. Saison, M. Hertog, B.M. Nicolaï, J. Lammertyn, Beer quality screening by FT-IR spectrometry: impact of measurement strategies, data pre-processings and variable selection algorithms, J. Food Eng. 106 (2011) 188–198.

[31] A. Urtubia, J.R. Pérez-Correa, F. Pizarro, E. Agosin, Exploring the applicability of MIR spectroscopy to detect early indications of wine fermentation problems, Food Control 19 (2008) 382–388.

[32] M. Ferreiro-González, A. Ruiz-Rodríguez, G.F. Barbero, J. Ayuso, C.G. Barroso, FT-IR, vis spectroscopy, color and multivariate analysis for the control of ageing processes in distinctive Spanish wines, Food Chem. 277 (2019) 6–11.

[33] G.P. Danezis, A.S. Tsagkaris, V. Brusic, C.A. Georgiou, Food authentication: state of the art and prospects, Curr. Opin. Food Sci. 10 (2016) 22–31.

[34] J. Engel, L. Blanchet, L.M.C. Buydens, G. Downey, Confirmation of brand identity of a Trappist beer by mid-infrared spectroscopy coupled with multivariate data analysis, Talanta 99 (2012) 426–432.

[35] T. Cajka, K. Riddellova, M. Tomaniova, J. Hajslova, Recognition of beer brand based on multivariate analysis of volatile fingerprint, J. Chromatogr. A 1217 (2010) 4195–4203.

[36] D. Cozzolino, W.U. Cynkar, N. Shah, P.A. Smith, Can spectroscopy geographically classify Sauvignon Blasnc wines from Australia and New Zealand? Food Chem. 126 (2) (2011) 673–678.

[37] D.A. Magdas, F. Guyon, I. Feher, S. Cinta-Pinzaru, Wine discrimination based on chemometric analysis of untargeted markers using FT-Raman spectroscopy, Food Control 85 (2018) 385–391.

[38] F. Zanuttin, E. Gurian, I. Ignat, S. Fornasaro, A. Calabretti, G. Bigot, A. Bonifacio, Characterization of white wines from north-eastern Italy with surface-enhanced Raman spectroscopy, Talanta 203 (2019) 99–105.

[39] D.A. Magdas, S. Cinta Pinzaru, F. Guyon, I. Feher, B.I. Cozar, Application of SERS technique in white wines discrimination, Food Control 92 (2018) 30–36.

Chapter 10

Some concluding remarks

Historically speaking, the fascinating world of vibrational spectroscopy started with the discovery of IR radiation by Sir William Herschel in 1800 and further blossomed with Raman effect by Sir Chandrasekhara Venkata Raman in 1928. The exemplary, up-to-date application of vibrational spectroscopy in the field of pharmaceutical, biomedical, and food analysis has been reviewed and discussed in this book.

Vibrational spectroscopy, IR and Raman, are multipurpose tools in combination with chemometrics that can offer both qualitative and quantitative methods typified by rapidity, simplicity, low-cost analysis, efficiency and requiring nondestructive minimal sample preparation. Due to its versatility, it is relatively easy to record spectra from samples in any state of matter (i.e., liquid, solid, gas). Undoubtedly, they have been extensively used in pharmaceutical, biomedical, and food analysis because they can give information about the sample under study through molecularly specific probe without using extrinsic label and being invasive.

In biomedical analysis, the greatest advantage of vibrational spectroscopy is related to its high molecular sensitivity and micrometer-level spatial resolution. Due to advances in instrumentation, vibrational spectroscopic imaging has been increasingly chosen for biomedical investigation, e.g., identification of a disease, lesion, tumor, or infection. This is because it requires no reagents and the methods could be automated. It can offer the clinicians with new diagnostic approaches that are better than some other approaches (e.g., pathological observations) in terms of sensitivity and specificity estimates, and the patients have an early diagnosis and/or prognosis for any illness. Nonetheless, vibrational spectromicroscopy has not been exploited yet in preclinical or clinical trials for the reason that there is a challenge in translating research evidence into validated applications in clinical practice suitably used in biomedical laboratories. This drawback may come, in part, from the limited background knowledge of clinicians and biomedical researchers in vibrational spectroscopy.

In pharmaceutical analysis, vibrational spectroscopy has successfully proved to be efficient for quantification and structural identification of both active ingredients and excipients in different types of dosage forms. Its capability as spectroscopic fingerprint features could be employed for discrimination and authentication of brand-name, generic and over-the-counter medications, and herbal remedies. Especially, with innovative handheld/portable Raman and NIR spectrophotometers the analysis of pharmaceutical products could be done in a timely manner for both quality control and in-process tests.

Vibrational Spectroscopy Applications in Biomedical, Pharmaceutical and Food Sciences
https://doi.org/10.1016/B978-0-12-818827-9.09991-1

© 2020 Elsevier Inc. All rights reserved.

In food analysis, the application of vibrational spectroscopy has markedly supported a better understanding of the sensory, functional, chemical, and textural characteristics of various foods such as alcoholic and nonalcoholic beverages, dairy products, and oils. It is being applied not only for determining beverage or food composition, adulteration or authentication, assessing and predicting quality and process-induced variation, but also for detecting microbiological or chemical contaminants associated with food safety.

The edition of this book is made with the hope of encouraging post-graduate students and researchers to further exploit the advantages of vibrational spectroscopy. Given the evidence-based biomedical, pharmaceutical, and food analysis, the following graphical abstract can be used not only to summarize the evolution of vibrational spectroscopy in the applied fields but also to replace the closing statements of this book.

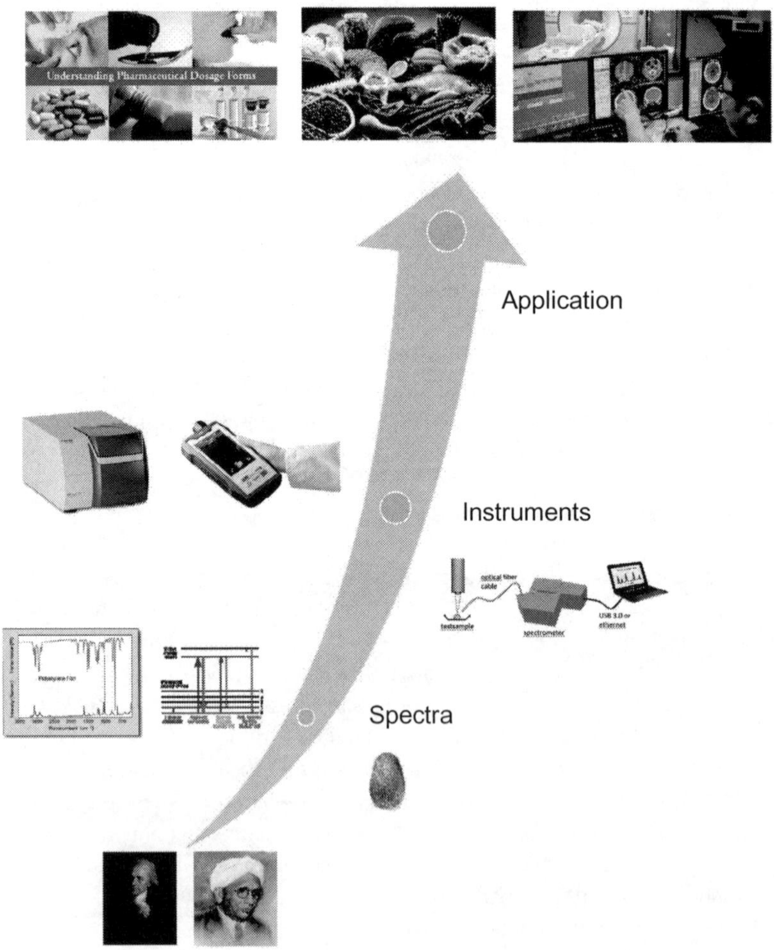

Appendix

Chemometric processing of spectroscopic data

Alessandra Biancolillo[a] and Federico Marini[b]

[a]*Department of Physical and Chemical Sciences, University of L'Aquila, L'Aquila, Italy,*
[b]*Department of Chemistry, University of Rome "La Sapienza", Rome, Italy*

1 Introduction: From the spectra to the data

Instrumental signals, and spectroscopic ones are not an exception, are usually collected to characterize samples in terms of similarities and dissimilarities, to highlight possible trends or groupings, to constitute the basis for the prediction of properties or responses, which cannot be measured directly or whose measurement may be costly, difficult and/or time consuming, or for their classification. In order for any of the aforementioned tasks to be accomplished as best as possible, it is not enough to look at the profiles (spectra) as they are produced by the instrument, but some sort of data processing should necessarily be involved. Chemometrics, as a discipline, deals with the use of mathematical, statistical, logical, and computer science tools for the extraction of the maximum relevant information from the data, and constitutes the link between the raw measurements and the final aims of the analysis, as briefly summarized earlier [1, 2]. Since there are already several papers and monographs dealing with the use of chemometrics for the analysis of spectral data, in this appendix only the main issues connected with the processing of spectroscopic data will be presented, together with a brief description of the tools most commonly used for the purpose; for a most detailed treatment of the different subjects, the reader is suggested to consult the referred bibliography [3–5].

Anyway, before getting into the different stages of data analysis, it is necessary to introduce the different ways spectroscopic data (and data in general) may be represented. The representation, which chemists and spectroscopists are more familiar with, is the one when, as a result of an experiment, a spectrum is collected: such spectrum is usually represented as a plot of the recorded light intensity (absorbed, transmitted, or reflected by the sample) as a function of the wavelength or frequency of the corresponding radiation (see Fig. A.1A). This representation is quite straightforward, as an experienced spectroscopist can relate the position of peaks to the presence of specific substances or functional

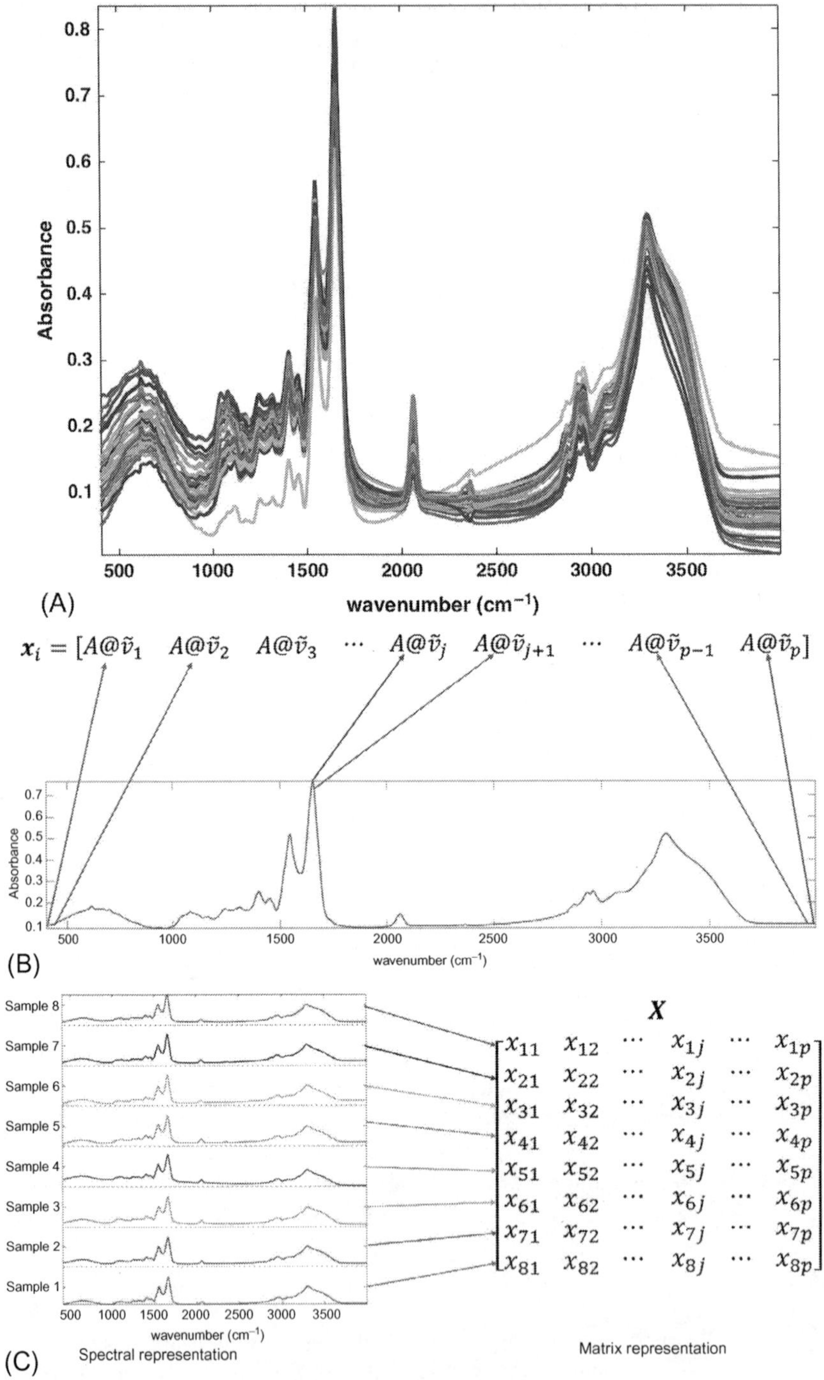

FIG. A.1 (A) "Usual" graphical representation of spectroscopic data, as plot of the intensity of the (absorbed) radiation as a function of wavelength (or, in the present case, wavenumber). (B) Equivalence between the spectroscopic profile of an individual sample and the corresponding vector of measurements $\boldsymbol{x}_i$. (C) Equivalence between the spectroscopic profiles of a set of samples and the corresponding data matrix $\boldsymbol{X}$.

groups and the peak areas to the corresponding relative or absolute concentrations. However, for the sake of further data processing, a different spectral representation is more effective, which is the one describing a profile as an ordered array of numbers, in particular a row vector, each element of which corresponds to the spectral intensity at a particular wavelength (see Fig. A.1B):

$$\boldsymbol{x}_i = \begin{bmatrix} I_{\lambda_1} & I_{\lambda_2} & I_{\lambda_3} & \cdots & I_{\lambda_{p-1}} & I_{\lambda_p} \end{bmatrix} = \begin{bmatrix} x_{i1} & x_{i2} & x_{i3} & \cdots & x_{ip-1} & x_{ip} \end{bmatrix} \quad \text{(A.1)}$$

In Eq. (A.1), $I\lambda_1$, $I\lambda_2$, ..., $I\lambda_p$ represent the intensity of the radiation (expressed as absorbance, transmittance, reflectance or, sometimes, even just as arbitrary units) collected at the different wavelengths λ_1, λ_2, ..., λ_p, p being the number of sampled frequencies (variables); since this vector (representing the spectrum of the ith sample) will here be referred to as $\boldsymbol{x}_i$, analogously its elements can also be indicated as x_{i1}, x_{i2}, ..., x_{ip}.

However, as also already mentioned in the beginning of this appendix, it is very rare that the problem to be solved may be dealt with just by recording the spectrum of a single sample. Under these circumstances, since each spectrum corresponds to a row vector of numbers, the profiles collected on the different samples can be organized in the form of a data matrix, each row of which corresponds to the signal recorded on a particular sample (Fig. A.1C):

$$\boldsymbol{X} = \begin{bmatrix} x_{11} & x_{12} & \cdots & x_{1j} & \cdots & x_{1p} \\ x_{21} & x_{22} & \cdots & x_{2j} & \cdots & x_{2p} \\ \vdots & \vdots & \ddots & \vdots & \ddots & \vdots \\ x_{i1} & x_{i2} & \cdots & x_{ij} & \cdots & x_{ip} \\ \vdots & \vdots & \ddots & \vdots & \ddots & \vdots \\ x_{n1} & x_{n2} & \cdots & x_{nj} & \cdots & x_{np} \end{bmatrix} \quad \text{(A.2)}$$

where x_{ij} indicates the generic element (corresponding to the light intensity collected on the ith sample at the jth wavelength) of the data matrix $\boldsymbol{X}$, which is made of n rows (as many as the number of samples) and p columns (as many as the number of variables/frequencies).

The matrix representation of the data is not only useful as it allows to quickly and straightforwardly calculate and express models, through the use of the appropriate algebra, but also because it is as an immediate geometric counterpart, which can allow graphical display. Indeed, each row of a data matrix (a sample) can be represented as a point in a multidimensional space whose axes are the p measured variables (row-space representation). Then, by looking at the distribution of the points in the multivariate space, it could be possible to visually appreciate similarities and differences between samples, based on the consideration that closeness in space can be directly related to similarity in the chemical/spectroscopic behavior, to identify clusters or groupings, to highlight possible trends and, in general, to characterize the system. On the other hand, relating the observed relationship among the samples to the directions along which they occur is the main tool for chemical interpretation.

2 Exploratory data analysis

All the possibilities related to the geometrical representation of a data matrix and sketched at the end of the previous section, are actually the aims of what is normally referred to as exploratory data analysis [6]. Indeed, exploratory data analysis focuses on trying to identify and summarize the main characteristics of a data set, relying as much as possible on graphical display. It was proposed originally by Tukey as a way of looking at the data with no prejudice, instead of just searching for the confirmation of premade hypotheses (e.g., by testing), and "let the data talk," so that, in case, hypotheses can be formulated ex post, grounded on the measurements.

Based on these ideas, the main aims of exploratory data analysis are maximizing insight into a data set, revealing underlying structures, identifying important variables, detecting outliers and anomalies, and building parsimonious models [7]. As already said, in principle, all these tasks should be accomplished by making extensive use of graphical and data display tools. However, most real-world applications, and especially those based on spectroscopic profiling, involve the measurement of a high number of variables (hundreds or thousands, if not even more), so that one should in principle be able to represent the data in such a high dimensional space, which is of course not feasible.

To overcome this limitation, exploratory data analysis makes extensive use of so-called projection techniques, which allow to condensate the relevant information in the data onto a parsimonious set of components (new directions in the multivariate space), which may also be used for graphical display. Indeed, a projection induces a mapping from the original p-dimensional space onto a lower-dimensional subspace, which can even be composed of a single direction only. If the axes of this subspace are chosen as to be relevant to describe the phenomena of interest, then data compression may be achieved without a significant loss of information.

2.1 PCA

Although alternative methods exist and are used in some application, in the context of exploratory data analysis, principal component analysis (PCA) [8, 9] is by far the projection method most frequently adopted, and this is not surprising, as it is based on finding the subspace, which best fits (approximates) the data in a least squares sense, i.e., by minimizing the sum of the squared distances (residuals) between the points in the original space and the same points after the projection. Moreover, the axes of this reduced space are chosen to be orthogonal by construction, so that any new direction brings new information, ensuring the model to be as parsimonious as possible.

Briefly, PCA proceeds as follows. The first direction to be selected as an axis for the reduced subspace (principal component) is identified as the one providing the best one-dimensional approximation of the data in a least squares sense. By calling loading vector $\boldsymbol{p}_1$, the unit norm vector identifying the direction of the component, such criterion can be expressed as:

$$\min_{p_1} \left\| \boldsymbol{X} - \boldsymbol{X}\boldsymbol{p}_1\boldsymbol{p}_1^T \right\|^2 \tag{A.3}$$

$\boldsymbol{X}$ being the data matrix. When projecting the data onto the principal component, the new coordinates of the samples are called scores and are collected in the corresponding vector $\boldsymbol{t}_1$. Then, it can be demonstrated that the least squares criterion expressed by Eq. (A.3) is equivalent to affirming that the first principal component is selected as the direction in space along which the variance of the corresponding scores is maximum:

$$\max_{p_1} \lambda_1 = \max_{p_1} \frac{\sum_{i=1}^{n} \boldsymbol{t}_1^T \boldsymbol{t}_1}{n-1} \tag{A.4}$$

where n is the number of samples and the variance along the principal component is usually called its eigenvalue and indicated as λ_1. If additional components are sought, they are identified according to the criteria described in Eqs. (A.3) or (A.4), provided that the search be restricted to all the possible directions orthogonal to the ones already extracted. Eventually, if F components are calculated, i.e., if the data are projected onto a F-dimensional subspace, the scores and the loadings for the various directions can be collected into matrices,

$$\begin{aligned} \boldsymbol{T} &= \begin{bmatrix} \boldsymbol{t}_1 & \boldsymbol{t}_2 & \cdots & \boldsymbol{t}_F \end{bmatrix} \\ \boldsymbol{P} &= \begin{bmatrix} \boldsymbol{p}_1 & \boldsymbol{p}_2 & \cdots & \boldsymbol{p}_F \end{bmatrix} \end{aligned} \tag{A.5}$$

which are related by

$$\boldsymbol{T} = \boldsymbol{X}\boldsymbol{P} \tag{A.6}$$

The scores matrix $\boldsymbol{T}$ collects the "new" coordinates of the samples onto the subspace spanned by the principal components and can be used to describe the individuals for the exploratory analysis (or even for further modeling stages). Two- or three-dimensional scatterplots of the sample scores along selected principal components can be used to infer information about possible trends in the data, similarity and dissimilarity among individuals, presence of clusters or of anomalous observation and/or suspect outliers. On the other hand, inspection of the loadings can support the chemical interpretation of the relations among samples found in the corresponding scores plot.

2.2 Explorative multiblock approaches

In the last years, the greater availability of instrumentations and the advancement thereof, has ensured complex systems can be analyzed by different analytical techniques, allowing a multiplatform investigation of the problems under study. When the same samples are analyzed by different analytical techniques and/or in diverse time points, the outcome is a *multiblock* data set.

How to handle these kind of data sets has been discussed into the literature, and different strategies have been proposed. The most common "nomenclature" for organizing multiblock methods is the one, which divides them into *Low-*, *Mid-*, and *High-Level Data Fusion* approaches, and it refers to the level information is joint [10, 11].

Very briefly, when original data from different platforms are merged and analyzed (by classical chemometric tools) the approach is a Low-level one. On the other hand, if features (for instance, scores) are extracted by each data block, and then modeled all together, the applied strategy belongs to the class of the Mid-level approaches. Finally, the High-Level data fusion consist of joining the results of diverse models into a unique outcome [11].

Other classifications of multiblock strategies take into account whether data structures are modeled simultaneously or in parallel, more details on this regard can be found in Ref. [11].

A number of explorative multiblock methods have been proposed into the literature, based on different principles. For instance, some, like Common Components and Specific Weights Analysis, are based on the extraction of the common variability among the different data matrices [12]; on the other hand, others put the accent on the distinctive information present in data, as Distinctive and common components with simultaneous-component analysis [13] does. All these approaches are very efficient and widely applied in different fields. Nevertheless, one of the most widely used explorative data fusion strategies is a natural extension of PCA to multitab arrays. In fact, handling a multiblock data set, it is possible to block-scale each data matrix (e.g., dividing it by its Euclidean norm), concatenate all of them, and finally perform PCA on the resulting structure. This strategy owns its diffusion to the fact it has a very reduced computational cost and, at the same time, it performs well in several situations. As a consequence, it is easy to find papers where this strategy is satisfactorily applied both to achieve exploratory analysis and as a starting point for classification (few examples [14–17]).

However, when problems are particularly complex, or some a priori knowledge on the data blocks requires different solutions, it could be more suitable to apply other data fusion approaches, developed to meet specific requirements. A brief list of explorative multiblock methods is reported in Table A.1.

3 Regression

Exploratory data analysis provides a number of advantages briefly described in the previous section. Nevertheless, it has a huge constraint: in its classical formulation, it is not suitable for correlating a set of independent variables (e.g., spectral signal) to the dependent ones.

TABLE A.1 Brief list of explorative multiblock methods.

Multiblock method	Level of fusion	References
Common components and specific weights analysis	Mid	[12]
Distinctive and common components with simultaneous-component analysis	Mid	[13]
Joint and individual variation explained	Mid	[18]
Hierarchical principal component analysis	Mid	[19]

Finding out the relation between observations and specific characteristics of the system under study is a common task encountered in different contexts; for instance, it is what one aims at when instrumental data (i.e., a set of *predictors*) are collected to estimate the concentration of a specific compound in a sample.

The characteristic, which is predicted on the basis of the collected data, is called *response* ($\boldsymbol{y}$) and it can be estimated by the general formula:

$$\boldsymbol{y} = \hat{\boldsymbol{y}} + \boldsymbol{e} = \boldsymbol{X}\boldsymbol{b} + \boldsymbol{e} \qquad \text{(A.7)}$$

where $\hat{\boldsymbol{y}}$ is the estimated response, $\boldsymbol{e}$ is the residuals ($\boldsymbol{y} - \hat{\boldsymbol{y}}$), $\boldsymbol{X}$ is the set of predictors, and $\boldsymbol{b}$ is the regression coefficients. Solving Eq. (A.7) corresponds to creating a multivariate regression model. Once $\boldsymbol{b}$ is estimated (i.e., once the calibration model is built), it can be exploited to draw predictions on new objects [4].

Into the literature, different regression methods have been proposed; among the others, one of the most well-known and widely applied is partial least squares (PLS) [20–22].

3.1 Partial least squares

PLS is a multivariate regression approach developed to correlate a data matrix of predictors ($\boldsymbol{X}$) to a response ($\boldsymbol{y}$) even in those cases where $\boldsymbol{X}$ is not invertible.

The starting point of the method is the projection of $\boldsymbol{X}$ onto a restricted novel space of *latent variables* called *scores* ($\boldsymbol{T}$). A brief description of the algorithm is reported in now.

3.1.1 Estimation of a response vector y by PLS

Plainly, taking into consideration the case where a set of measurements $\boldsymbol{X}_{(N \times M)}$ is used to predict the response vector $\boldsymbol{y}_{(N \times 1)}$ through the PLS algorithm, the first required step is to estimate a set of orthonormal *weights* $\boldsymbol{w}_1$ ($\boldsymbol{w}_1 = \boldsymbol{X}\boldsymbol{t}_1$) suitable for maximizing the covariance between the first score ($\boldsymbol{t}_1$) and $\boldsymbol{y}$:

$$\max_{w_1}\left[\operatorname{cov}\left(t_1,y\right)\right]=\max\left(t_1^T y\right) \tag{A.8}$$

The relation between $\boldsymbol{y}$ and $\boldsymbol{t}_1$ (so-called *inner relation*) is defined by the coefficient $\boldsymbol{c}_1 = \frac{\boldsymbol{y}^T\boldsymbol{t}_1}{\boldsymbol{t}_1^T\boldsymbol{t}_1}$.

Once Eq. (A.8) is solved, it is possible to extract further PLS scores, but after *deflation* (i.e., removal of the modeled variance) of $\boldsymbol{X}$.

This is achieved by means of a set of coefficients called *loadings* ($\boldsymbol{p}_1 = \frac{\boldsymbol{X}^T\boldsymbol{t}_1}{\boldsymbol{t}_1^T\boldsymbol{t}_1}$):

$$\boldsymbol{E}_{X,1} = \boldsymbol{X} - \boldsymbol{t}_1\boldsymbol{p}_1^T \tag{A.9}$$

And it leads to the matrix $\boldsymbol{E}_{X,1}$, which is $\boldsymbol{X}$ deflated of the variance accounted into the model through the first component.

In principle, also $\boldsymbol{y}$ could be deflated, but it is not necessary (on the contrary, as it will be exposed later, this is needed for the estimation of a response matrix $\boldsymbol{Y}$) [20].

The desired number of components (F) can be extracted repeating all the aforementioned steps (for instance, for the second component, $\boldsymbol{w}_2$ will maximize the covariance between $\boldsymbol{E}_{X,1}$ and $\boldsymbol{t}_2$, new loadings $\boldsymbol{p}_2$ will be calculated and $\boldsymbol{E}_{X,1}$ will be deflated into $\boldsymbol{E}_{X,2}$) and, eventually, the global model can be expressed as a function of the PLS components ($\boldsymbol{T}$):

$$\boldsymbol{T} = \boldsymbol{X}\boldsymbol{R} \tag{A.10a}$$

with

$$\boldsymbol{R} = \boldsymbol{W}\left(\boldsymbol{P}^T\boldsymbol{W}\right)^{-1} \tag{A.10b}$$

where $\boldsymbol{W} = [\boldsymbol{w}_1\ \boldsymbol{w}_2 \ldots \boldsymbol{w}_F]$ and $\boldsymbol{P} = [\boldsymbol{p}_1\ \boldsymbol{p}_2 \ldots \boldsymbol{p}_F]$.

Then, the response can be estimated by

$$\boldsymbol{y} = \hat{\boldsymbol{y}} + \boldsymbol{e} = \boldsymbol{X}\boldsymbol{R}\boldsymbol{c} + \boldsymbol{e} = \boldsymbol{X}\,\boldsymbol{W}\left(\boldsymbol{P}^T\boldsymbol{W}\right)^{-1}\boldsymbol{c} \tag{A.11}$$

where $\boldsymbol{C}$ are the inner relation coefficients ($\boldsymbol{C} = [\boldsymbol{c}_1\ \boldsymbol{c}_2 \ldots \boldsymbol{c}_F]$).

Finally, Eq. (A.7) can be expressed as:

$$\boldsymbol{y} = \boldsymbol{X}\boldsymbol{b}_{PLS} + \boldsymbol{e} \tag{A.12}$$

Once the calibration model is built, i.e., when $\boldsymbol{b}_{PLS}$ is estimated, it is possible to make predictions on new observations ($\boldsymbol{X}_{new}$):

$$\hat{\boldsymbol{y}} = \boldsymbol{X}_{new}\boldsymbol{b}_{PLS} \tag{A.13}$$

3.1.2 Estimation of a response matrix *Y* by PLS (PLS2)

The procedure applied to estimate a response matrix $\boldsymbol{Y}$ is similar to the one exposed earlier for $\boldsymbol{y}$ with slight changes. The first one, is that it requires two sets of latent variables, the $\boldsymbol{X}$-scores (i.e., the $\boldsymbol{T}$ mentioned above) and the $\boldsymbol{Y}$-scores ($\boldsymbol{U}$). In fact, in the PLS2 algorithm, both $\boldsymbol{X}$ and $\boldsymbol{Y}$ are projected onto the same latent variable subspace; this is achieved, for the first component, by finding the weights $\boldsymbol{w}_1$, and the loadings $\boldsymbol{q}_1$, such as:

$$\max_{\boldsymbol{w}_1,\boldsymbol{q}_1}\left[\operatorname{cov}\left(\boldsymbol{t}_1,\boldsymbol{u}_1\right)\right] \tag{A.14}$$

where $\boldsymbol{t}_1 = \boldsymbol{X}\boldsymbol{w}_1$ and $\boldsymbol{u}_1 = \dfrac{\boldsymbol{Y}^T\boldsymbol{q}_1}{\boldsymbol{q}_1^T\boldsymbol{q}_1}$

At this point of the algorithm, the inner relation (for the first component) can be formulated:

$$\boldsymbol{u}_1 = c_1\boldsymbol{t}_1 \tag{A.15}$$

With $c_1 = \dfrac{\boldsymbol{t}_1^T\boldsymbol{u}_1}{\boldsymbol{t}_1^T\boldsymbol{t}_1}$

As mentioned earlier, prior the extraction of the further components, it is necessary to deflate both $\boldsymbol{X}$ and $\boldsymbol{Y}$ of the variance accounted (so far) by the model. For $\boldsymbol{X}$ this is accomplished by Eq. (A.9); for $\boldsymbol{Y}$ by Eq. (A.16):

$$\boldsymbol{E}_{Y,1} = \boldsymbol{Y} - \boldsymbol{u}_1\boldsymbol{q}_1^T \tag{A.16}$$

At this point, it is possible to repeat all the steps (estimation of $\boldsymbol{w}$ and $\boldsymbol{q}$, of the scores and the deflation of the blocks) as long as the desired number of components is extracted.

Eventually, global scores $\boldsymbol{T}$ and $\boldsymbol{U}$, and loadings $\boldsymbol{Q}$ are collected into matrices by concatenation of the individual vectors:

$$\boldsymbol{T} = \left[\boldsymbol{t}_1\,\boldsymbol{t}_2\ldots\boldsymbol{t}_F\right] \tag{A.17a}$$

$$\boldsymbol{U} = \left[\boldsymbol{u}_1\,\boldsymbol{u}_2\ldots\boldsymbol{u}_F\right] \tag{A.17b}$$

$$\boldsymbol{Q} = \left[\boldsymbol{q}_1\,\boldsymbol{q}_2\ldots\boldsymbol{q}_F\right] \tag{A.17c}$$

The matrix $\boldsymbol{C}$ of the inner relation is organized in the form reported in Eq. (A.18):

$$\boldsymbol{C} = \begin{bmatrix} c_1 & \cdots & 0 \\ \vdots & \ddots & \vdots \\ 0 & \cdots & c_F \end{bmatrix} \tag{A.18}$$

And the final regression model is calculated:

$$\boldsymbol{Y} = \hat{\boldsymbol{Y}} + \boldsymbol{E}_Y = \boldsymbol{U}\boldsymbol{Q}^T + \boldsymbol{E}_Y = \boldsymbol{T}\boldsymbol{C}\boldsymbol{Q}^T + \boldsymbol{E}_Y = \boldsymbol{X}\boldsymbol{B}_{PLS} \tag{A.19}$$

Once the calibration model is built and $\boldsymbol{B}_{PLS}$ estimated, it is possible to make predictions on new observations ($\boldsymbol{X}_{new}$):

$$\hat{\boldsymbol{Y}} = \boldsymbol{X}_{new}\boldsymbol{B}_{PLS} \quad \text{(A.20)}$$

3.1.3 Estimation of the goodness of a fit

Once a regression model is built, it is important to check how reliable it is.

The goodness of the fit can be investigated through different figures of merit; for instance, the root mean square error (RMSE):

$$\text{RMSE} = \sqrt{\frac{\sum_{i=1}^{N}(\hat{y}_i - y_i)^2}{N}} \quad \text{(A.21)}$$

Plainly, RMSE is calculated by taking the square root of the sum of squares of the residuals normalized by the number of observations. Consequently, it provides an indication of the predictive ability of the model. The crossvalidated counterpart of this entity (the Root Mean Square Error in Cross-Validation) is particularly useful because it can be used for defining model parameters; a wider discussion about this aspect can be found in Section 5.

Similar to the RMSE is the BIAS; this additional figure of merit allows estimating the average error made by the model:

$$\text{BIAS} = \frac{\sum_{i=1}^{N}(\hat{y}_i - y_i)^2}{N} \quad \text{(A.22)}$$

Together with the RMSE and the BIAS, further indications about the fit are provided by the *coefficient of determination* or R^2. This additional feature reveals how much of the total variance is explained by the model. This coefficient is calculated by the following formula:

$$R^2 = \frac{\sum_{i=1}^{N}(\hat{y}_i - y_i)^2}{\sum_{i=1}^{N}(\hat{y}_i - \overline{y}_i)^2} \quad \text{(A.23)}$$

Due to its mathematical formulation, R^2 is positive-valued and upper-limited to 1. Customarily, the closest R^2 is to 1, the most the model is considered reliable. This is due to the fact that $R^2 = 1$ corresponds to the case where the explained variance $\left(\sum_{i=1}^{N}(\hat{y}_i - y_i)^2\right)$ is equal to the total variance $\left(\sum_{i=1}^{N}(\hat{y}_i - \overline{y}_i)^2\right)$. Nevertheless, this index naturally increases with the number of the observations; consequently, its interpretation should be carefully handled.

3.2 Multiblock regression approaches

Several multiblock regression approaches have been proposed into the literature and their application spaces on different fields [11, 23]. Among the Low-Level data fusion methods, one of the most commonly used is Multi-Block-PLS (MB-PLS) [24–27]. This approach is quite widely applied, and it represents a natural way of handling multiplatform data. In its easiest formulation [28], each block of data is block-scaled and then they are concatenated into a unique matrix, which is lately modeled by PLS.

Also Mid-level data fusion approaches are widely used; as a consequence, several approaches belonging to this category are available. Some of these, like Predictive Com Dim [29] or Parallel Orthogonal-PLS [30], focus the attention on the common variability among the blocks; others, like Sequential and Orthogonalized-PLS (SO-PLS) [31] or Sequential and Orthogonalized-Covariance Selection (SO-CovSel) [32], sequentially extract information from the different data matrices exploiting some tricks to get rid of redundant information. A wider overview of the most commonly applied multiblock regression methods is reported in Table A.2.

TABLE A.2 Brief list of multiblock regression methods.

Multiblock method	Level of fusion	References
Multiblock-PLS	Low	[24–27]
Predictive-ComDim	Mid	[29]
Parallel orthogonalized-PLS	Mid	[30]
Sequential and orthogonalized-PLS	Mid	[31]
Sequential and orthogonalized-covariance selection	Mid	[32]
Hierarchical-PLS	Mid	[33]
Sequential and orthogonalized N-PLS	Mid	[34]
On-PLS	Mid	[35]
Multiblock redundancy analysis	Mid	[36, 37]

4 Classification

As previously described, inspecting a complex system, the aim, and, consequently, the tools applied during the investigation, are conceptually different. The explorative analysis described allows recognizing the main characteristics of the data, and it provides indication about possible outliers. Nevertheless, it is not suitable for investigating the relation between the collected measures and a specific dependent variable. This latter task is admirably accomplished by regression approaches, although they do not directly provide indications about similarities/dissimilarities among samples (as PCA, for instance, does).

When a complex system is investigated with the aim of finding out recursive patterns in samples, classification methods are the most suitable tools. In particular, *supervised* pattern recognition approaches could represent an efficient way of highlighting grouping tendencies among objects.

Classifiers are commonly applied for solving problems in various fields with different aims; for instance, in food science, to trace or authenticate raw materials [38–40] and food products [41–43], in forensic and pharmaceutical science [44–46], in case-control studies, to analyze clinical data [47, 48], or in data mining [49], to investigate consumers' preferences.

Although a wide number of supervised classifiers have been proposed into the literature, all of them can be divided into two macrocategories: *discriminant* and *class-modeling* approaches. A brief general description of these methods with focus on some specific approaches is provided later [4, 50].

4.1 Discriminant analysis

Different classifiers belong to the category "discriminant methods," nevertheless, they obviously present some common characteristics. For instance, all of them base the classification outcome on the dissimilarities among objects belonging to different classes. In fact, all the discriminant approaches divide the entire multivariate space of samples into as many regions as the number of categories present into the training set. The main consequence of this is that each object will be assigned to solely one category and it cannot be unsigned. One of the first discriminant classifiers proposed is Fisher's Linear Discriminant Analysis [51]. This method is still widely used because very effective; nevertheless, it presents some strictly constrains, which prevent its utilization in different contexts. In fact, it is applicable only on invertible data matrices; condition rarely met, in particular handling instrumental data, where the number of variables is often higher than the numerosity of the investigated objects. Consequently, LDA is often-used after a feature extraction approach, in order to decrease the number of variables. Additionally, some modification/extensions have been proposed, in order to overcome this limitation. Among these, one of the most diffused is partial least square discriminant analysis (PLS-DA).

4.1.1 PLS-DA

PLS-DA [52, 53] has been developed combining PLS and discriminant analysis. Plainly, the algorithm exploits PLS to solve the classification problem transforming it into a regression one. The keystone, which makes it possible is the so-called *Y dummy,* which is a matrix encoding class belongings for each sample [54]. For example, taking into account a three-category case, given N samples and G classes, Y will have dimensions $N \times G$, i.e., it will be made of N binary row vectors of different forms (depending on the category each specific sample belongs to) and of as many columns as the number of the categories present in the training set. Class-belonging is expressed putting 1s in the column correspondent to the membership of each observation, and 0s in all the other positions. For instance, samples belonging to Class 1, Class 2, and Class 3 will be represented by the vector in Eqs. (A.24a), (A.24b), (A.24c), respectively.

$$y_1 = [100] \tag{A.24a}$$

$$y_2 = [010] \tag{A.24b}$$

$$y_3 = [001] \tag{A.24c}$$

Once the Y-dummy is created, the classification problem can be handled solving Eq. (A.7) by PLS. Also in this case, $\boldsymbol{X}$ is the data matrix of observations, whereas $\boldsymbol{Y}$ is the dummy. The calibration model built in this way can be used to classify new observations. Nevertheless, the predicted $\boldsymbol{Y}$ $(\hat{\boldsymbol{Y}})$ will be composed of continuous variables rather than 0s and 1s. Different solutions have been proposed to correlate this response and class membership of new samples [55, 56]. One easy option is to assign the sample to the class corresponding to the column where the highest value is present; for instance, given a ith $\hat{\boldsymbol{Y}}$-vector:

$$\hat{y}_i = [0.5 - 0.1 \; 0.8] \tag{A.25}$$

the ith sample will be assigned to Class 3, because the highest value is in the third column.

When only two categories are involved, the $\boldsymbol{Y}$-dummy can be represented as a category vector where 1s indicate samples belonging to Class 1 and 0s objects appertaining to Class 2. Also in the two-category case, $\hat{\boldsymbol{y}}$ is not binary, but constituted of continuous values; when $\hat{y}_i < 0.5$ the object will be assigned to class 1, otherwise ($\hat{y}_i > 0.5$) to Class 2.

4.2 Class modeling

As described earlier, discriminant methods classify samples according to the intraclass dissimilarities. On the other hand, class-modeling techniques tend to classify objects on the basis of their intercategory similarity. As a consequence, these approaches are often used facing *asymmetric classification* problems, i.e., all those cases where, for some reasons, there is a specific class of interest with respect to the others.

Contrary to discriminant methods, these approaches do not divide the entire space of samples into categories, but they define the class region of each specific category. The main consequence of this is that a sample could be assigned to one or more classes, or it could even be rejected by all the class models and it will not be assigned to any of them [57].

The difference between the diverse class-modeling approaches belongs on the law applied for the calculation of the class regions. Among the others, the most commonly used method is called Soft Independent Modeling of Class Analogies (SIMCA) [58, 59], which is briefly described now.

4.2.1 SIMCA

Considering the rationale exposed earlier, it is straightforward SIMCA individually models each class of the training set. Its algorithm is based on the assumption the variability present in each category can be represented by a bilinear model. In other words, a PCA model is built on samples belonging to the same class and then further objects are assigned/rejected on the basis of their spatial position into the PCA space. For an *i*th sample, one conventional rule for acceptance/rejection into a modeled class is based on the distance d_i, calculated by Eq. (A.26):

$$d_i = \sqrt{\left(\frac{T_i^2}{T_{0.95}^2}\right)^2 + \left(\frac{Q_i}{Q_{0.95}}\right)^2} = \sqrt{\left(T_{i,\mathrm{red}}^2\right)^2 + \left(Q_{i,\mathrm{red}}\right)^2} : \quad \text{(A.26)}$$

where T^2 (the Mahalanobis distance of the sample from the center of the class space) and Q (the orthogonal distance of the *i*th sample from the model) are two statistical entities used to define whether an object is "close enough" to those belonging to the class to be accepted or not. As can be seen from Eq. (A.26), T^2 and Q are normalized by ${T_{0.95}}^2$ and $Q_{0.95}$, i.e., the 95th percentiles of their distributions, leading their reduced counterparts (T_{red}^2 and Q_{red}) [60].

Due to the normalization, the "critical" value of T_{red}^2 and Q_{red} becomes 1 for both statistics; accordingly, decision about acceptance or rejection of an unknown sample is based on setting a distance threshold equal to $\sqrt{2}$: an object falling at $d > \sqrt{2}$ will be rejected by the model; otherwise, it is accepted.

4.3 Multiblock approaches for classification

As mentioned in Section 3.2, it is more convenient to handle multiblock data sets by data fusion approaches; this consideration is valid also in classification contexts. For this reason, several of the multiblock methods already described have been extended to the classification field. Multiblock classifiers are widely used for solving problems in pharmaceutical analysis, in particular for checking the quality of raw materials and/or final products [61–63], in food science, for authenticating and/or tracing [11, 64–67] aliments, to analyze clinical data, in case-control studies [68, 69], and in several other fields [11].

A common procedure is to use data fusion methods to reduce features and then apply a discriminant classifier (for instance, LDA).

In general, this strategy is constituted of two main steps:

1. A explorative/regression multiblock model is built.
2. The classification model is calculated on the $\hat{\boldsymbol{Y}}$.

Whether wanted, any suitable classifier can be also applied on the extracted features, since they span the same space as the predicted response.

This procedure has been discussed for SO-PLS (i.e., SO-PLS-LDA [70, 71], SO-N-PLS [34]) and SO-CovSel (i.e., SO-CovSel-LDA) [32] and application of this strategy (i.e., multiblock method coupled with classifier) can be found for further data fusion approaches, both low- [72–74] and mid-level [73, 75–77].

5 Validation

All the techniques described in the present appendix constitute examples of how chemometrics translates the information in the collected experimental data into models whose aim is to approximate the system under study or to allow predictions of selected responses. This means that, whenever a specific problem is concerned, there is usually more than a single model, which can be computed on the available data. Accordingly, given the empirical (data-driven) nature of this approach, not all the models, which could in principle be calculated on a data set, may result in the same quality due to various concurrent factors, such as the number and distribution of samples and whether the experiments were properly designed or not, the inherent properties of the algorithm, and so on [78]. Therefore, whenever a model is computed, it is necessary to verify whether it can allow drawing reliable considerations on the system under investigation and/or whether accurate predictions can be formulated on new (unknown) samples. This fundamental process is called validation [79], and its main aims are to check how appropriate the model is, how reliable its predictions are, and whether any resulting interpretation properly explains the phenomena under study and can be generalizable [80].

In practice, to evaluate the quality of possible models, which can be built on the data, the validation process relies on the definition of suitable diagnostics, which are normally based on the inspection of the model parameters (e.g., loadings, regression coefficients, and so on) or, most frequently, on the residuals. Then, since most of the methods are based on some sort of least squares criterion or, anyway, may involve an iterative procedure where model parameters are calculated as the one leading to the smallest residuals, it is apparent that, in order to have a proper validation and to avoid overoptimism, it is fundamental not to use the same data exploited for calculating the model parameters (*training set*) to estimate such a diagnostics, as the corresponding residuals would rarely be representative of the ones one could expect on applying the model on new data.

Accordingly, proper validation should be carried out by applying the model to a completely new set of data (test set), and the corresponding results, in particular the residuals, should be used to calculate the required diagnostics. Ideally, this new set of data should be collected in conditions, which resemble as closely as possible the ones under which the model should be routinely used, covering as exhaustively as possible the range of future measurements. When collecting brand-new data is not possible, one can still split all the available samples into a training and a test set, to proceed with validation as described earlier. In such cases, it is of utmost importance that both sets are representative: to this purpose, various criteria have been proposed in the literature, which select individuals to be placed in either group of samples, so to span as uniformly as possible the variable space. Examples of such approaches are Kennard-Stone [81] or duplex [82] algorithms, the use of D-optimal designs [83–85] or Kohonen neural networks [86, 87].

When the number of samples is not sufficient for building a training and a test set, which are both representative, an alternative strategy could be to make use of resampling approaches, i.e., approaches in which subsets of individuals are repeatedly drawn from the overall data set to constitute the training and the test set, and the results on the individual subsets are then pooled together to evaluate the model performances (and to calculate the model diagnostics). In this framework, the most commonly used resampling method is crossvalidation [88]. However, it should be stressed that, although particularly efficient for small data sets (and, if needed, to evaluate the distribution of model parameters), resampling strategies may provide a biased estimate of model quality, since the same observations are used alternatively for model building and model validation.

Accordingly, crossvalidation (or another resampling strategy) is preferably used only during the stages of model selection and/or model comparison, e.g., when models with varying number of components or pretreated in a different way are fitted to the same data and the one performing best has to be selected, or when the results of applying different techniques on the same matrix(es) are used to choose the best algorithm. In such cases, resampling is applied only to the training set: eventually, for the evaluation of the reliability of the final (chosen) model, test set validation should always be carried out, whenever possible.

References

[1] S. Wold, Chemometrics; what do we mean with it, and what do we want from it? Chemom. Intell. Lab. Syst. 30 (1995) 109–115.

[2] R.G. Brereton, J. Jansen, J. Lopes, F. Marini, A. Pomerantsev, O. Rodionova, J.-M. Roger, B. Walczak, R. Tauler, Chemometrics in analytical chemistry-part I: history, experimental design and data analysis tools, Anal. Bioanal. Chem. 409 (2017) 5891–5899.

[3] H. Mark, J. Workman Jr., Chemometrics in Spectroscopy, second ed., Associated Press, London, UK, 2018.

[4] A. Biancolillo, F. Marini, Chemometrics applied to plant spectral analysis, in: J. Lopes, C. Sousa (Eds.), Vibrational Spectroscopy for Plant Varieties and Cultivars Characterization, Comprehensive Analytical Chemistry, vol. 80, Elsevier, Amsterdam, 2018, pp. 69–104 (Chapter 4).

[5] A. Biancolillo, F. Marini, Chemometric methods for spectroscopy-based pharmaceutical analysis, Front. Chem. 6 (2018) 576.

[6] J.W. Tukey, Exploratory Data Analysis, Addison-Wesley, Reading, MA, 1977.

[7] A. Smilde, R. Bro, P. Geladi, Multi-Way Analysis, John Wiley & Sons, Chichester, 2004.

[8] S. Wold, K. Esbensen, P. Geldadi, Principal component analysis, Chemom. Intell. Lab. Syst. 2 (1987) 37–52.

[9] R. Bro, A.K. Smilde, Principal component analysis, Anal. Methods 6 (2014) 2812–2831.

[10] M. Cocchi, Chapter 1—Introduction: ways and means to deal with data from multiple sources, in: M. Cocchi (Ed.), Data Handling in Science and Technology, vol. 31, Elsevier, Amsterdam, 2019, pp. 1–26.

[11] A. Biancolillo, R. Boqué, M. Cocchi, F. Marini, Data fusion strategies in food analysis, in: M. Cocchi (Ed.), Data Handling in Science and Technology, vol. 31, Elsevier, Amsterdam, 2019, pp. 271–310.

[12] G. Mazerolles, M. Hanafi, E. Dufour, D. Bertrand, E.M. Qannari, Common components and specific weights analysis: a chemometric method for dealing with complexity of food products, Chemom. Intell. Lab. Syst. 81 (2006) 41–49.

[13] M. Schouteden, K. Van Deun, S. Pattyn, I. Van Mechelen, SCA with rotation to distinguish common and distinctive information in linked data, Behav. Res. Methods 45 (2013) 822–833.

[14] M. Cocchi, M. Li Vigni, C. Durante, Chemometrics-bioinformatics, in: A.G. Constatinos, G.P. Danezis (Eds.), Food Authentication, Management, Analysis and Regulation, Wiley Blackwell, Hoboken, NJ, 2017, pp. 483–520.

[15] L. Li, S. Xie, J. Ning, Q. Chen, Z. Zhang, Evaluating green tea quality based on multisensory data fusion combining hyperspectral imaging and olfactory visualization systems, J. Sci. Food Agric. 99 (2018) 1787–1794.

[16] A. Biancolillo, M. Tomassetti, R. Bucci, S. Izzo, F. Candilio, F. Marini, Ancient human bones studied and compared by near infrared spectroscopy, thermogravimetry and chemometrics, J. Near Infrared Spectrosc. 27 (2018) 6–14.

[17] C. Alamprese, M. Casale, N. Sinelli, S. Lanteri, E. Casiraghi, Detection of minced beef adulteration with Turkey meat by UV-vis, NIR and MIR spectroscopy, LWT Food Sci. Technol. 53 (2013) 225–232.

[18] E.F. Lock, K.A. Hoadley, J.S. Marron, A.B. Nobel, Joint and individual variation explained (JIVE) for integrated analysis of multiple data types, Ann. Appl. Stat. 7 (2013) 523–542.

[19] M. Hanafi, A. Kohler, M. Qannari, Shedding new light on hierarchical principal component analysis, J. Chemom. 24 (2010) 703–709.

[20] H. Martens, T. Naes, Multivariate Calibration, John Wiley & Sons, New York, NY, 1991.

[21] S. Wold, H. Martens, H. Wold, The multivariate calibration problem in chemistry solved by the PLS methods, in: A. Ruhe, B. Kagstrøm (Eds.), Matrix Pencils: Proceedings of a Conference Held at Pite Havsbad, Sweden, March 22–24, 1982, Springer Verlag, Heidelberg, Germany, 1983, pp. 286–293.

[22] P. Geladi, B.R. Kowalski, Partial least squares regression: a tutorial, Anal. Chim. Acta 185 (1986) 1–17.

[23] M. Cocchi, Data Handling in Science and Technology, vol. 31, Elsevier, Amsterdam, 2019.

[24] I.E. Frank, J. Feikema, N. Constantine, B.R. Kowalski, J. Chem. Inf. Comput. Sci. 24 (1984) 20–24.

[25] I.E. Frank, B.R. Kowalski, Prediction of wine quality and geographic origin from chemical measurements by partial least-squares regression modeling, Anal. Chim. Acta 162 (1984) 241–251.

[26] L.E. Wangen, B.R. Kowalski, A multiblock partial least squares algorithm for investigating complex chemical systems, J. Chemom. 3 (1988) 3–20.

[27] J.A. Westerhuis, A.K. Smilde, Deflation in multiblock PLS, J. Chemom. 15 (2001) 485–493.

[28] S.J. Qin, S. Valle, M.J. Piovoso, On unifying multiblock analysis with application to decentralized process monitoring, J. Chemom. 15 (2001) 715–742.

[29] A. El Ghaziri, V. Cariou, D.N. Rutledge, M. Qannari, Analysis of multiblock datasets using ComDim: overview and extension to the analysis of (K + 1) datasets, J. Chemom. 30 (2016) 420–429.

[30] I. Måge, E. Menichelli, T. Næs, Preference mapping by PO-PLS: separating common and unique information in several data blocks, Food Qual. Prefer. 24 (2012) 8–16.

[31] T. Næs, O. Tomic, B.H. Mevik, H. Martens, Path modelling by sequential PLS regression, J. Chemom. 25 (2011) 28–40.

[32] A. Biancolillo, F. Marini, J.-M. Roger, SO-CovSel: a novel method for variable selection in a multiblock framework, J. Chemom. (2019) e3120, https://doi.org/10.1002/cem.3120.

[33] S. Wold, N. Kettaneh, K. Tjessem, Hierarchical multiblock PLS and PC models for easier model interpretation and as an alternative to variable selection, J. Chemom. 10 (1996) 463–482.

[34] A. Biancolillo, T. Næs, R. Bro, I. Måge, Extension of SO-PLS to multi-way arrays: SO-N-PLS, Chemom. Intell. Lab. Syst. 164 (2017) 113–126.

[35] T. Löfstedt, J. Trygg, OnPLS—a novel multiblock method for the modelling of predictive and orthogonal variation, J. Chemom. 25 (2011) 441–455.

[36] S. Bougeard, M. Qannari, N. Rose, Multiblock redundancy analysis: interpretation tools and application in epidemiology, J. Chemom. 25 (2011) 467–475.

[37] S. Bougeard, M. Qannari, C. Lupo, M. Hanafi, From multiblock partial least squares to multiblock redundancy analysis. A continuum approach, Informatica 22 (2011) 11–26.

[38] M.A. Quelal-Vásconez, E. Pérez-Esteve, A. Arnau-Bonachera, J.M. Barat, P. Talens, Rapid fraud detection of cocoa powder with carob flour using near infrared spectroscopy, Food Control 92 (2018) 183–189.

[39] S. De Luca, E. Ciotoli, A. Biancolillo, A.D. Magrì, F. Marini, Simultaneous quantification of caffeine and chlorogenic acid in coffee green beans and varietal classification of the samples by HPLC-DAD coupled with chemometrics, Environ. Sci. Pollut. Res. 25 (2018) 28748–28759.

[40] A. De Girolamo, S. Cervellieri, M. Cortese, A.C.R. Porricelli, M. Pascale, F. Longobardi, C. von Holst, L. Ciaccheri, V. Lippolis, Fourier transform near-infrared and mid-infrared spectroscopy as efficient tools for rapid screening of deoxynivalenol contamination in wheat bran, J. Sci. Food Agric. 99 (2019) 1946–1953.

[41] A. Biancolillo, S. De Luca, S. Bassi, L. Roudier, R. Bucci, A.D. Magrì, F. Marini, Authentication of an Italian PDO hazelnut ("Nocciola Romana") by NIR spectroscopy, Environ. Sci. Pollut. Res. 25 (2018) 28780–28786.

[42] C. Márquez, I. Ruisánchez, M.P. Callao, Qualitative and quantitative multivariate strategies for determining paprika adulteration with SUDAN I and II dyes, Microchem. J. 145 (2019) 686–692.

[43] P. Firmani, S. De Luca, R. Bucci, F. Marini, A. Biancolillo, Near Infrared (NIR) spectroscopy-based classification for the authentication of Darjeeling black tea, Food Control 100 (2019) 292–299.

[44] E. Deconinck, R. Van Campenhout, C. Aouadi, M. Canfyn, J.L. Bothy, L. Gremeaux, P. Blanckaert, P. Courselle, Combining attenuated total reflectance- infrared spectroscopy and chemometrics for the identification and the dosage estimation of MDMA tablets, Talanta 195 (2019) 142–151.

[45] D. Custers, S. Vandemoortele, J.-L. Bothy, J.O. De Beer, P. Courselle, S. Apers, E. Deconinck, Physical profiling and IR spectroscopy: simple and effective methods to discriminate between genuine and counterfeit samples of Viagra® and Cialis®, Drug Test. Anal. 8 (2016) 378–387.

[46] X. Chen, X. Sun, H. Hua, Y. Yi, H. Li, C. Chen, Quality evaluation of decoction pieces of Rhizoma Atractylodis Macrocephalae by near infrared spectroscopy coupled with chemometrics, Spectrochim. Acta A 221 (2019) 117169.

[47] R. Calvani, A. Picca, F. Marini, A. Biancolillo, J. Gervasoni, S. Persichilli, A. Primiano, H.J. Coelho-Junior, M. Bossola, A. Urbani, F. Landi, R. Bernabei, E. Marzetti, A distinct pattern of circulating amino acids characterizes older persons with physical frailty and sarcopenia: results from the BIOSPHERE study, Nutrients 10 (2018) e1691.

[48] H.G. Park, K.S. Jang, H.M. Park, W.S. Song, Y.Y. Jeong, D.H. Ahn, S.M. Kim, Y.H. Yang, Y.G. Kim, MALDI-TOF MS-based total serum protein fingerprinting for liver cancer diagnosis, Analyst 144 (2019) 2231–2238.

[49] K. Christensen, K.H. Liland, K. Kvaal, E. Risvik, A. Biancolillo, J. Scholderer, S. Nørskov, T. Næs, Mining online community data: the nature of ideas in online communities, Food Qual. Prefer. 62 (2017) 246–256.

[50] C. Albano, W. Dunn III, U. Edlund, E. Johansson, B. Norden, M. Sjöström, S. Wold, Four levels of pattern recognition, Anal. Chim. Acta 103 (1978) 429–443.

[51] R.A. Fisher, The use of multiple measurements in taxonomic problems, Ann. Eugen. 7 (1936) 179–188.

[52] M. Sjöström, S. Wold, B. Söderström, PLS discriminant plots, in: E.S. Gelsema, L.N. Kanal (Eds.), Pattern Recognition in Practice, Elsevier, Amsterdam, 1986, pp. 461–470.

[53] L. Ståle, S. Wold, Partial least squares analysis with cross-validation for the two-class problem: a Monte Carlo study, J. Chemom. 1 (1987) 185–196.

[54] M. Barker, W. Rayens, Partial least squares for discrimination, J. Chemom. 17 (2003) 166–173.

[55] U.G. Indahl, H. Martens, T. Naes, From dummy regression to prior probabilities in PLS-DA, J. Chemom. 21 (2007) 529–536.

[56] H. Nocairi, E.M. Qannari, E. Vigneau, D. Bertrand, Discrimination on latent components with respect to patterns. Application to multicollinear data, Comput. Stat. Data Anal. 48 (2004) 139–147.

[57] P. Oliveri, G. Downey, Multivariate class modeling for the verification of food-authenticity claims, TrAC Trend Anal. Chem. 35 (2012) 74–86.

[58] S. Wold, M. Sjöström, SIMCA: a method for analysing chemical data in terms of similarity and analogy, in: B.R. Kowalski (Ed.), Chemometrics, Theory and Application, American Chemical Society Symposium Series No. 52, American Chemical Society, Washington, DC, 1977, pp. 243–282.

[59] S. Wold, M. Sjöström, Comments on a recent evaluation of the SIMCA method, J. Chemom. 1 (1987) 243–245.

[60] H.H. Yue, S.J. Qin, Reconstruction-based fault identification based on a combined index, Ind. Eng. Chem. Res. 40 (2001) 4403–4414.

[61] Y. Zhang, B. Xu, X. Wang, S. Dai, X. Shi, Y. Qiao, Optimal selection of incoming materials from the inventory for achieving the target drug release profile of high drug load sustained-release matrix tablet, AAPS PharmSciTech 20 (2019) 76.

[62] T. Casian, A. Farkas, K. Ilyés, B. Démuth, E. Borbás, L. Madarász, Z. Rapi, B. Farkas, A. Balogh, A. Domokos, G. Marosi, I. Tomută, Z.K. Nagy, Data fusion strategies for performance improvement of a process analytical technology platform consisting of four instruments: an electrospinning case study, Int. J Pharm. 567 (2019) 118473.

[63] X.M. Wu, Q.Z. Zhang, Y.Z. Wang, Traceability of wild Paris polyphylla Smith var. yunnanensis based on data fusion strategy of FT-MIR and UV–Vis combined with SVM and random forest, Spectrochim. Acta A 205 (2018) 479–488.

[64] Y.B. Monakhova, R. Godelmann, A. Hermann, T. Kuballa, C. Cannet, H. Schäfer, M. Spraul, D.N. Rutledge, Synergistic effect of the simultaneous chemometric analysis of 1H NMR spectroscopic and stable isotope (SNIF-NMR, 18O, 13C) data: application to wine analysis, Anal. Chim. Acta 833 (2014) 29–39.

[65] M. Silvestri, L. Bertacchini, C. Durante, A. Marchetti, E. Salvatore, M. Cocchi, Application of data fusion techniques to direct geographical traceability indicators, Anal. Chim. Acta 769 (2013) 1–9.

[66] S.A. Drivelos, K. Higgins, J.H. Kalivas, S.A. Haroutounian, C.A. Georgiou, Data fusion for food authentication. Combining rare earth elements and trace metals to discriminate "Fava Santorinis" from other yellow split peas using chemometric tools, Food Chem. 165 (2014) 316–322.

[67] P. Firmani, A. Nardecchia, F. Nocente, L. Gazza, F. Marini, A. Biancolillo, Multi-block classification of Italian semolina based on near infrared spectroscopy (NIR) analysis and alveographic indices, Food Chem. 309 (2020) 125677.

[68] R. Calvani, A. Picca, F. Marini, A. Biancolillo, M. Cesari, V. Pesce, A.M.S. Lezza, M. Bossola, C. Leeuwenburgh, R. Bernabei, F. Landi, E. Marzetti, The "BIOmarkers associated with Sarcopenia and PHysical frailty in EldeRly pErsons" (BIOSPHERE) study: rationale, design and methods, Eur. J. Intern. Med. 56 (2018) 19–25.

[69] M.L.M. Andersen, M.A. Rasmussen, S. Pörksen, J. Svensson, J. Vikre-Jørgensen, J. Thomsen, N.T. Hertel, J. Johannesen, F. Pociot, J.S. Petersen, L. Hansen, H.B. Mortensen, L.B. Nielsen, Complex multi-block analysis identifies new immunologic and genetic disease progression patterns associated with the residual β-cell function 1 year after diagnosis of type 1 diabetes, PLoS One 8 (2013) e64632.

[70] A. Biancolillo, I. Måge, T. Næs, Combining SO-PLS and linear discriminant analysis for multi-block classification, Chemom. Intell. Lab. Syst. 141 (2015) 58–67.

[71] A. Biancolillo, T. Naes, The sequential and orthogonalised PLS regression (SO-PLS) for multi-block regression: theory, examples and extensions, in: M. Cocchi (Ed.), Data Handling in Science and Technology, vol. 31, Elsevier, Amsterdam, 2019, pp. 157–177 (Chapter 6).

[72] E. Borras, J. Ferré, R. Boqué, M. Mestres, L. Acena, A. Calvo, O. Busto, Olive oil sensory defects classification with data fusion of instrumental techniques and multivariate analysis (PLS-DA), Food Chem. 203 (2016) 314–322.

[73] W. Wang, C. Li, A multimodal quality inspection system based on 3D, hyperspectral, and X-ray imaging for onions, in: ASABE and CSBE/SCGAB Annual International Meeting, 2014.

[74] A. Biancolillo, R. Bucci, A.L. Magrì, A.D. Magrì, A.D.F. Marini, Data-fusion for multiplatform characterization of an Italian craft beer aimed at its authentication, Anal. Chim. Acta 820 (2014) 23–31.

[75] D.D. Marshall, S. Lei, B. Worley, Y. Huang, A. Garcia-Garcia, R. Franco, E.D. Dodds, R. Powers, Combining DI-ESI–MS and NMR datasets for metabolic profiling, Metabolomics 11 (2015) 391–402.

[76] A. Rudnitskaya, D. Kirsanov, A. Legin, K. Beullens, J. Lammertyn, B.M. Nicolaı, J. Irudayaraj, Analysis of apples varieties e comparison of electronic tongue with different analytical techniques, Sensors Actuators B Chem. 116 (2006) 23–28.

[77] A. Biancolillo, K.H. Liland, I. Måge, T. Næs, R. Bro, Variable selection in multi-block regression, Chemom. Intell. Lab. Syst. 156 (2016) 89–101.

[78] R.G. Brereton, J. Jansen, J. Lopes, F. Marini, A. Pomerantsev, O. Rodionova, J.M. Roger, B. Walczak, R. Tauler, Chemometrics in analytical chemistry—part II: modeling, validation, and applications, Anal. Bioanal. Chem. 410 (2018) 6691–6704.

[79] F. Westad, F. Marini, Validation of chemometric models—a tutorial, Anal. Chim. Acta 893 (2015) 14–24.

[80] R. Harshman, How can I know if it's real? A catalogue of diagnostics for use with three-mode factor analysis and multidimensional scaling, in: H.G. Law, C.W. Snyder Jr., J. Hattie, R.P. Mc Donald (Eds.), Research Methods for Multimode Data Analysis, Praeger, New York, 1984, pp. 566–591.

[81] R.W. Kennard, L.A. Stone, Computer aided design of experiments, Technometrics 11 (1969) 137–148.

[82] R.D. Snee, Validation of regression models: methods and examples, Technometrics 19 (1977) 415–428.

[83] W. Wu, B. Walczak, D.L. Massart, S. Heuerding, F. Erni, I.R. Last, K.A. Prebble, Artificial neural networks in classification of NIR spectral data: design of the training set, Chemom. Intell. Lab. Syst. 33 (1996) 35–46.

[84] I.-M. Olsson, J. Gottfries, S. Wold, D-optimal onion designs in statistical molecular design, Chemom. Intell. Lab. Syst. 73 (2004) 37–46.

[85] L. Eriksson, T. Arnhold, B. Beck, T. Fox, E. Johansson, J.M. Kriegl, Onion design and its application to a pharmaceutical QSAR problem, J. Chemom. 18 (2004) 188–202.

[86] K. Rajer-Kanduč, J. Zupan, N. Majcen, Separation of data on the training and test set for modeling: a case study for modeling of five colours properties of a white pigment, Chemom. Intell. Lab. Syst. 65 (2003) 221–229.

[87] F. Marini, A.L. Magrì, R. Bucci, A.D. Magrì, Use of different artificial neural networks to resolve binary blends of monocultivar Italian olive oils, Anal. Chim. Acta 599 (2007) 232–240.

[88] M. Stone, Cross-validatory choice and assessment of statistical prediction, J. R. Stat. Soc. Ser. B 36 (1974) 111–147.

Index

Note: Page numbers followed by *f* indicate figures and *t* indicate tables.

A

V

W

Made in the USA
Monee, IL
13 September 2026

62991934R00146